Bio-Nano-Geo Sciences

The Future Challenge

Bio-Nano-Geo Sciences

The Future Challenge

Edited by

Prof. Dr. Alok Srivastava

Department of Chemistry
Punjab University
Chandigarh

&

Prof. Dr. Ipsita Roy

Department of Biotechnology
National Institute of Pharmaceutical Education and Research (NIPER)
S.A.S. Nagar, Punjab.

Ane Books Pvt. Ltd.

New Delhi ♦ Chennai

Bio -Nano-Geo Sciences : The Future Challenge

Alok Srivastava and Ipsita Roy

Published by

Ane Books Pvt. Ltd.

4821, Parwana Bhawan, 1st Floor, 24 Ansari Road, Darya Ganj,
New Delhi - 110 002, Tel.: +91(011) 23276843-44, Fax: +91(011) 23276863
e-mail: kapoor@anebooks.com, Website: www.anebooks.com

Branch Office:

- Avantika Niwas, 1st Floor, 19 Doraiswamy Road, T. Nagar,
  Chennai - 600 017, Tel.: +91(044) 28141554, 28141209
  e-mail: anebookschennai@gmail.com, rathinam@anebooks.com

ISBN : 978-81-80521-80-5

Printed at : R.K. Offset Process, New Delhi.

FOREWORD

Dr. Georg Schütte
Secretary General, AvH Foundation

The German Alexander von Humboldt Foundation (AvH) sponsors highly-qualified scientists and scholars from all over the world who come to Germany on research visits. One of its aims is to strengthen networking amongst the academics it sponsors, on both a regional and disciplinary basis, as well as to secure the sustainability of its contacts worldwide. Awakening the interest of junior researchers in the Humboldt Foundation and in the research location Germany is a further aspect. To achieve its goals, the AvH offers a comprehensive palette of attractive sponsorship programmes. In the context of alumni sponsorship, it supports Humboldt Associations/Humboldt Academies and individual Humboldtians in running regional and specialist conferences on current topics.

In March 2006, a so-called Humboldt Kolleg took place in Palampur entitled, "Bio-Nano-Geo Sciences: The Future Challenge" which was attended by academics from India, Germany, Greece, USA and Bangladesh. This meeting was vastly more than just a specialist congress – it was a gathering of distinguished academics transcending the borders of disciplines and countries. Humboldtians are highly respected throughout the world. Academic excellence is the only key to becoming a member of the international Humboldt Network. Once an Humboldtian, always an Humboldtian. The alumni network is the Foundation's greatest capital.

In the long term, this network of excellence can only be maintained and extended if highly-qualified young scientists and scholars are introduced to the sponsorship opportunities the AvH offers. Consequently, junior researchers were also included in the Humboldt Kolleg at Palampur. They gained an impression of the excellence of the Humboldt alumni and the advantages of being a member of this international network. Furthermore, they had an opportunity to make personal contact with partners in Germany participating in the Humboldt Kolleg.

Of the 23,000 Humboldt alumni in 135 countries, including 40 Nobel Prize winners, approximately 1,000 currently live in India alone. They have got together to form 14 Humboldt Chapters and Humboldt Academies in various regions of India and they cultivate academic and social contacts inside and outside the country.

The results of the Humboldt Kolleg "Bio-Nano-Geo Sciences: The Future Challenge" in Palampur are partly documented in this book. The contributions selected demonstrate the diversity of the topics, all of which are of relevance to the future of mankind. They range from ethical issues via climate change to nanotechnology.

The Humboldt Kolleg at Palampur was mainly organised and run by the Humboldt Academy of Chandigarh, under the direction of Professor Alok Srivastava. Prof. Srivastava and the members of the Academy had already organised highly regarded academic and social meetings in previous years. I should like to take this opportunity to thank both him and the members of the Academy very much indeed for such a successful event.

Dr. Georg Schütte

Secretary General

Alexander von Humboldt Foundation, Germany

PREFACE

When the idea of holding a symposium (Kolleg) covering a broad area under the aegis of Alexander von Humboldt Foundation, Germany first came up, there was a lot of debate as to the defined field that the event would cover. The topic "Bio-Nano-Geo Sciences: The Future Challenge" was the result. We had been fortunate enough to attract scientists and research workers, who are leaders in their respective fields not only from India but abroad too. Some of them presented their work formally in the various sessions of the symposium, many more in informal sessions which were held late in the evenings. The pleasant environment of the Institute of Himalayan Bioresource Technology (IHBT), Palampur, in the midst of the Dhauladhar range, provided the perfect backdrop. We are extremely grateful to Dr. P. S. Ahuja, Director, and Dr. A. A. Zaidi, Scientist, at IHBT for making available excellent facilities for the smooth functioning of the symposium.

This book, which is a compilation of invited articles, the authenticity of the data therein for which we are not responsible is based on formal and informal presentations made at the said symposium, has been peer-reviewed by eminent researchers. We are thankful to them as well as the authors who agreed to write up the manuscripts and have waited patiently with us to see this book in a printed form. Though the reviewers can not be thanked by name, we take this opportunity to express our gratitude to them for taking time out from their busy schedule and going through the manuscripts painstakingly. We are also thankful to the research scholars who participated enthusiastically in the poster sessions. About 30 posters were presented and judged by a panel of eminent scientists.

This Kolleg was organized jointly by the Chandigarh and Kanpur Academies of the Humboldt Foundation in India. We are indebted to all the members of the Chandigarh Academy, especially Prof. R. K. Gupta, Prof. J. K. Gupta, Prof. L. M. Gupta, Prof. B. K. Das, Prof. V. K. Jindal, Dr. R. P. Bajpai, Dr. B. K. Handa, Dr. P. V. Bharatam, Dr. M. N. Ravi Kumar and Prof. M. L. Garg, for their encouragement and help at the symposium. We are also grateful to Prof. Amitabha Ghosh, Prof. S. Sarkar and Prof. Sandeep Verma of the Kanpur Academy, who mobilized a huge contingent and ensured an enthusiastic participation of academecians/research scientists not only from IIT Kanpur but from other Institutes in India and abroad too.

The help provided by Shri. S. Saxena, Shri. J. R. Kapoor and the staff of Ane Books Pvt. Ltd. in the form of designing of cover and printing the book in a short time is gratefully acknowledged.

Last but not the least, the Alexander von Humboldt Foundation at Bonn, Germany has always been a source of moral and financial support. This endeavour has also received support from the past President of the Foundation Prof. Dr. Wolfgang Fruehwald and the present President Prof. Dr. Helmut Schwarz. We are indeed appreciative of the gesture of the Secretary General Dr. Georg Schuette, Dr. Klaus Manderla, Dr. G. Limberg, Frau Ulrika Holdefleiss-Walter and Frau Anne Heming and look forward to continued interaction with the Foundation. *Vielen dank*!

Finally we would like to place in record our heartfelt thanks to Prof. Ashok Sahni the President of the Humboldt Academy of Chandigarh who was a constant source of encouragement and was always there whenever his help was required.

Alok Srivastava
Ipsita Roy

CONTENTS

PLATE 1

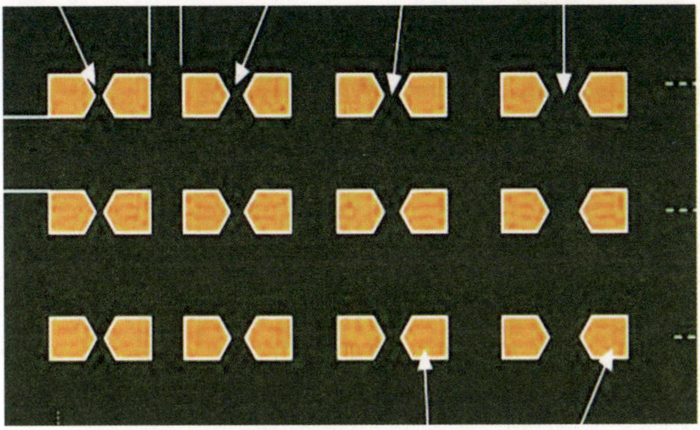

Microelectrodes with varying spacing from 5-25 micron,
(For details refer p 10)

Binding of DNA in between the electrodes, (For details refer p 11)

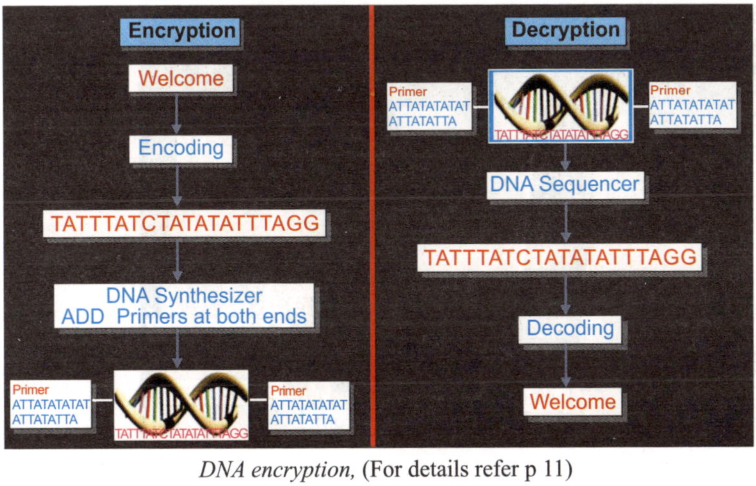

DNA encryption, (For details refer p 11)

PLATE 2

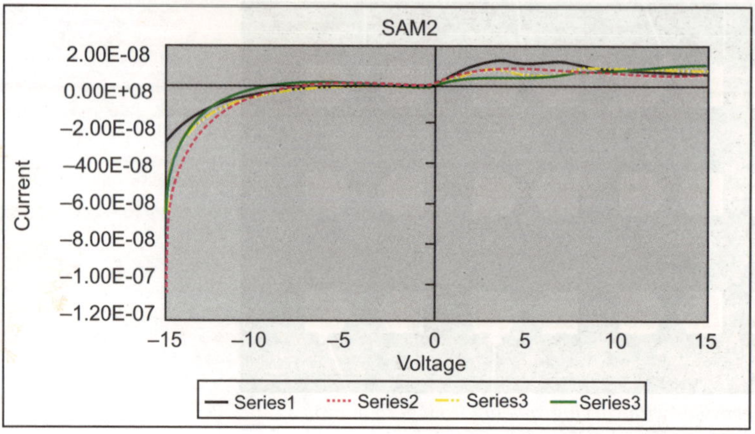

Current-voltage plot of DNA between microelectrodes, (For details refer p 11)

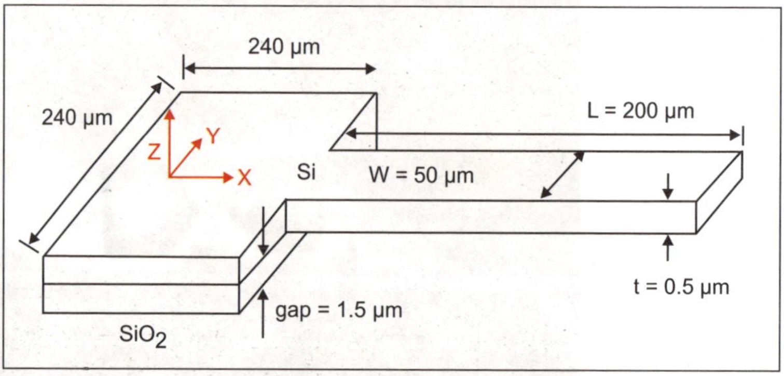

Schematic of cantilever, (For details refer p 12)

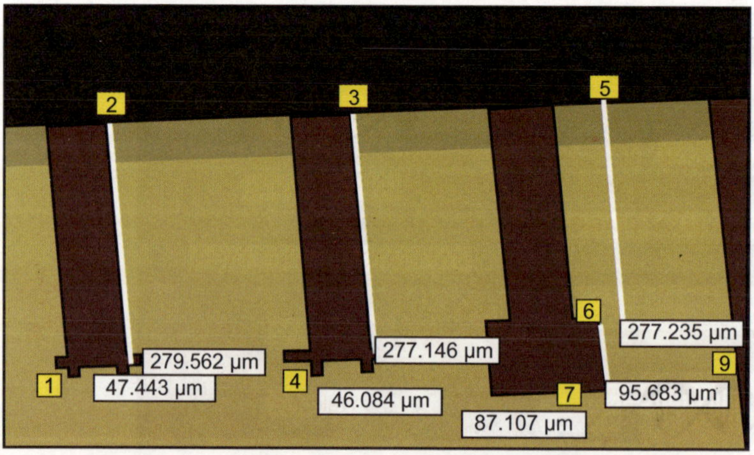

Cantilever fabricated at SCL, (For details refer p 13)

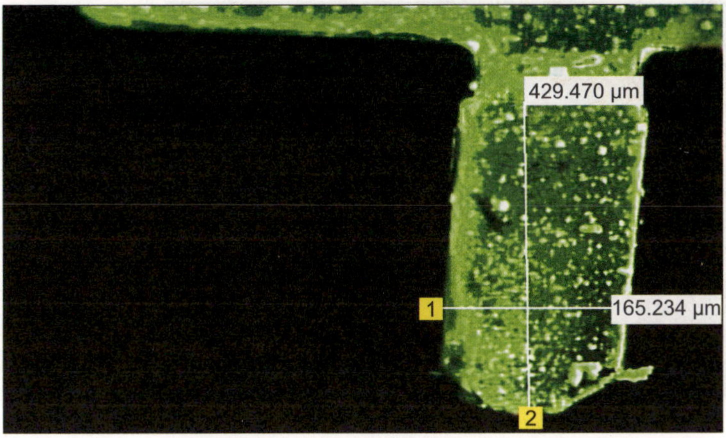

Antibody coated cantilever, (For details refer p 13)

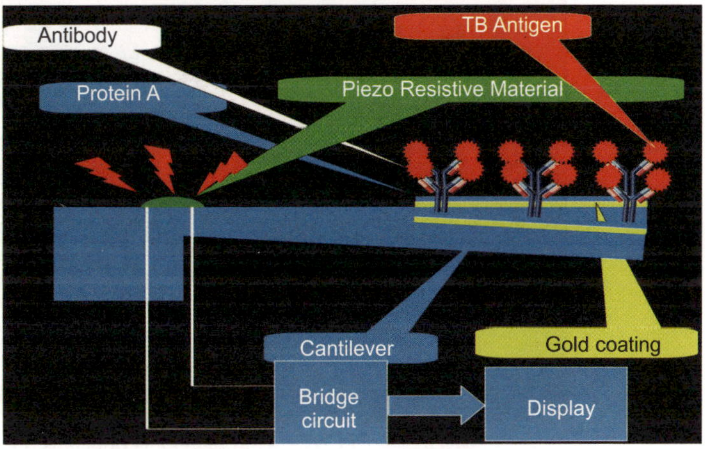

Working principle of cantilever, (For details refer p 13)

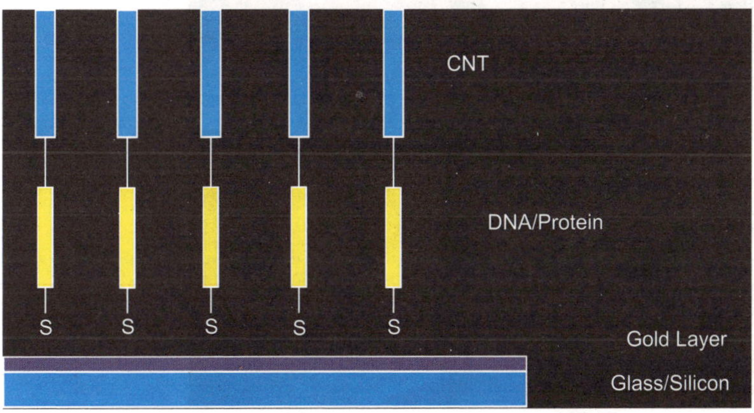

Alignment of CNTs, (For details refer p 14)

PLATE 4

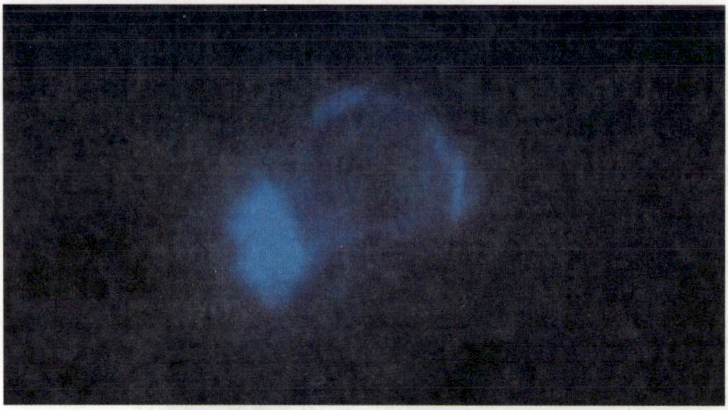

DNA-CNT binding (fluorescent image), (For details refer p 14)

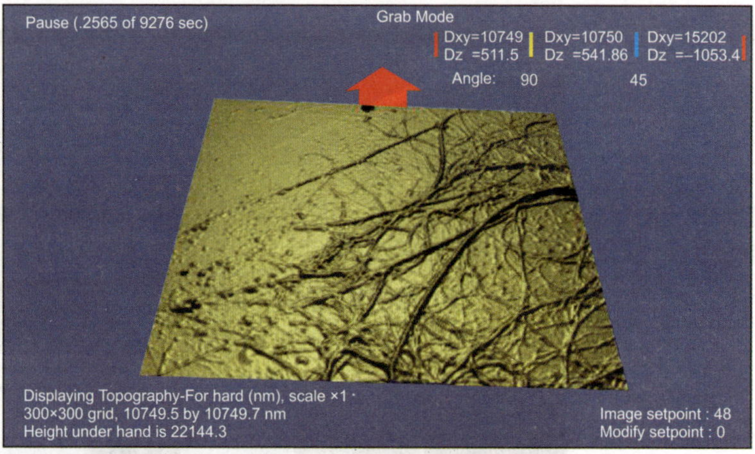

Dispersed CNTs (AFM image), (For details refer p 15)

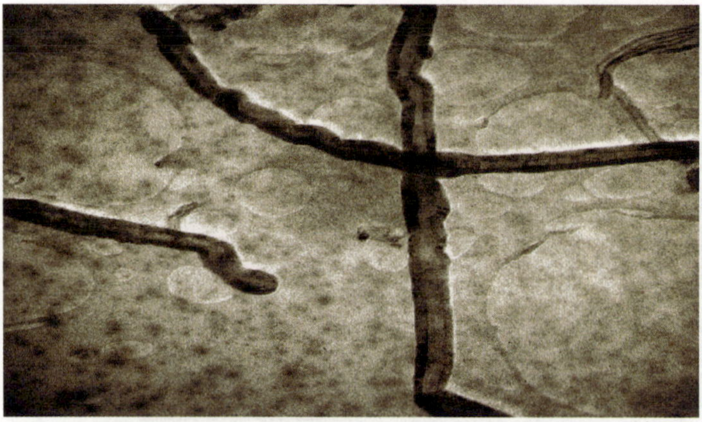

Dispersed CNTs (TEM image), (For details refer p 15)

PLATE 5

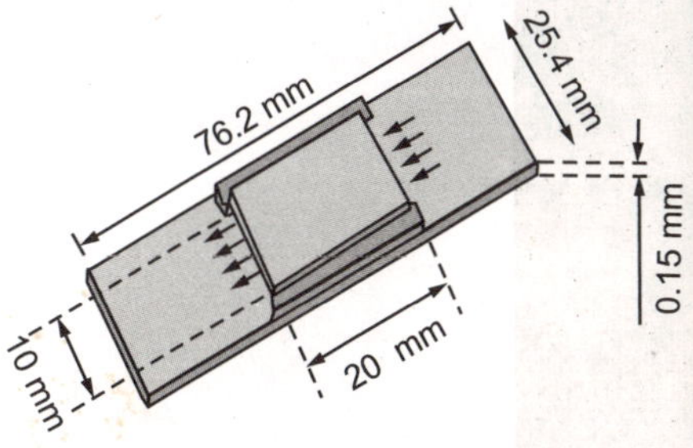

Schematic of flow cell for molecular motors, (For details refer p 16)

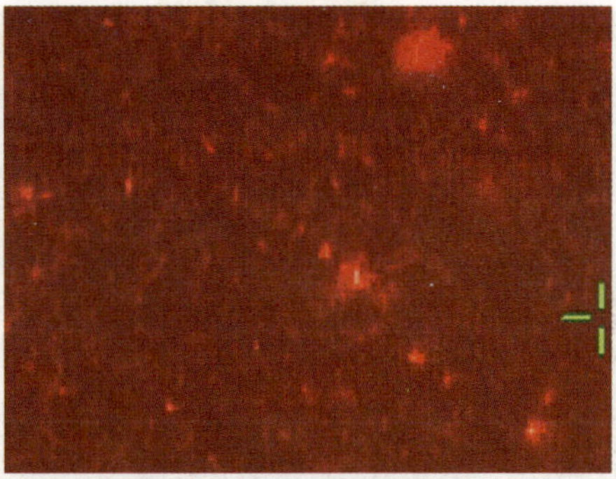

Motility of actin filament, (For details refer p 16)

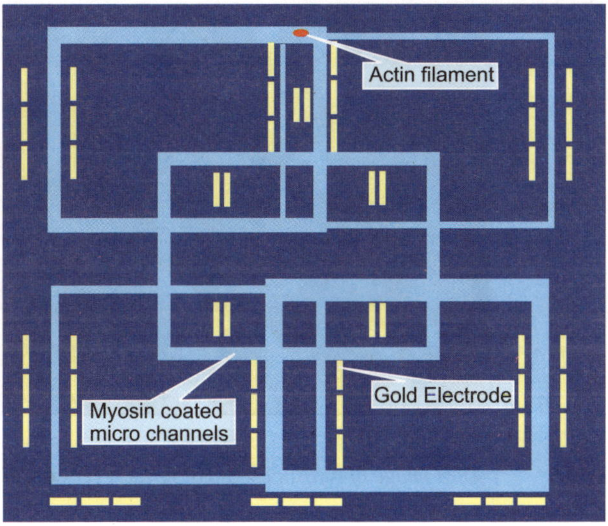

In-vitro control of molecular motors, (For details refer p 16)

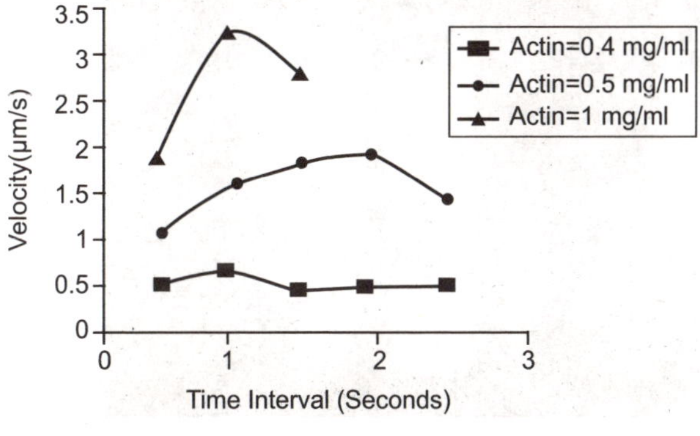

Motility of action filament Pranam 65(5) 967, (2005), (For details refer p 16)

PLATE 6

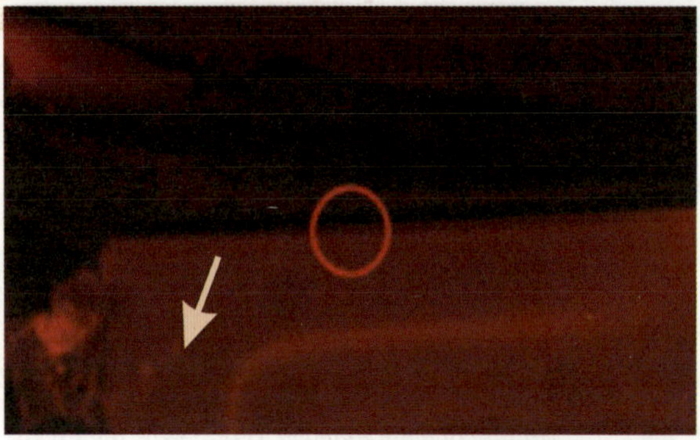

Controlled movement of actin filament in Microchannel,
(For details refer p 17)

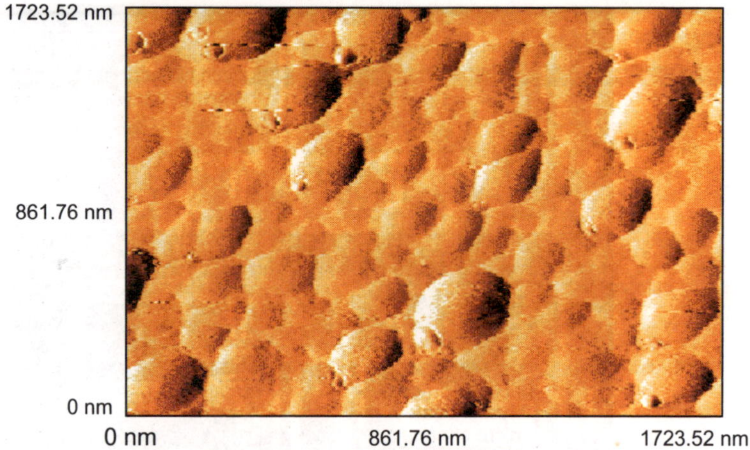

AFM image of myosin on Glass Surface, (For details refer p 17)

Fig. 4 *Simple setup for fullerene production used by students at PU.,* (For details refer p 34)

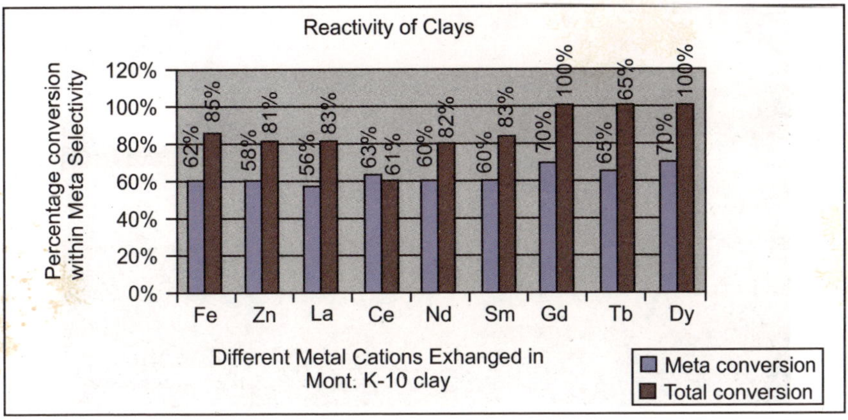

Fig. 4 *Reactivity of various cation exchanged clays,* (For details refer p 65)

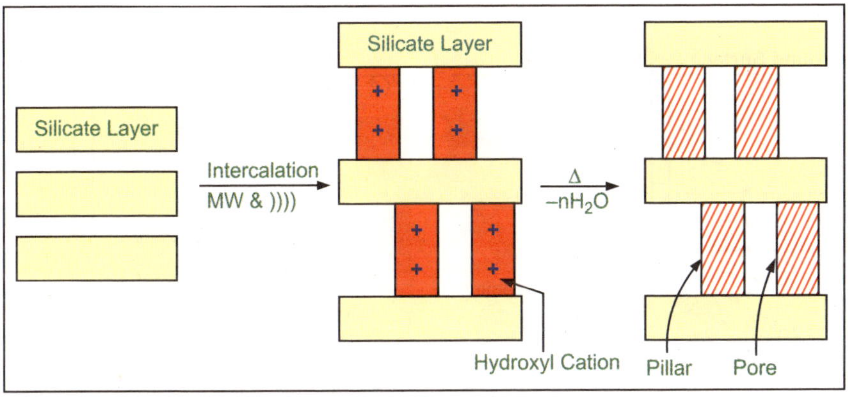

Fig. 5 *ZrO_2-pillared clay prepared by ultrasonic and microwave activation* (For details refer p 65)

Fig. 1a *Mansar Lake, situated in Lower Siwalik Formation,* (For details refer p 118)

PLATE 8

Fig. 1b *A panoramic view of Mansar lake, showing vegetation along the banks,* (For details refer p 118)

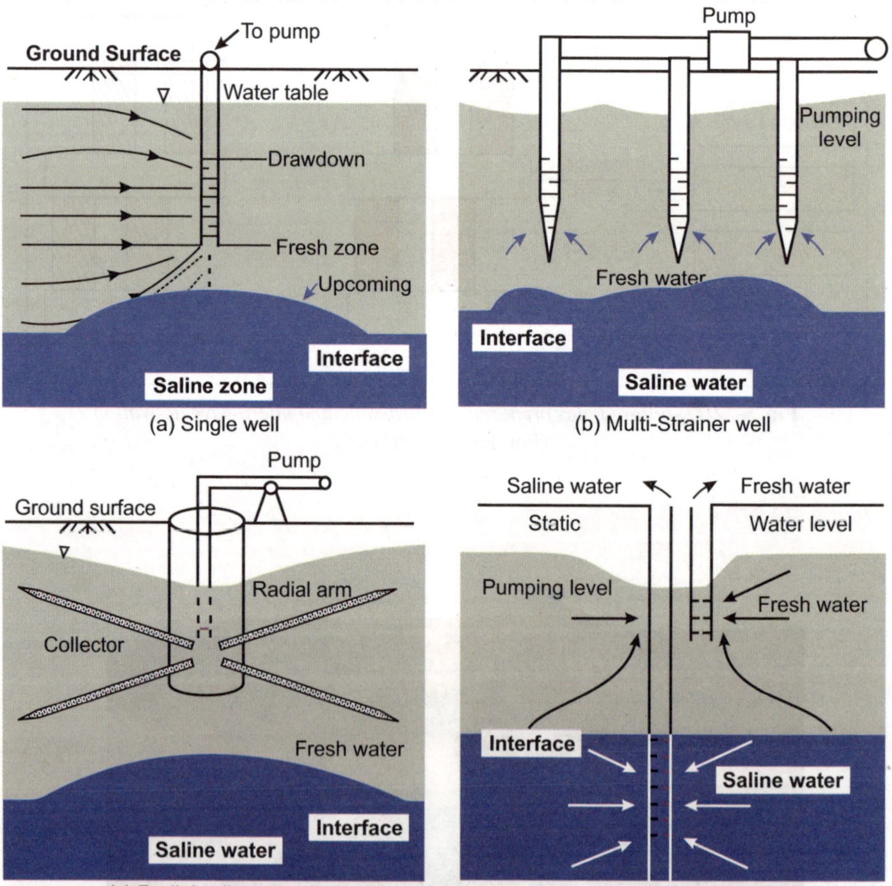

Fig. 1 *Different types of fresh water skimming wells,* (For details refer p 125)

PLATE 9

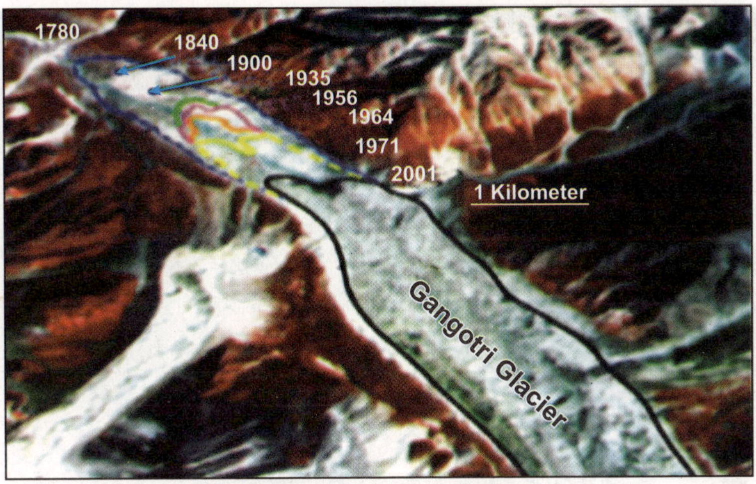

Fig. 2 *Retreat in Gangotri glacier in the last 220 years (source, Jeff Kargel, USGS),* (For details refer p 143)

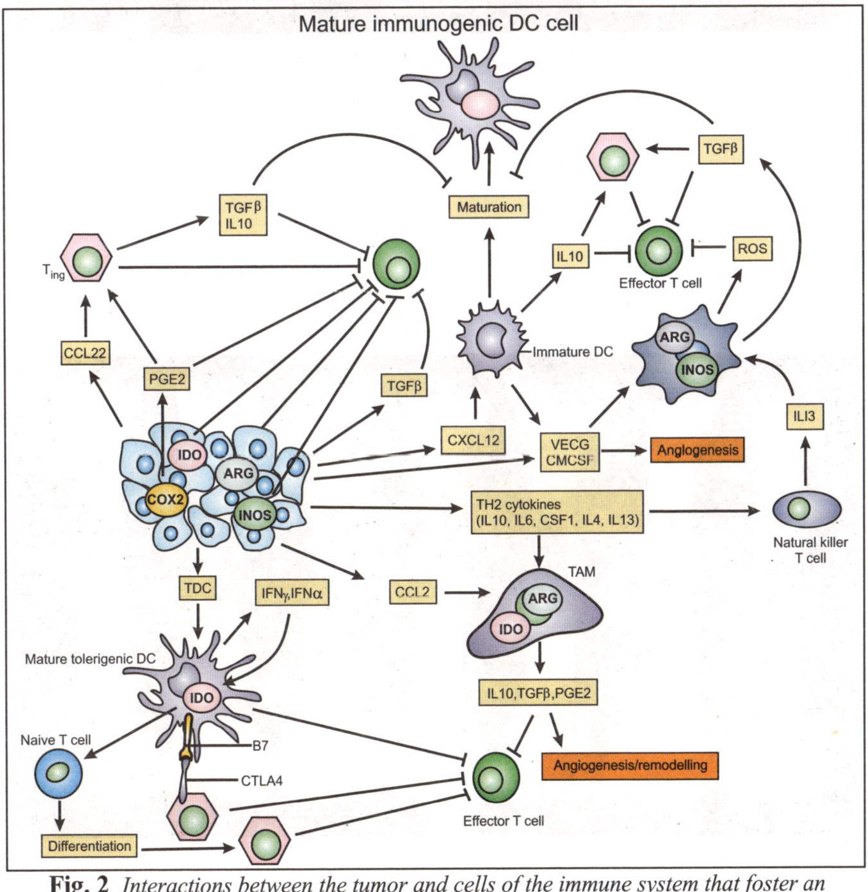

Fig. 2 *Interactions between the tumor and cells of the immune system that foster an immunosuppressive microenvironment Figure ta en from Ref No 2,* (For details refer p 151)

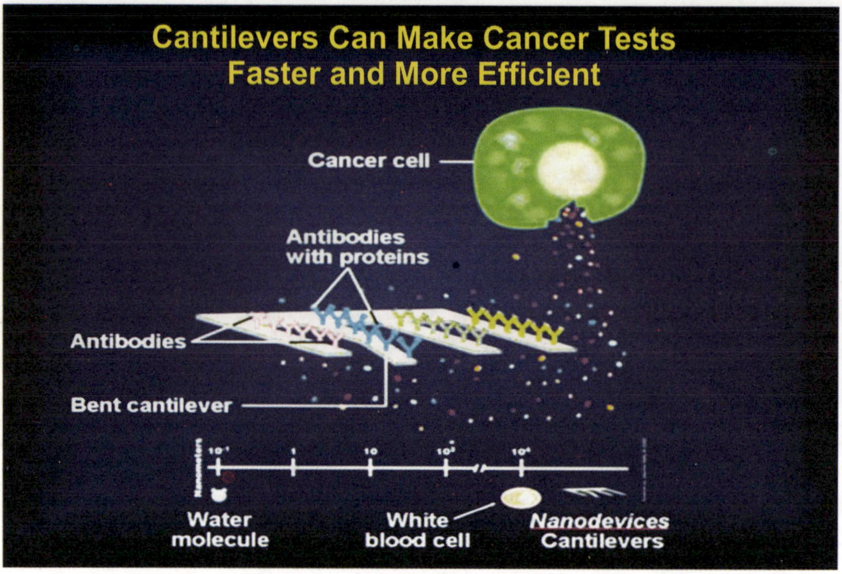

Fig. 3 *Role of Nanocantlevers in detecting cancer cells,* (For details refer p 152)

Fig. 4 *Fluorescence induced by exposure to ultraviolet light in vials containing various sized Cadmium selenide (CdSe) quantum dots,* (For details refer p 153)

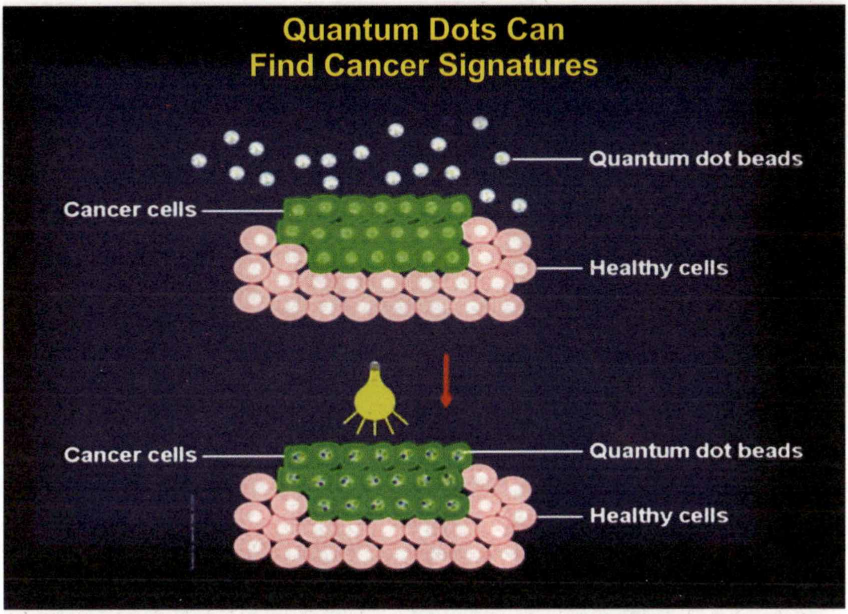

Fig. 5 *Quantum Dots can be used in the body, eliminating the need for biopsy,* (For details refer p 153)

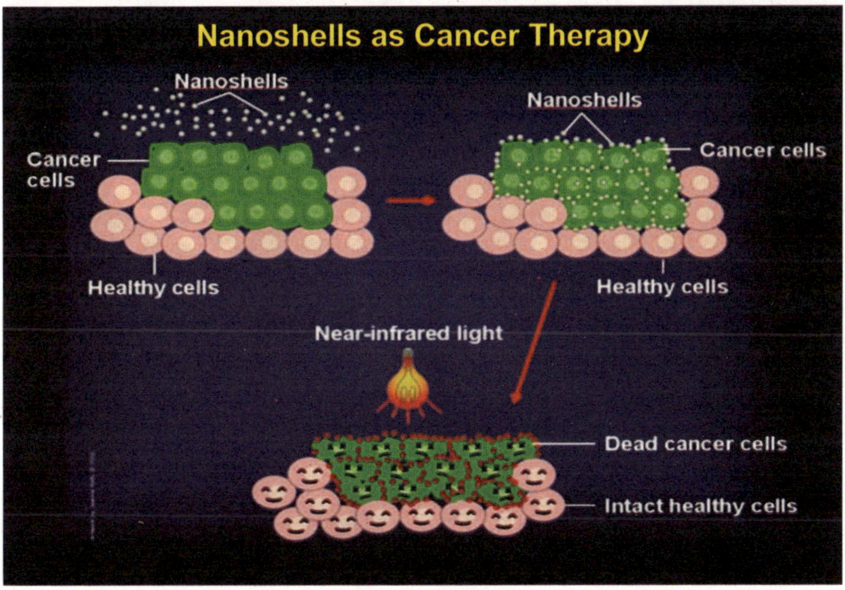

Fig. 6 *The use of nanoshells in cancer therapy,* (For details refer p 154)

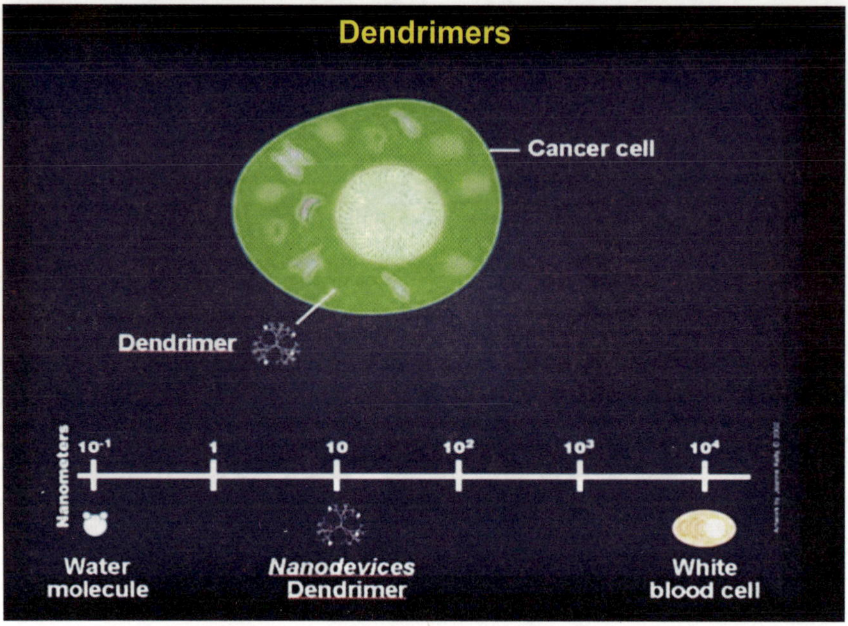

Fig. 7 *Comparison between the size of cancer cell and a dendrimer,* (For details refer p 154)

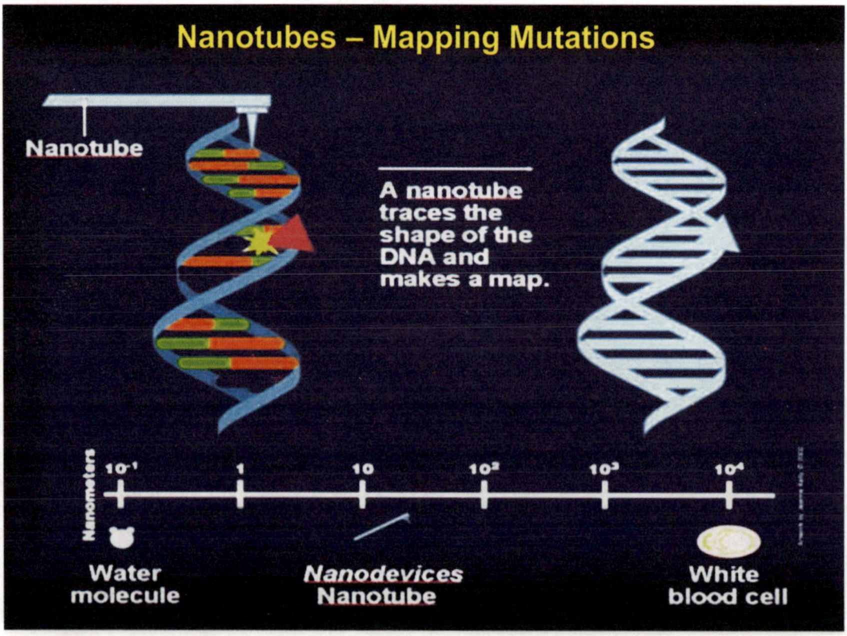

Fig. 8 *Identification of mutations in the DNA by nanotube,* (For details refer p 155)

PLATE 13

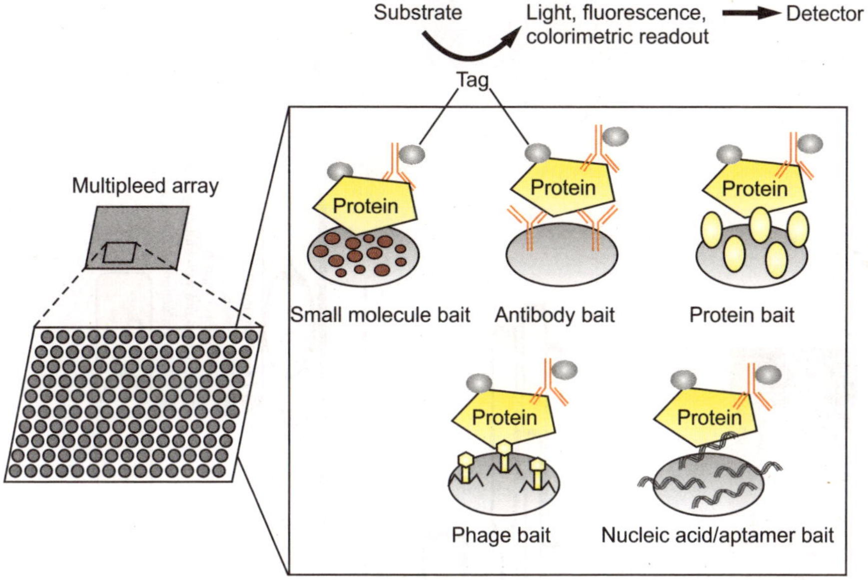

Fig. 9 *Enlarged image of the Protein Microarray system,* (For details refer p 157)

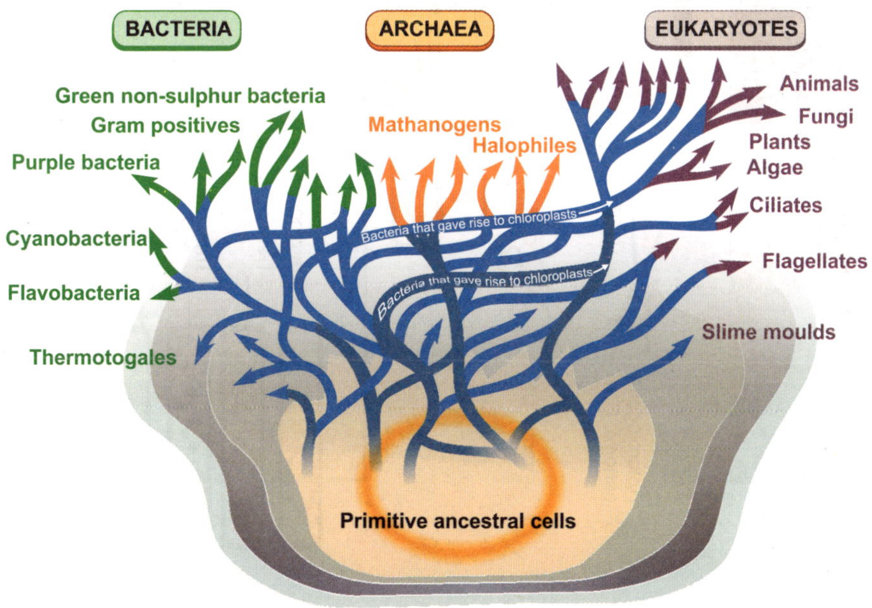

Fig. 1 *Microbial diversity,* (For details refer p 162)

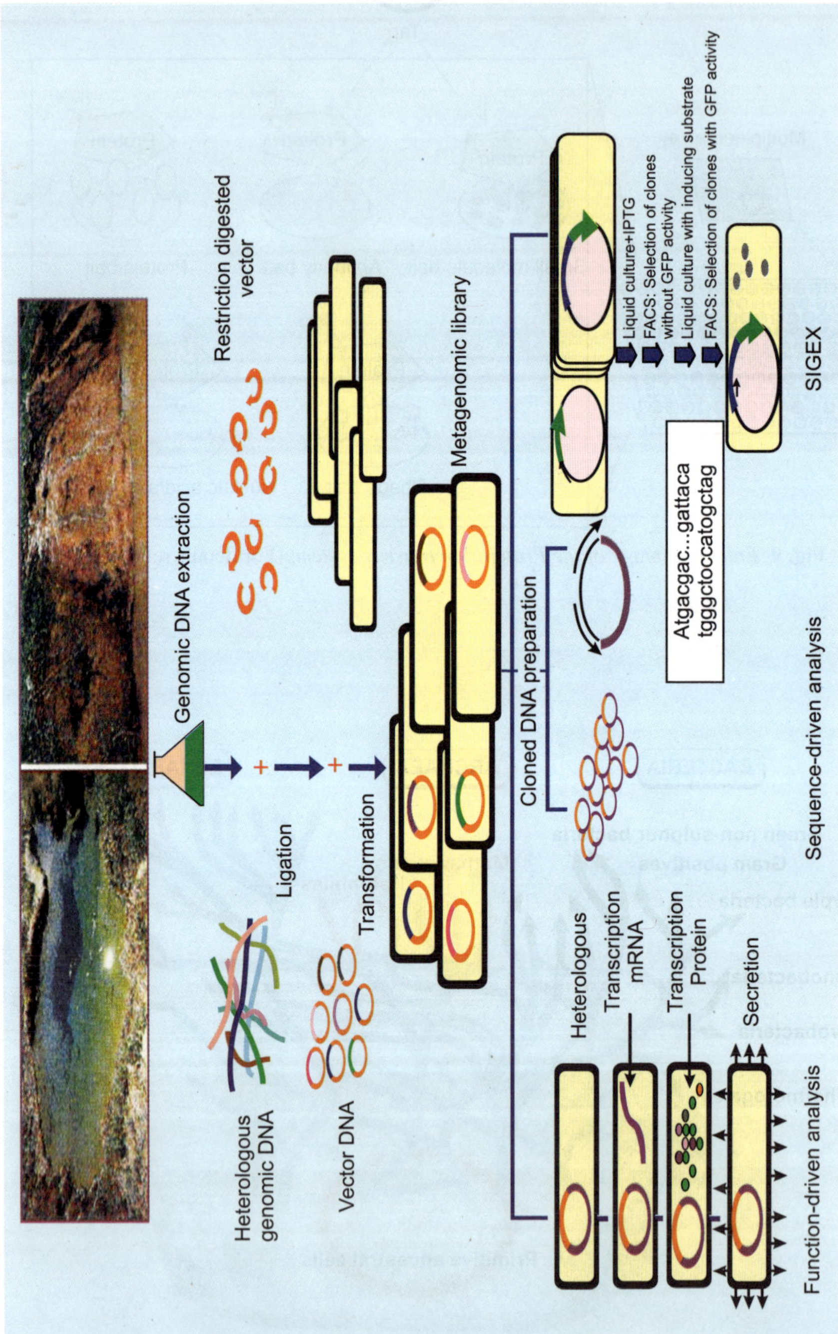

Fig. 2 *Metagenomic libraries,* (For details refer p 166)

1

NATURAL NANO-MACHINES OF LIFE: A BIOLOGICAL ROUTE TO NANOTECHNOLOGY

Debashish Chowdhury

ABSTRACT

An eukaryotic cell is filled with machines, each typically a few tens of nanometers long, which have specific tasks and operate in a well coordinated manner. I present examples of linear and rotary motors as well as pumps that drive various processes essential for sustaining life. Then, I summarize the methods of studying the materials and mechanisms of these machines emphasizing the crucial differences between the natural nano-machines and their macroscopic counterparts. Finally, I highlight some potential applications of the fundamental research on molecular machines in biomedical science and nano-technology.

1. INTRODUCTION

Three centuries ago Marcello Malpighi conjectured "Nature, in order to carry out the marvelous operations in animals and plants, has been pleased to construct their organized bodies with a very large number of machines, which are of necessity made up of extremely minute parts so shaped and situated, such as to form a marvelous organ, the composition of which are usually invisible to the naked eye, without the aid of the microscope." (As quoted by Marco Piccolino [1]).

A machine is a device, having a unique purpose, that augments or replaces human or animal effort for the accomplishment of physical tasks. The operation of a machine may involve the transformation of chemical, thermal, electrical, or nuclear energy into mechanical energy or vice versa. All machines have an input, an output, and a transforming (or, modifying) and a transmitting device. A man-made complex macroscopic device is usually an assembly of simpler components each of which is designed to achieve a specific function. For example, a hair dryer consists of a heater, a fan and a switch suitably connected by electric wires and assembled in a compact framework; combined and coordinated operation of these components gives rise to the function of the hair dryer- blowing hot wind.

Naturally, to understand how a machine works, one has to understand the structure and functions of its components. So, the belief in the role of organic machines in sustaining life led to intense anatomical investigations in the seventeenth century. Curiously, the emphasis shifted away from organic machines during the eighteenth and nineteenth centuries; a critical analysis for plausible reasons has been carried out by Piccolino [1]. With the triumph of the molecular theory of matter in the beginning of the twentieth century molecule became the prime object of investigation in chemistry as well as in biology. Interestingly, already in 1906, Loeb (as quoted in [2]) referred to the cell as a "chemical machine" (see also [3]).

Does a cell really function as a machine? Indeed, each of the cells in our body is like a 'factory' made up of an enormously large number of molecular machines [4,5]. However, unlike man-made machines, these are products of Nature's *evolutionary design* over billions of years and the mechanisms underlying their function are often very different from their macroscopic counterparts in spite of many superficial similarities. The typical sizes of the molecular machines in our body cells range from tens to a few *nano-meters*. Therefore, one of the biggest challenges of modern science is to determine the principles behind the design of these natural nano-machines and to understand their dynamics which help in elucidating the mechanisms of their functions in terms of the fundamental principles of physics.

2. EXAMPLES OF NATURAL NANO-MACHINES

The *cytoskeleton* of each cell is a dual-purpose network made of filamentary proteins some of which are like girders while others are like cables and ropes; it gives strength to the architecture of the cell and, at the same time, also forms the back bone of the intra-cellular transportation network [5]. These filamentary proteins are dynamic and are, often, targets of anti-cancer drugs as these play crucial roles in cell division [6].

The molecular motors do not move smoothly; their jerky motion is similar to that of a person trying to move straight in a blizzard. The molecular motors moving along the cables and struts have no wheels. They look more like two-legged humans than the four-wheeled vehicles we are familar with. Each carries a heavy bundle of cargo on their head and move like porters. *Kinesin* and *dynein* motor proteins are like trains of a rapid mass transit system that ran shuttle service from the city centre to the peripheral suburbs. The stiff struts that forms the tracks for these motors are called *microtubules*. The shuttle buses that carry the passengers in these suburbs are mostly *myosins* and they move on more flexible cables called *actin*.

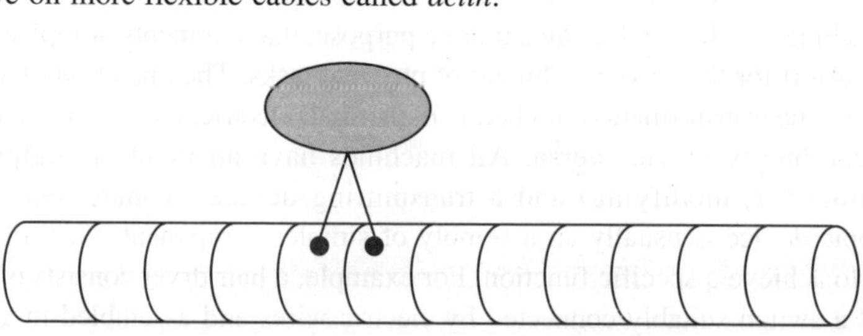

Fig. 1 *A cartoon of a two-headed molecular motor moving on a filamentary track.(Copyright: Indrani Chowdhury; reproduced with permission).*

The molecular motors consume "chemical fuels". Each of the motors utilize the chemical energy released by a chemical reaction. The fuels, namely molecules of a compound called ATP, are abundant everywhere in the cell and the motors use the energy released by their hydrolysis to generate the mechanical energy required for their directed movement. All of the spent fuel is recycled in the cell; the spent fuel is recharged to manufacture fresh fuel. Some other machines in the cell directly use sunlight and, therefore, is very eco-friendly as no waste product is generated by those machines. Moreover, another advantage of the machines running on light energy is that light can be switched on and off easily and rapidly.

Inside the cell nucleus, a member of a special class of machines, called *DNA helicase*, walk along a DNA fiber unwinding the two strands. This process resembles disentangling of two strands of find thread that are snarled together. The local opening of the double stranded DNA is, effectively, an unzipping process that cuts the weak bonds which hold the two strands together. Such openings form one step in the overall processes of DNA repair or polymerisation of DNA and proteins.

RNA polymerase are also nucleotide-associated motors that move on single-stranded DNA and polymerise the so-called messenger RNA (mRNA) by joining nucleotides that are selected on the basis of the templates formed by the single-strand of DNA. After moving out of the nucleus the mRNA itself serves as a template for the synthesis of a protein by another set of machines. Little ball-like machines, called *ribosomes*, move along the mRNA strand stiching amino acids and releasing the polypeptides as fresh proteins.

A mitochondrion, a power house that recharges the spent fuel ADP into fresh fuel ATP, looks very different from the chemical factories in the macro world. There is a tiny machine, called *ATP synthase,* stuck on the wall of the mitochondrion. This rotary motor currently holds the world record: it is the smallest rotary motor. This is, at least superficially, very similar to the motor of a hair dryer. This is not the only rotary motor that exists in the sub-cellular world. Bacteria, which are uni-cellular organism, also have a slightly larger rotary motor fixed in their skin. This motor rotates a helical filament called a flagellum.

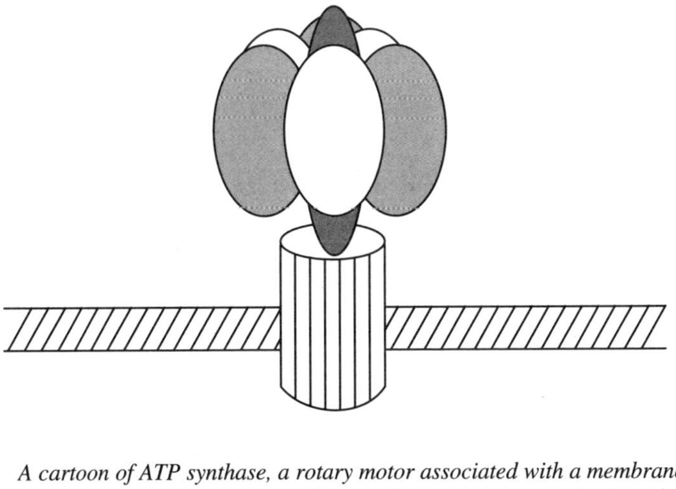

Fig. 2 *A cartoon of ATP synthase, a rotary motor associated with a membrane. The cylindrical part is called the F_0 motor while the other part represents the F_1 motor. Both F_0 and F_1 are reversible and are tightly coupled to each other. (Copyright: Indrani Chowdhury; reproduced with permission).*

Ion pumps are also molecular machines. These are fitted on the membranous wall of some cells and these are capable of pumping sodium ions from inside to outside while simultaneously, pumping potassium ions in the reverse direction. These pumps consume energy to carry the ions against concentration gradient, i.e., opposite to the direction of the spontaneous flow, just as a water pump consumes electrical energy to draw water upward against gravity.

3. PHYSICAL MODELS AND METHODS FOR MATERIALS AND MECHANISMS

Let me briefly mention our methods of investigation, namely, experimental, and theoretical techniques, for studying the common fundamental physical principles that govern the generic features of the structure and dynamics of the biological nano-machines as well as the corresponding underlying mechanisms of their operations. This is a frontier area of current interdisciplinary research that involves molecular cell biology, (bio-)chemistry, physics as well as engineering.

3.1 Materials: Soft Matter and Weak Forces

The durable parts of macroscopic machines are normally "hard" so as to survive the regular wear and tear. In contrast, the structural elements of the cell, e.g., filaments and membranes, are "soft".

These materials can be deformed easily because their conformations are determined by weak forces like, for example, Van der Waals forces, hydrogen bonding, etc. Moreover, since the thermal energy k_BT available at a temperature T (k_B being the Boltzamann constant) is comparable to these weaker forces, thermal fluctuations can also lead to conformational changes. Furthermore, straightening of a filament or flattening of a membrane reduces its entropy and the corresponding restoring force is of entropic origin. Therefore,the free energy of soft matter are often dominated by entropy which gives rise to phenomena that are not observed in hard materials. Therefore, biomolecular machines operate in a domain where the appropriate units of length, time, force and energy are *nano-meter*, *milli-second*, *pico-Newton* and, k_BT, respectively.

3.2 Experimental Methods: Imaging vs. Manipulation

Seeing is believing. However, until late twentieth century, scientists could not see individual molecules although everybody believed in its existence. We got our first glimpse of the macromolecules via X-ray diffraction and, then, electron microscopy. But, these were static pictures. Towards the end of the twentieth-century a series of novel imaging techniques emerged which allowed us to monitor the dynamics of single-molecules [7]. Confocal fluorescence microscopy is, perhaps, the most versatile among these methods.

The recently developed single-molecule techniques can be broadly classified into two groups: (i) Methods of imaging, and (ii) methods of manipulation. For example, green (and red) fluorescence microscopy enables us to monitor the dynamics of single motors just as ecologists use "radio collars" to track individual animals. But, for a clear understandng of the mechanism of the molecular machines it is not enough merely to watch them passively; we must be able also to manipulate them. Such manipulations have been possible over the last decade because of the availability of sophisticated techniques like, for example, optical tweezers and scanning force microscopes (SFM).

3.3 Theoretical Methods

By "theoretical methods" we mean analytical (mathematical) treatment of theoretical models. Since such an analysis can be accomplished exactly only in rare cases, one has to make sensible approximations so as to get results as accurate as possible.

3.3.1 Theoretical Models

In biology, often the simplest among a family of objects is called a model system for the purpose of experimental investigations. For example, the fruit fly *Drosophila* and the worm *C-elegans* are often called model organisms because of their simpler features among all multi-cellular organisms. However, by the term model we shall almost exclusively mean *theoretical* model. A theoretical model is an abstract representation of the real system. Biologists often develop *qualitative* models to explain empirical observations. In contrast, physicists work with *quantitative* models, formulated in terms of mathematical symbols, not only to interpret experimental data but also to make new predictions that can be tested through experiments.

3.3.2 Mechanisms-Conceptual Challenges

Are the mechanisms of molecular machines identical to their macroscopic counterparts except that the machines are nano-meter sized objects and the forces are in the range of pico-Newons? NO, there are crucial differences between the molecular machines and their macroscopic counterparts.

The molecular machines get bombarded from all sides by the randomly moving water molecules. The masses of molecular machines are so small that these *random thermal forces* have strong infuence on their movement. It is because of these random forces that the molecular motors inside the cell move in a manner that resembles the movement of a human being in a strong blizzard. However, unlike passive Brownian particles (like, for example, pollen grains in water), the molecular machines consume (free-) energy. At least some of these machines are physical realizations of Brownian ratchet, a device that exhibits directed, albeit noisy, movement in spite of being subjected to random Brownian forces as they operate in conditions far from thermodynamic equilibrium [8].

Interestingly, the nano-motors can come to a halt practically instantaneously as soon as they put on brake. This is a consequence of low Reynold's number which is defined as the ratio of the inertial and viscous forces. In order to appreciate the difficulties of swimming at low Reynold's number, it may be worth pointing out that if a bacterium tried to swim in an aqueous medium adopting the swimming styles of humans, the difficulties it would face will be very similar to those that you and I would face if we ever tried to swim in honey.

Most of the kinesin and dynein motors can walk over a long distance on their tracks before getting detached; these "porter"-like motors are called processive. In contrast, the non-processive motors work collectively in a fashion that resembles a group of "rowers". It is generally believed that close coordination of the two heads of kinesin motors is essential for their processive movements. If that is true, how can we explain the experimentally observed fact that single-headed kinesin motors KIF1A are also processive? An interesting mechanism, based on the principle of ratchet-and-pawl device (called Brownian ratchet) has been suggested recently for KIF1A [9]. Moreover, for the ribosome, the largest and the most complex molecular machine, very recently a theoretical model has been developed that accounts for the detailed mechano-chemistry involved in translation of the genetic message stored in mRNA into proteins [10].

4. IMPORTANCE OF RESEARCH ON MOLECULAR MACHINES

4.1 Importance in Fundamental Research-Understanding Life

Interest in molecular machines blossomed with the realization that much of cellular architecture, dynamics and function depend on those of molecular machines. So far as the molecular machines are concerned, some of the fundamental questions are as follows:

(i) **Machine components**: what are the *moving parts* of the motor and what are their molecular constituents?

(ii) **Fuel**: what is the *fuel* that supplies the (free-)energy input for the machine ?

(iii) **Mechanisms of energy transduction**: How does the *transduction* of energy take place?

(iv) **Efficiency**: What is the most appropriate definition of efficiency of the motor and how to estimate that efficiency?

(v) **Regulation and control**: How is the operation of the machine regulated? For example, how is the machine switched on and off? In addition to these general questions, several other fundamental questions are addressed in the context of specific molecular machines.

4.2 Importance in Biomedical Research-Improving Quality of Life

Just as occasional disruption of work in any department of a factory can bring entire factory to a standstill, defective molecular machines can cause diseases [11,12]. If we understand how motors work at the molecular level we will be able to fix them when they malfunction and perhaps manipulate them to improve human health and fitness. We might also devise ways to selectively either arrest sub-cellular processes that cause diseases like cancer or slow down metabolism of organisms that invade cells causing other types of disease.

4.3 Importance in Research on Nano-Technology-Mimicking Life

Although quest for ever-smaller size of devices began long ago, Feynman's famous talk of 1959 [13] is accepted by the majority of physicists as the defining moment of nano-technology. Feynman first attracted the attention of the physics community to the unlimited possibilities of manipulating and controlling things on the scale of nano-meters. He indicated the potential advantages of nano-technology.

From the perspective of applied research, the natural molecular machines have opened up a new frontier of nano-technology. The miniaturization of components for the fabrication of useful devices, which are essential for modern technology, is currently being pursued by engineers following mostly a top-down (from larger to smaller) approach. On the other hand, an alternative approach, pursued mostly by chemists, is a bottom-up (from smaller to larger) approach [14]. Insight into natural nanomachines will give us clue as to the possible design principles that can be utilized to synthesize artificial nanomachines. Significant progress has already been made in this direction. For example, cytoskeletal motor proteins are being utilized to synthesize molecular shuttles [15] and vehicles for drug delivery [16].

Summary and Conclusion

In recent years it has become clear that a cell is more like a factory where active biological processes are driven by coordinated operation of a large variety of molecular machines. One of the challenges of fundamental research on these machines is to understand the mechanisms of their operation in terms of the *physical principles* governing their structure and dynamics. The results of these investigations are likely to have significant impact on applied research in biomedical sciences and to open novel approaches to nano-technology.

REFERENCES

1. Piccolini, M. (2000) Biological machines: from mills to molecules. *Nature Rev. Mol. Cell Biol.* **1**, 149-152.
2. Taylor, E. W. (2001) 1999 E.B. Wilson lecture: the cell as molecular machine. Mol. Biol. Cell. **12**, 251-254.
3. Alberts, B. (1998) The cell as a collection of protein machines: preparing the next generation of molecular biologists. *Cell* **92**, 291-294.
4. Mavroidis, C., Dubey, A. and Yarmush, M. L. (2004) Molecular machines. *Annu. Rev. Biomed. Eng.* **6**, 363-395.
5. Howard, J. (2001) *Mechanics of Motor Proteins and the Cytoskeleton*. Sinauer Associates, Massachusetts, USA.
6. Mishra, P. K., Kunwar, A., Mukherji, S. and Chowdhury, D. (2005) Dynamic instability of microtubules: effect of catastrophe-suppressing drugs. *Phys. Rev. E* **72**, 051914.
7. Imhof, J. M. and Bout, D. A. V. (2003) Resource letter: Laser-based modern optical microscopy. *Am. J. Phys.* **71**, 429-436.
8. Juelicher, F. Ajdari, A. Prost, J. (1997) Modeling molecular motors. *Rev. Mod. Phys.* **69**, 1269-1281.
9. Nishinari, K., Okada, Y., Schadschneider, A. and Chowdhury, D. (2005) Intracellular transport of single-headed molecular motors KIF1A. *Phys. Rev. Lett.* **95**, 118101-.
10. Basu, A. and Chowdhury, D. (2006) communicated for publication.

11. Aridor, M. and Hannan, L. A. (2000) Traffic jam: A compendium of human diseases that affect intracellular transport processes. **1**, 836-851 (2000); Traffic jams II: An update of diseases of itracellular tansport (2002) **3**, 781-790.
12. Hirokawa, N. and Takemura, R. (2003) Biochemical and molecular characterization of diseases linked to motor proteins. *Trends Biochem. Sci.* **28**, 558-565.
13. Feynman, R. P. (1959), included also in: *The pleasure of finding things out*, (Perseus Books, Cambridge, Massachusetts, 1999), chapter 5.
14. Balzani, V., Venturi, M. and Credi, A. (2003) *Molecular devices and machines: a journey into the nano-world*, Wiley-VCH, Weinheim, Germany.
15. Hess, H. and Vogel, V. (2001) Molecular shuttles made from motor proteins: active transport in non-biological environments, *Rev. Mol. Biotech.* **82**, 67-85.
16. Cohen, R. N., Rashkin, M. J., Wen, X. and Szoka Jr., F. C. (2005) Molecular motors as drug delivery vehicles. Drug Discovery Today: Technologies, **2**, 111-118.

2

HYBRID BIO-NANODEVICES FOR ENGINEERING AND HEALTHCARE

R. P. Bajpai and Lalit M. Bharadwaj

ABSTRACT

Development of biomolecular and nanodevices for engineering and medical applications has been described in this article. These include: biomolecular electronics, disease diagnosis and targeted drug delivery based on molecular devices working in living systems. The group is taking multi-disciplinary integrated approach to various R & D activities related to broad areas such as DNA nanowire, Bio-MEMS for disease diagnosis, biomolecular motors for nano electronic switching & nanorobotics etc. Studies on Carbon Nanotubes (CNTs) interfaced with biomolecules is targeted towards new sensor development.

1. INTRODUCTION

Nature has created large variety of molecular devices which are working in living system ranging from micro-organism to human beings. Nature creates these devices by self-assembly process and does not require billions of dollar lithography based fabrication facilities used in VLSI devices. The life processes and materials have been perfected by nature in millions of years. These devices will find applications in electronics, life sciences, medicine, space, defense, bio-warfare, security, etc. Thus, the best way is to directly utilize large number of molecular devices working in the living systems for engineering applications by interfacing with non-biological materials and devices. These hybrid devices will have higher performance and perfection as molecular devices, such as, DNA, biomolecular motors, proteins, enzymes have well established dedicated functions to perform under given set of environmental conditions. At the same time, these molecular devices can be isolated from the living system or synthesized in simple molecular biology laboratory without huge investments as required in semiconductor industry. The system that works inside our bodies, in our cells, and in the nature that surrounds us is full of examples of nanoscale systems. Whether it's our genetic makeup that works for storing and transferring of information about our looks, personality or susceptibility to diseases or increased understanding of the calcium carbonate nanostructures used in seashell construction. Nanotechnology thus holds massive potential for the health care and life sciences market. Nanodevices enable drug delivery mechanisms that can send biologically active materials directly to the location where they are beneficial and reduce side effects to make pharmaceuticals more effective. Some important healthcare areas are self-assembly, biointerface, implants, nano-robots, tissue engineering. Nucleic Acid Diagnostics/Molecular Diagnostic Assays, Point-of-care testing, Genomics, Proteomics, DNA biochips, and microarrays, regenerative medicine, the healthcare environment, drug discovery tools, liquid handling, separation and filtration, nanopore sequencing and separations etc.. Bionanotechnology will bring new kind of sensors for applications in biology and engineering based on nanoparticle tagging, nanotube and nanowire, quantum dots, piezoelectric acoustic assays/rupture event scanning, enzymatic electrochemical detection, single-molecule sequencing and analysis, Lab-on-a-tape, nanoarrays, dip pen nanolithography, nano mass spectrometry, Nanogrid etc. These devices will play important role in early warning and treating life-threatening conditions, including cancer and heart disease in the form of body implants to deliver smart drugs or carry new cells to repair damaged tissue. Bionanotechnology also provides new tools/sensors for agriculture for improving the efficiency of crop production, food processing, food safety and environmental consequences, storage and distribution, remote control etc.

DNA Nanowire based Devices: DNA is the best molecular electronic device produced by nature which can store, process and provide information for growth and maintenance of any living system. Single cell production carries with it all requisite information and processing capability to produce living species by taking materials from environment. DNA can be synthesized in any desired sequence of four bases A, T, G, C to act as information storage. Encryption and decryption techniques can be applied which qualify this to be secure device. With large possibilities available with four bases, hence large data storage can be possible.

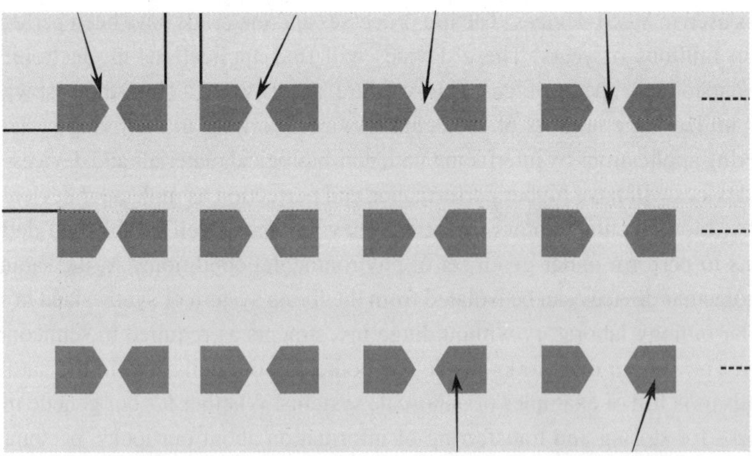

Microelectrodes with varying spacing from 5-25 micron,
(For color illustration refer p xi)

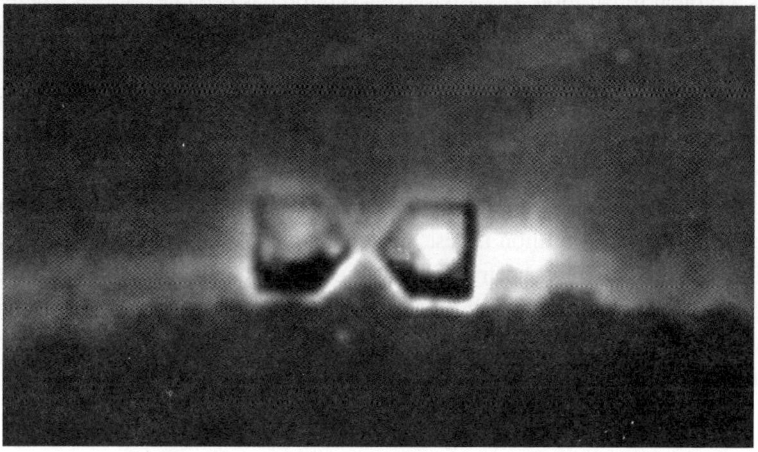

Binding of DNA in between the electrodes, (For color illustration refer p xi)

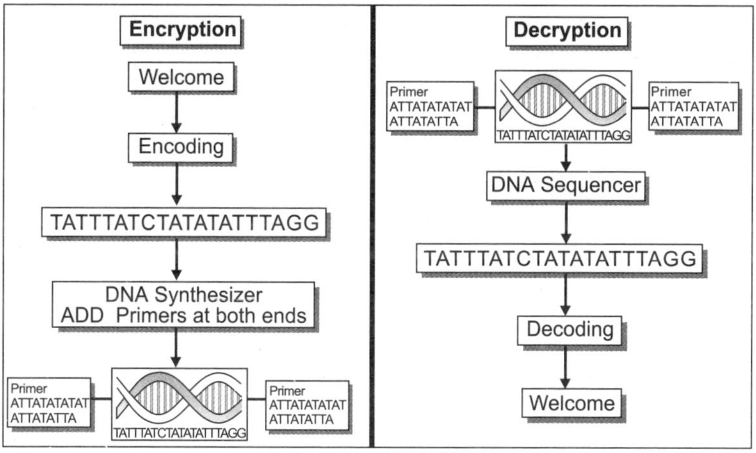

DNA encryption, (For color illustration refer p xi)

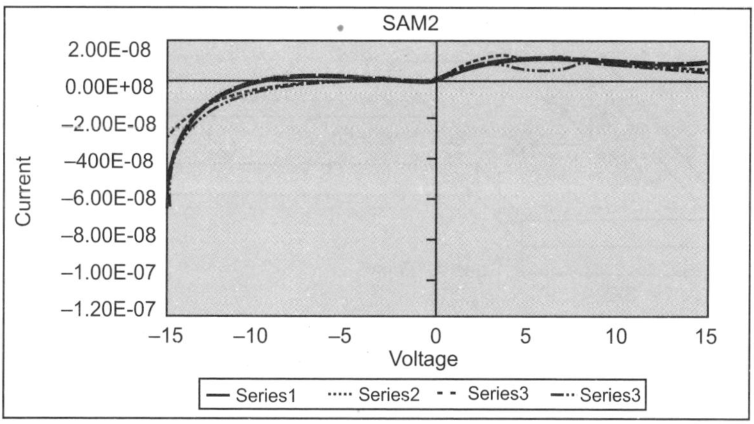

Current-voltage plot of DNA between microelectrodes, (For color illustration refer p xii)

As a typical illustration, DNA based encryption has been developed where all the 256 ASCII characters are defined in terms of four bases (A,T,G,C) using a dedicated software developed for this. Basic number systems with arithmetical operations in terms of DNA sequences were also attempted which can handle up to rational numbers. Work on complex number is in progress. This will lead towards molecular ultra-high density memory devices. For example, National Flag can be coded in terms of a DNA sequence of 7924 bases i.e. in our single genome (containing 6.4 billion bases) tricolor can be copied almost a million times. Space occupied by a genome is only few nm^3. Work is on towards studying electrical behavior of various segment of DNA to understand their information processing capability using femto range (10^{-15}) IV-CV measurements. This will help in sensor development for very low concentrations. DNA sequences can be bound on micro/nano electrodes to study such characteristics.

Bio-MEMS based Microdiagnostic Kits

The Bio-MEMS based sensors are the most recent development that provide specific molecular direction. The high specificity of the sensors is provided by the biomolecular recognition characteristics of antigen-antibody, probe DNA-target DNA, protein-receptor and enzyme-substrate. The change in mass on the microcantilever surface due to the binding of the analyte molecules is directly proportional to the deflection of the microcantilever.

Significant work has been done towards development of disposable microdiagnostic kits for diseases by developing innovative technique for binding antibodies on cantilever with surface coverage greater than 65%. With this achievement it will be possible to have sensitive device based on 0.5 micron thick cantilever design. This concept has been validated by theoretical simulation and demonstrated the relative bending of cantilever (500 × 150 × 1 micron fabricated at Semiconductor Complex Limited, Mohali) using AFM

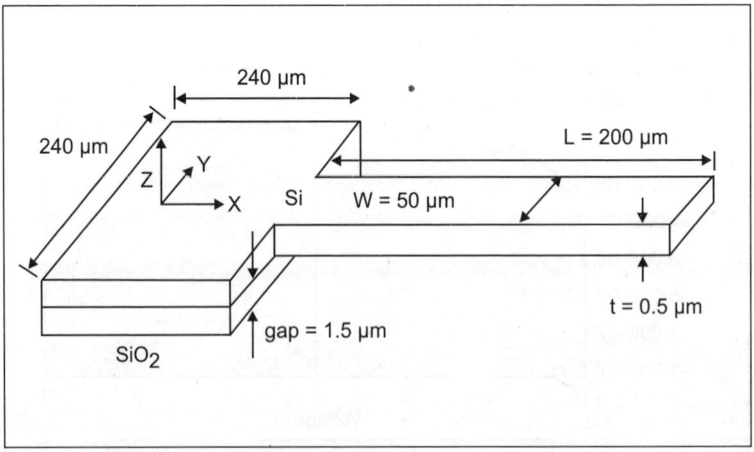

Schematic of cantilever, (For color illustration refer p xii)

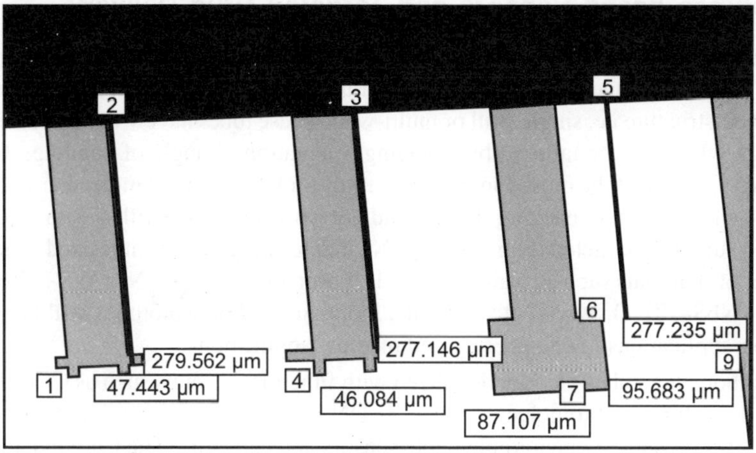

Cantilever fabricated at SCL, (For color illustration refer p xii)

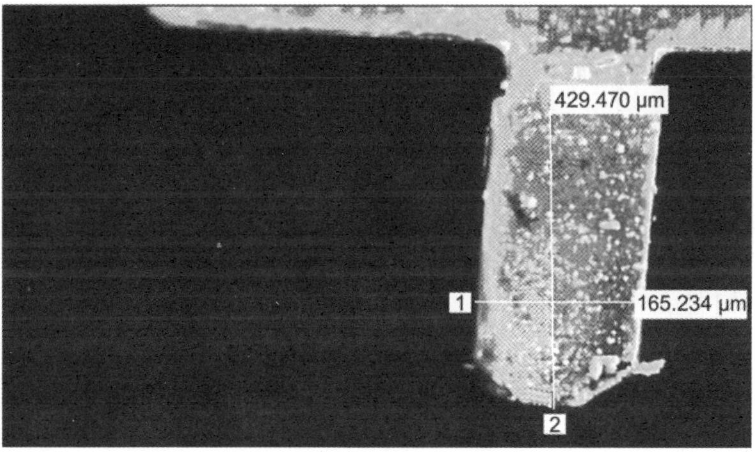

Antibody coated cantilever, (For color illustration refer p xiii)

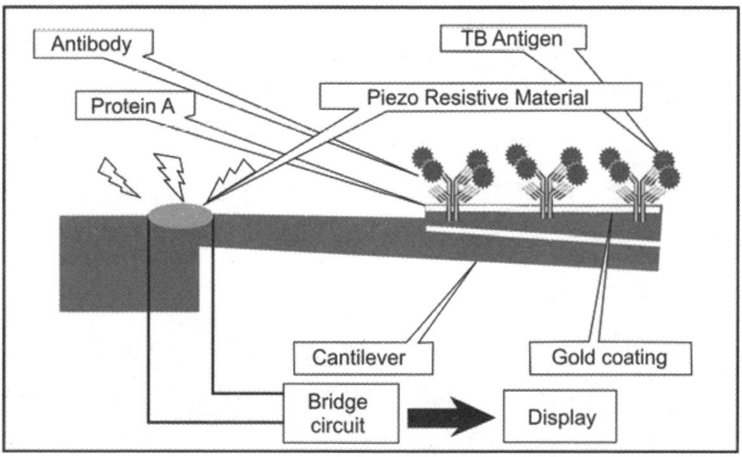

Working principle of cantilever, (For color illustration refer p xiii)

Nanotubes based Sensors and Targeted Drug Delivery

Carbon Nanotubes (CNT) are tiny molecular tubes made up of hexagonal carbon rings joined together. Dimensions of these tubes range from 1 nm to tens of nm depending upon their structure i.e. single wall or multi-wall. These tubes have wonderful electrical behavior which can be tailored by changing orientation or angle of bending. Carbon nanotubes are one of the most dominant materials for fabrication of molecular electronic devices with ultra high packing density and performance. CNTs filled with medicines are well suited for targeted drug delivery. Nanotubes have been synthesized from large number of materials such as vanadium oxide, boron nitride, $B_xC_yN_x$, WS_2, Rossette, $NiCl_2$, NbS_2, Ti_2O, MoS_2, etc. Organic, inorganic and bionanotubes will be studied for their applications in engineering and medical applications.

Carbon Nanotubes are functionalized with functional groups such as COOH, NH_2 and OH for binding biomolecules such as DNA, antibody-antigen, bioreceptors and proteins for developing nanoimplantable sensors and actuators. The Carbon nanotubes can be filled with drug and encapsulated using disease receptors based biomembranes for drug delivery applications.

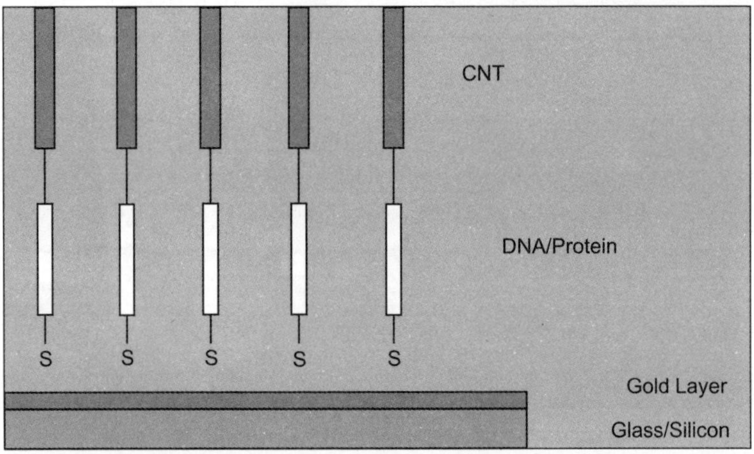

Alignment of CNTs, (For color illustration refer p xiii)

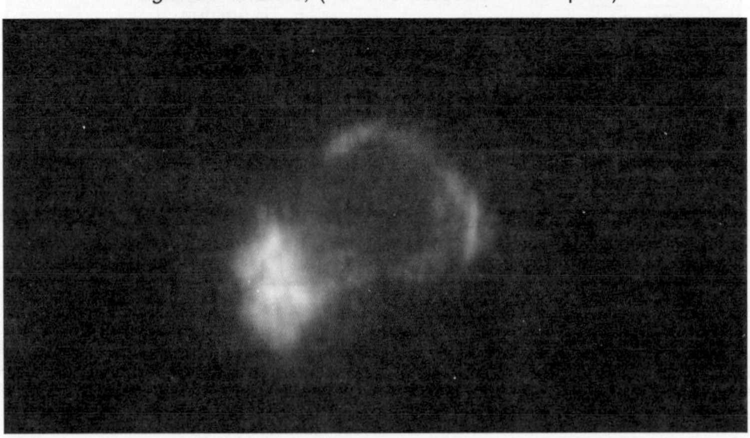

DNA-CNT binding (fluorescent image), (For color illustration refer p xiv)

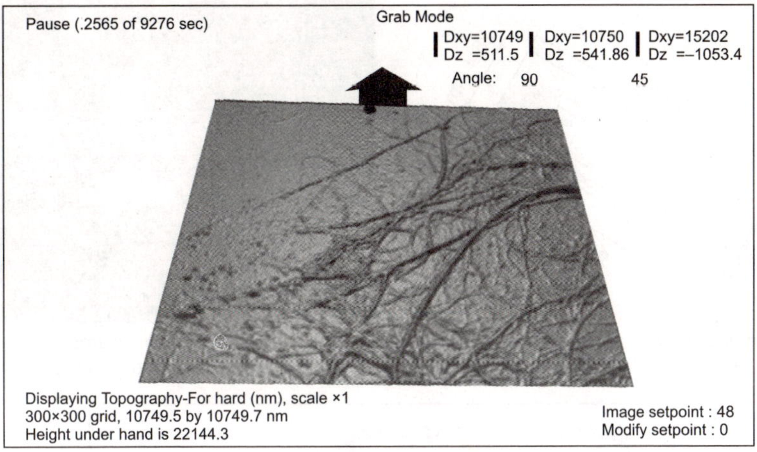

Dispersed CNTs (AFM image), (For color illustration refer p xiv)

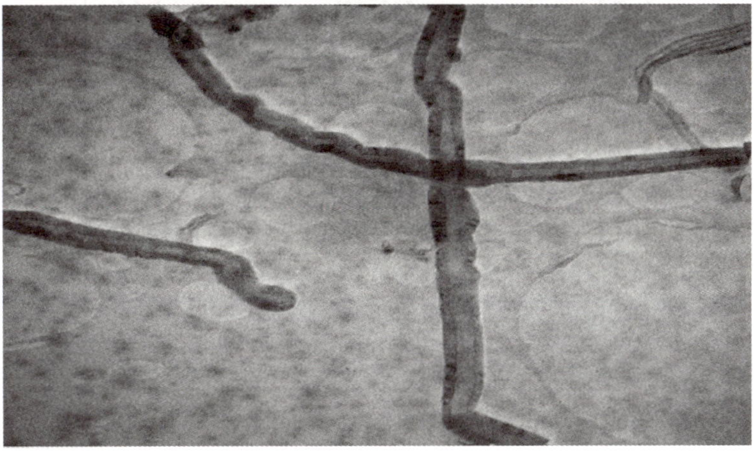

Dispersed CNTs (TEM image), (For color illustration refer p xiv)

Interfacing biomolecules with carbon nanotube will be of immense importance in development of highly sensitive and selective miniaturized biocompatible sensors and actuators with dimensions in micro and nano range. These will revolutionize in early predictive diagnosis.

BioMolecular Motors: In living system various kinds of movements of vital organs and muscles take place through different systems of molecular motors, such as myosin-actin, microtubules-kinesin, microtubules-dynein, microtubules-bacterial flagella motor etc. It is possible to extract motors from their native environment, prepare assemblies of them on a plane surface. For these motors to be useful in nanofabrication, it is essential that external control be applied to the following properties: (a) translocation speed; (b) translocation direction; (c) activation and arrest. These characteristics can be controlled by altering the chemical and/or the physical environment of the motors.

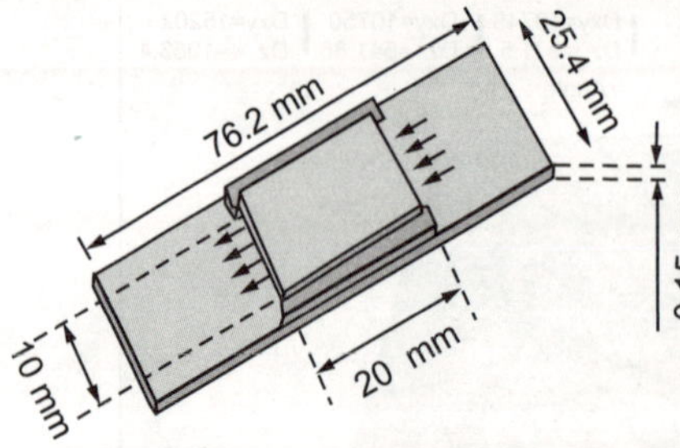

Schematic of flow cell for molecular motors,
(For color illustration refer p xv)

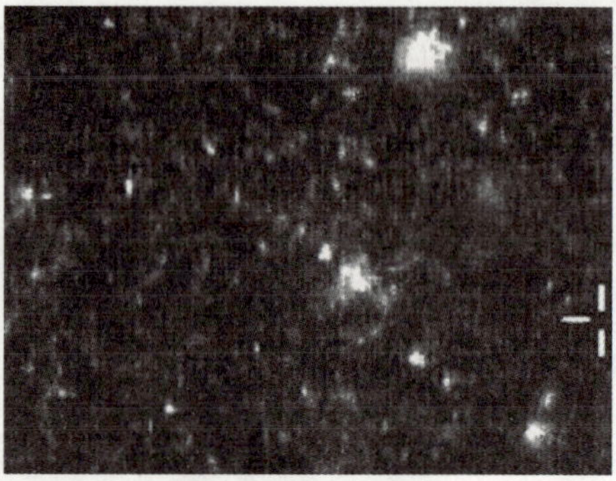

Motility of actin filament,
(For color illustration refer p xv)

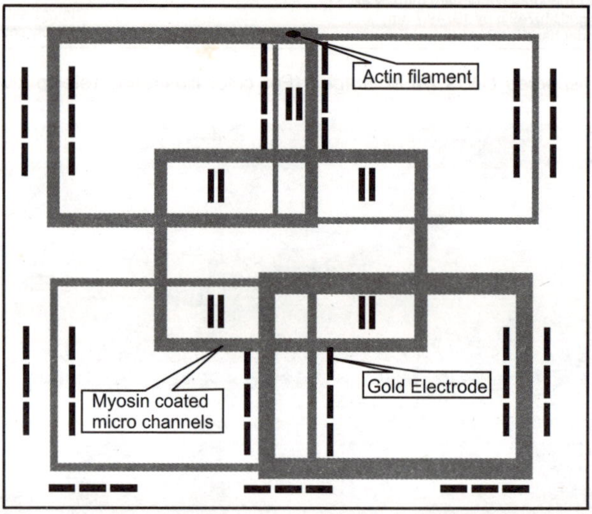

In-vitro control of molecular motors, (For color illustration refer p xv)

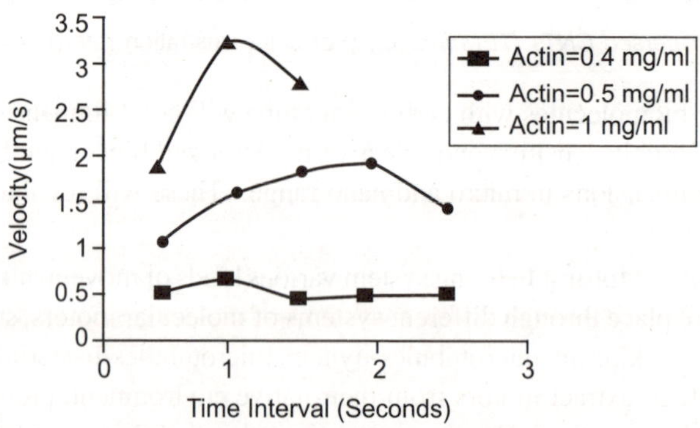

Motility of action filament Pranam 65(5) 967, (2005), (For color illustration refer p xv)

In-vitro studies have been made on myosin actin systems which are responsible for our muscular movements to control the movement of actin against myosin in predictable and reproducible manner using myosin coated channels on silicon/glass wafer. Study of this system under various electromagnetic field, electrolyte, pH, etc. is of importance in development of:

- Targeted drug delivery
- Designing molecular switches with ultra-high density, using low cost fabrication technique of molecular biology
- Understanding muscular movement and their diagnosis and cure
- Nanorobotics

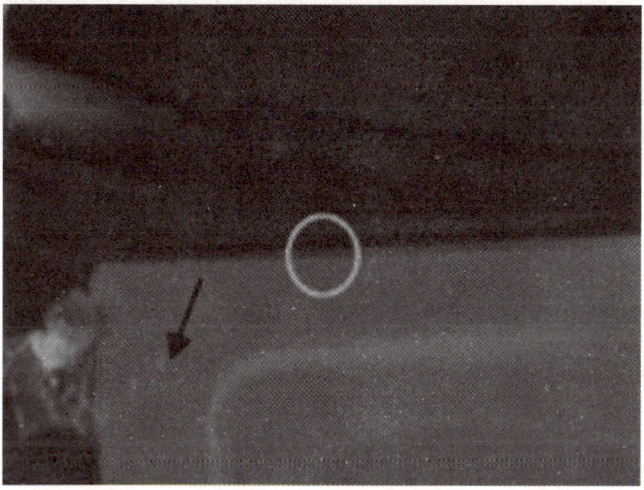

Controlled movement of actin filament in Microchannel, (For color illustration refer p xvi)

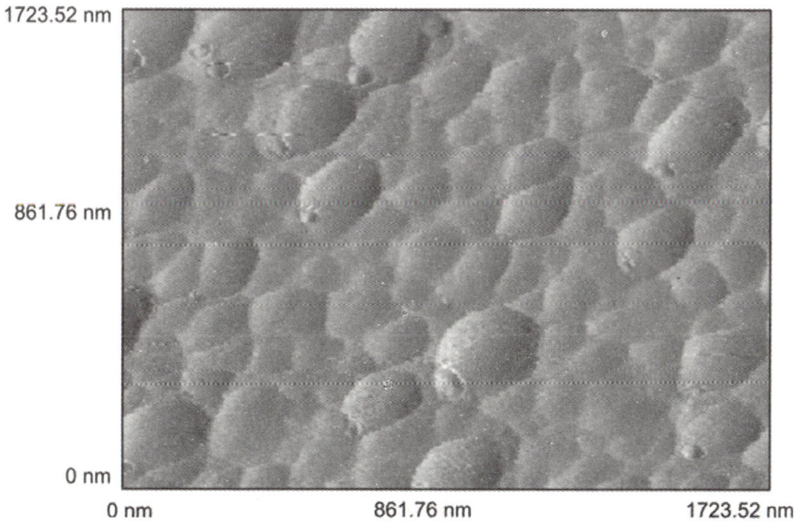

AFM image of myosin on Glass Surface, (For color illustration refer p xvi)

Myosin was immobilized on different solid substrates like gold-coated glass slide, mica, silicon which involve both covalent binding using cross linkers like EDC, APTES, silanes and non-covalent binding employing Protein A, surface adsorption. Atomic force microscope was used to investigate topological characteristics of different surfaces.

The velocity of actin filaments sliding over myosin heads is calculated using tracking program developed in MATLAB. Velocity of actin filament on myosin tracks immobilized on gold coated glass surface is found to be 1.8 μm/sec on mica surface is 2.99μm/sec and on glass slide cleaned with 0.1M KOH in ethanol is 3.65 μm/sec. In-vitro studies show that average velocity of the actin molecules was ranging from 1.0-1.5 μm/sec at pH 8.0. Microchannels are fabricated on gold coated glass surface by ablating the surface using laser micro dissection system to control the position of the actin filaments

Nanorobotic application of the system has been demonstrated by transportation of carboxylated polystyrene microbeads attached to actin filament.

Experiments are in progress to control directionality as well as the movement of actin filaments, by varying environmental conditions in terms of electrical and magnetic field, pH, temperature etc.

Future Program at CSIO

The potential of hybrid devices based on biomolecules and micro/nano-structures has been demonstrated and initial studies are quite encouraging. Further focused research is required for the development of bionanosensors for health care and agriculture applications. This includes development of early diagnostic techniques, targeted drug delivery, bionanosensors for pathogens, nanorobotics, etc. These devices will find applications besides healthcare and agriculture in defense, space, environment etc. also.

Acknowledgement

The article highlights the work carried out by the Scientists of Biomolecular Electronics and Nanotechnology Division of CSIO. Thanks are due to all government agencies such as DRDO, DST, DBT and DIT for providing financial support and to the Director, CSIO for encouragement and facilities.

Further Reading

- Bhalla, V, Bajpai, R.P. and Bharadwaj, L.M. (2004) DNA electronics. *EMBO Rep.* **4**, 1-4.
- Kumar, R., Shukla, A.K., Bagga, E., Kumari, S., Bajpai, R.P. and Bharadwaj, L.M. (2005) 1-Ethyl-3-(3-Dimethylaminopropyl) Carbodiimide interference with Lowry method. *Anal. Biochem.* **336**, 132-134.
- Vashist, S.K., Raiteri, R., Tewari, R., Bajpai, R.P. and Bharadwaj, L.M. (2006) Quantification of human immunoglobulin G immobilised on gold-coated silicon chip for biosensing applications. *J. Phys. Conf. Series* **34**, 806-811.
- Ajore, R., Kaur, I., Sobti, R.S. and Bharadwaj, L.M. (2007) DNA immobilization chemical interference due to aggregates study by dip and drop approach. *J. Biochem. Biophys. Methods* (*accepted for publication*)

- Kumar, S., Jindal, V.K. and Bharadwaj, L.M. (2007) Immobilization of single walled carbon nanotubes on glass surface. *Mater. Lett.* (*accepted for publication*)
- Kumar, S., Kumar, R., Shukla, A. and Bharadwaj, L.M. (2007) Microelectrodes fabrication using laser scissors. *Mater. Lett.* (*in press*)
- Madan, R., Kumar, S., Bagga, E., Bajpai, R.P. and Bharadwaj, L.M. (2004) Smart sensor chip based on BioMEMS. *BioMEMS Nanotechnol.* **5275**, 197-203.
- Bharadwaj, L.M. and Bhalla, V. (2007) Carbon nanotubes: Role in healthcare. In: *Nanoparticles for Pharmaceutical Applications*, pp. 177-190, American Scientific Publisher, USA.
- Singh, A., Bharadwaj, L.M. and Singh H. (2005) DNA and quantum based algorithms for VLSI circuits testing. *Natural Comput.* **4**, 53-72.
- Singh, A., Gupta, U., Singh, B. and Bharadwaj, L.M. (2004) A DNA based evolutionary approach for VLSI circuit partitioning. *Int. J. Comput. Intelligence* **2**, 105-110.
- Kumar, S., Kumar, R., Singh, R., Kumar, R., Shukla, A.K., Jindal, V.K. and Bharadwaj, L.M. (2006) Carbon nanotubes dispersed by optical tweezers on silicon surface. *Azo Nano* **2**, 1-10.
- Bagga, E., Kumari, S., Kumar, R., Kumar, R., Bajpai, R.P. and Bharadwaj, L.M. (2005) Covalent immobilization of myosin for *in-vitro* motility of actin. *Pranam* **65**, 967-972.
- Singh, A. and Bharadwaj, L.M. (2007) Quantum and DNA computing application to automated test pattern generation for VLSI circuits. Science Publication: An International Journal, New York, USA. (*accepted for publication*)
- Bharadwaj, L.M. (2005) Biomolecular electronics and nanotechnology. In: International Conference on Advances in Network Sciences, June 29-July 1, 2005, National University of Singapore, Singapore.
- Bharadwaj, L.M., Bhondekar, A.P., Shukla, A.K., Bhalla, V. and Bajpai, R.P. (2002) DNA based high density memory devices and biomolecular electronics at CSIO. *Proc SPIE* **4937**, 319-325.
- Bharadwaj, L.M., Kaur, I., Kumar, R. and Bajpai, R.P. (2002) Design simulation of DNA based electronic components. *Proc SPIE* **4937**, 226-230.
- Dhawan, I., Kumar, R., Bharadwaj, L.M. and Bajpai , R.P. (2003) DNA based biomolecular electronics-theoretical aspects of charge transmission through DNA. *Proc SPIE* **5118**,
- Shukla, A.K., Bhondekar, A.P., Kumar, R., Bajpai, R.P. and Bharadwaj, L.M. (2003) A novel digital steganographic technique. STEG '03': Pacific Rom Workshop on Digital Steganography held on July 3-4, 2003 at KIT, Kitakyushu, Japan.

3

PHARMACOSCINTIGRAPHIC EVALUATION OF NANOPARTICULATE DRUG DELIVERY SYSTEMS

Rakesh Kumar Sharma, Sheetal Soni, Anil Kumar Babbar and Amarnath Maitra

ABSTRACT

Drug delivery to the brain is often restricted by the blood-brain barrier (BBB), which regulates the passage of substances including drugs, between peripheral circulation and CNS. One approach to overcome this barrier has been using appropriately modified polymeric nanoparticles. Pharmacoscintigraphy is the application of nuclear medicine techniques for tracing the radiolabeled ingredient of the active component of drug/device or the formulation excipient. We have used surface modified chitosan nanoparticles as brain delivery carriers and evaluated their efficacy by pharmacoscintigraphy. The chitosan nanoparticles were radiolabeled with Technetium-99m, and the in vivo brain uptake profiles were studied in Swiss Albino Strain 'A' mice. The biodistribution of the preparation in the brains of New Zealand rabbits was also studied with gamma scintigraphy. Using Polysorbate-80 (1% w/w) coated nanoparticles, we have demonstrated a nearly five-fold increase in the brain accumulation compared with uncoated nanoparticles. Our results indicate that surface modification of chitosan nanoparticles by coating with polysorbate-80 is effective in drug delivery through BBB.

1. ARTICLE

Most of the drugs introduced to clinical medicine exert their effects by interactive interference with the cell or cell membrane and function through concentration dependent reversible interactions at specific receptor site(s). Obviously, to obtain a desirable therapeutic response, the correct amount of drug should be transported and delivered to the site of action with subsequent control of drug input rate. Its distribution to other tissues therefore seems unnecessary and wasteful and it is a potential cause of toxicity. The developments over past decades indicate explicit progress in the area of controlled and targeted drug delivery. The drug delivery technology has certainly infused new interests in seemingly traditional old drugs by providing them new life specifically through their therapeutic targets. It is appreciated that, target oriented drug administration with improvements in therapeutic efficacy, reduction in side effects and optimized dosing regimen, shall be the leading trends in the area of therapeutics. Targeting is signified if the target component is distinguished from other compartments, where toxicity may occur, also if the active drug could be placed predominantly in the proximity of target site. The restricted distribution of the parent drug to the non-target sites with effective accessibility to the target site could maximize the benefits of targeted drug delivery Fig. 1.

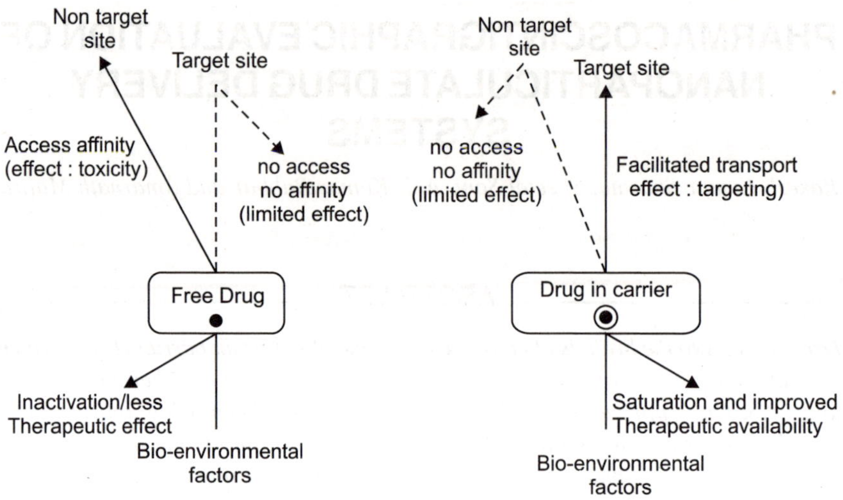

Fig. 1 *Principle and Rationale of Drug Targeting*

An ideal drug carrier engineered as a targetable device should have the following features:

- It must be able to cross anatomical barriers and in case of tumor chemotherapy tumor vasculature.
- It must be recognized specifically and selectively by the target cells and must maintain the avidity and specificity of the surface ligands.
- The linkage of the drug and the directing unit (ligand) should be stable in plasma, interstitial and other biofluids.
- Carrier should be non toxic, non-immunogenic and biodegradable particulate or macromolecule and after recognition, and internalization, the carrier system should release the drug moiety inside the target organs, tissues or cells.
- The biomolecules used for carrier navigation and site recognition should not be ubiquitous otherwise it may crossover the sites, defeating the concept of targeting.

Several carriers appended as pilot molecules to selectively deliver the drug to the intended cell lines have been reported (Table 1). Based on the nature of origin they are categorized as endogenous (low density lipoprotein, high density lipoprotein, serum albumin, erythrocytes) and exogenous (microparticulates, soluble polymeric and biodegradable polymeric drug carriers). Microparticulates have been investigated intensively during the last decade to explore and exploit intrinsic targeting potential of the plain or surface manipulated carriers. A variety of natural and synthetic biodegradable polymers have been used to prepare microparticles. For example human serum albumin (HSA) has been used to deliver more than 40 different drugs. Glutraldehyde crosslinked gelatin microspheres were prepared to deliver interferon to macrophages. Poly(lactide-co-glycolide) microspheres loaded with antibiotics including ampicilin, gentamycin, polymyxin B and chloramphenicol were investigated for sustained release. Poly(ε-caprolactone-co-lactic acid) has been used to prepare microspheres loaded with a narcotic antagonist, L-methadone, using solvent evaporation method. Polyanhydride, PCPP-SA(50:50) has been used to prepare microspheres loaded with insulin. The microspheres were prepared by hot-melt microencapsulation technique.

Table 1 *Carrier Systems Used for Targeting Drug Delivery*

1. **Colloidal carriers**
   (a) Vesicular systems
   Liposomes; Niosomes; Pharmacosomes; Virosomes; Immunoliposomes
   (b) Microparticulate systems
   Microparticles; Nanoparticles; Magnetic microspheres; Albumin microspheres; nanocapsules
2. **Cellular carriers**
   Resealed erythrocytes; serum albumin; antibodies; platelets; leukocytes
3. **Supramolecular delivery systems**
   Micelles; reverse micelles; mixed micelles; polymeric micelles; Liquid crystals; Lipoproteins (chylomicron; VLDL; LDL)
   Synthetic LDL mimicking particles (Supramolecule biovector systems)
4. **Polymer based systems**
   Signal sensitive; mucoadhesive; biodegradable; bioerodible; soluble synthetic polymeric carriers
5. **Macromolecular carrier**
   (a) Proteins, glycoproteins, neo glycoproteins and artificial viral envelopes (AVE)
   (b) Glycosylated water soluble polymers (poly-L-Lysine)
   (c) Mabs, Immunological Fabs fragments; antibody-enzyme complex and biospecific Abs
   (d) Toxins; Immunotoxins and rCD4 toxin conjugates
   (e) Lectins, polysaccharides

Colloidal drug carriers such as liposomes can also be used to improve the therapeutic index of both established and new drugs by modifying their distribution, and thus increasing their efficacy and/or reducing their toxicity. Liposomes are self-assembled vesicles composed of a lipid bilayer, which forms a closed shell surrounding an internal aqueous phase. Because of their structural characteristics, liposomes can entrap both hydrophilic and hydrophobic molecules. Liposomes have been used as carriers for many kinds of molecules. For the main purposes of reducing side effects and targeting, many anticancer drugs, such as doxorubicin, paclitaxel, annamycin, cisplatin derivatives, vincristine, daunorubicin, campothecin derivatives, 5-fluorouracil derivatives and so on have been encapsulated in the liposomes. Liposomes encapsulating anti-bacterial agents, such as amikacin and clofazimine and fungal agents and anti-viral agents are used for the treatment of infectious diseases. Although liposomes have unique advantages, such as the ability to protect the drugs from degradation, target the drugs to site of action and reduce the toxicity and side effects but the development work on liposomes have been limited due to inherent problems like low loading efficiency, rapid drug leakage, poor stability and production difficulty of the carrier.

Advent of nanotechnology has infused new dimensions into the target oriented delivery through self-assembling supramolecular based nanostructures. The colloidal carriers based on biodegradable and biocompatible polymeric systems have largely influenced the controlled and targeted drug delivery concepts. It was realized that the nanoparticles loaded bioactives could not only deliver drugs to specific organs within the body but delivery rate in addition could be controlled as being bystanders, burst, controlled, pulsatile or modulated. The possibilities and potentials further prompted the work and as a result a great deal of related information covering preparation methodologies, characterization, engineering, bio-fate and toxicology has been gathered. The understanding that relates to the biodistribution in particular has propelled and motivated the development of functionally designed nanoparticulates. The present and near-future applications for nanoparticles administered to human body can be grouped under the following categories:

- Drug Delivery
- Gene Delivery and vaccines
- Imaging Agents
- Artificial Blood (respirocytes)
- Detoxifying Agents (nanobullets)

It is apparent that the polymers, the building blocks of nanoparticulate composites, belong to natural or synthetic origins. Some of them have already been exploited for their biomedical applications. Obviously, the literature is abound with data concerning their safety, toxicology and biodegradation consideration.

Nanoparticles are sub-nanosized colloidal structures composed of synthetic or semi synthetic polymers. The continual quest and maneuvering towards physical stability improvization of liposomes resulted into development of solid core nanoparticles in eighties as an alternative drug carrier. The first reported nanoparticles were based on non-biodegradable polymeric systems (polyacrylamide, polymethylmethacrylate, polystyrene etc.). The Development of new drug delivery systems is a lucrative option as an alternative to drug development and discovery, as it requires a shorter time frame, less financial involvement and is associated with lesser degree of uncertainty and risk. Possibilities of chronic toxicity due to tissue and immunological response towards non-degradable polymeric burden, their use of systemic administration however could not be considered. Soon the biodegradable polymers were taken up and nanoparticles based on poly(cyanoacrylate) were extensively studied. The polymeric nanoparticles can carry drugs or proteinaceous substances, i.e. antigens. These bioactives are entrapped in the polymer matrix as particulates enmesh or solid solution or may be bound to the particle surface by physical adsorption or chemical conjugation. The drugs may be added during preparation of nanoparticles or to the previously prepared nanoparticles. The term particulate is suggestively general and doesn't account for the morphological and structural organization of the system. Thus they could be nanospheres, nanocapsules, nanocrystals or nanoparticles. The physical chemistry of these systems remains to be the same as of typical colloidal dispersions. The surface charges, dispersibility, density, hydrophobicity and hydrophilicity are critical factors which determine the absolute stability characteristics of a system vis-a-vis its in vivo disposition.

Nanoparticles mediated drug delivery therefore, makes it possible for drugs to permeate through cell walls, which is of critical importance to the expected growth of genetic medicine over the next few years. Living organisms are built of cells that are typically 10μm across. However, the cell parts are much smaller and are in the submicron size domain. Even smaller are the proteins with a typical size of just 5nm, which is comparable with the dimensions of smallest manmade nanoparticles. This simple size comparison gives an idea of using nanoparticles as very small probes that would allow us to spy at the cellular machinery without introducing too much interference. Various inorganic and organic nanoparticles exhibit all these attractive properties besides being biocompatible, biodegradable and non-antigenic and therefore, are thought to have potential as novel intravascular probes for diagnostics (e.g. imaging) and therapeutics (e.g. drug delivery).

Nowadays, ceramic nanostructures that interface with biological systems have recently attracted widespread interest in biology and medicine. These nanostructured ceramic materials both void and organically doped promise unique properties, with potential applications in a wide range of technologies including electronics, catalysis, sensors, biosensors, electrochemistry, immunochemistry, drug delivery, enzyme therapy, gene therapy and materials for use in environmental sciences. The term ceramic encompasses a wide variety of compounds including silicates, oxides, carbides, nitrides, phosphates, sulfides and borides. These ceramics have excellent mechanical properties under heavy stress, outstanding electrical and optical properties and exceptional resistance to extreme environmental conditions (temperature, pH, corrosive materials, microbes etc.). Some of them are non-toxic and biodegradable. Besides relatively mild conditions can be used for the preparation of these particles, which is advantageous in terms of entrapment of biomolecules like plasmid DNA and enzymes. One of the rapidly emerging areas of nanobiotechnology, where ceramic nanoparticles entrapped biomolecules can play a significant role, is gene delivery. The relevance of gene therapy for therapeutic purpose has come to be recognized as an important and exciting development in the arena of medical treatment. By insertion of plasmid DNA bearing therapeutic gene, it may be possible to rectify genetic disorders. Therapeutic agents in the form of proteins, peptides or even antigenic material may be produced in situ to treat diseased conditions or stimulate the immune system.

Intravenous administration of nanoparticles produces a distribution throughout the body different from that of conventional drugs. Nanoparticles are rapidly conditioned (or coated) by the element of circulation such as plasma proteins and glycoproteins by the process of opsonisation, which are then removed from blood circulation by phagocytosis. Fate of nanoparticles is largely dependent on their size and surface properties, charge, hydrophobicity and chemical composition. The process of particle recognition and capture can be modulated by particle size and surface properties of nanoparticles. Nanoparticles with size < 100 nm in diameter and having a hydrophilic surface are able to have higher residency time in blood and hence possess a greater ability to get targeted to the desired site. Rapid escape of nanoparticles from the degradative endo-lysosomal compartment and their sustained intracellular retention suggest that nanoparticles containing encapsulated therapeutic agent could serve as an efficient sustained release drug delivery system. Negatively charged and hydrophobic carriers undergo extensive opsonisation and distribute into the reticulo-endothelial system (RES) organs. Positively charged nanoparticles act as long-circulating carriers in blood and can also be useful for cancer therapy. They also distribute to brain at relatively higher concentrations and could be useful to deliver drugs to brain for therapeutic purposes. Particles less than 100 nm and having a hydrophobic surface are efficiently removed from the circulation but hydrophilic nanoparticles below can resist this coating process and consequently cleared more slowly. 'Tailored' nanoparticles can be exploited to control or avoid, in a predictable manner, their tropism for RES, particularly the tissues of the mononuclear phagocyte system (MPS), mainly macrophages of liver and spleen. Nanoparticles can be used to extend the time window of bioavailability along with various sustained and controlled release applications and to protect the drug from chemical and enzymatic decomposition. This can result in reduction of peripheral side effects of drugs by decreasing the overall dose of drugs in the body.

Drug delivery to the brain is often restricted by the blood-brain barrier (BBB), which regulates the passage of substances including drugs, between peripheral circulation and CNS. The BBB is an anatomic barrier formed by the tight endothelial cell junctions of the capillaries within the brain. One approach to overcoming this barrier has been using appropriately modified polymeric nanoparticles. However, a major obstacle to drug targeting with these particulate systems is non-specific uptake of injected carriers by the RES. Coating with amphiphilic surfactants significantly alter the pharmacokinetics and body distribution of nanoparticles. Since polysorbate-80 is known to facilitate uptake of nanoparticles by brain endothelial cells, nanoparticles coated with polysorbate-80 hold great promise for transport of agents across the BBB.

There is a definite need to gain further in vivo information about drug pharmacology in a man. This is especially important for certain agents with low therapeutic indices, such as anticancer drugs, where consequences of therapeutic failure and toxicity are significant. Besides, newer strategies such as antibody directed enzyme prodrug therapy and gene directed enzyme prodrug therapy are designed to specific tissues, where plasma monitoring may be irrelevant. The availability of non-invasive techniques such as gamma scintigraphy, positron emission tomography (PET), magnetic resonance imaging (MRI) and spectroscopy (MRS) for monitoring drug pharmacology has been a significant advance in furthering our knowledge. The assessment of drug pharmacology by nuclear medicine methods begins with the design and synthesis of labeled drug, followed by imaging, deteremination of parameters of interest and finally data interpretation. Medical diagnostic modalities currently in use include the application of gamma emitting radioactive materials, such as Technetium-99m (^{99m}Tc), Indium-111 (^{111}I), Iodine-125 (^{125}I), Iodine-131 (^{131}I) and Gallium-67 (^{67}Ga).

Radiopharmaceuticals have been defined as radioactive drugs that, when used for the purpose of diagnosis or therapy, typically elicit no physiological response from the patient. The design of these compounds is based solely upon physiological function of the target organ. Unlike radiographic procedures, which depend almost entirely on tissue density differences, external imaging of radiopharmaceutical is essentially independent of the density of the target organ. The mechanism of localization of a radiopharmaceutical in a particular target organ can be as simple as the physical trapping of particles or as sophisticated as an antigen-antibody reaction or chemisorption of an inorganic phosphate on the hydroxyapatite crystals deposited in an acute myocardial infection. Nearly 80% of all radiopharmaceuticals used in nuclear medicine are

^{99m}Tc-labeled compounds. In terms of physical properties, ^{99m}Tc is the radionuclide of choice for diagnosis in nuclear medicine.

Following properties of ^{99m}Tc makes it a suitable radionuclide for diagnosis:

- Short but reasonable half life of six hours.
- High photon yield (88%) of 140 keV gives good tissue penetration and is easily collimated and stopped by thc detector.
- No β-radiation results in a low radiation dose.
- It is readily available in a sterile, pyrogen free and carrier free state from generator for in hospital use.
- Chemically reactive yielding other chemical forms.

Pharmacoscintigraphy is the application of nuclear medicine techniques for tracing the radiolabeled ingredient of the active component of drug/device or the formulation excipient. It can provide vital information regarding the extent, rate, site, and mode of drug release in animals. In the context of evaluation of NDDS, pharmacoscintigraphy technique appear very promising as it is non-invasive, permits repeated measurements and allows use of same organism as its own pre-treatment control. It is possible to extend pharmacoscintigraphy studies in human patients also, thereby making it a pre-clinical screening modality of choice.

We have used surface modified chitosan nanoparticles as brain delivery carriers and evaluated their efficacy by pharmacoscintigraphy. The Chitosan nanoparticles were radiolabeled with Technetium-99m, and the in vivo brain uptake profiles were studied in Swiss Albino Strain 'A' mice. The biodistribution of the preparation in the brains of New Zealand rabbits was also studied with gamma scintigraphy.

Using Polysorbate-80 (1% w/w) coated nanoparticles we have demonstrated a nearly five-fold increase in the brain accumulation compared with uncoated nanoparticles. When these particles are coated with Polysorbate-80, the hydrophobic tail of the surfactant is protruded towards the outer surface of the particles. As a result, the nanoparticles acquire partial surface hydrophobicity and, thus, are expected to become susceptible to more rapid blood clearance and enhanced RES uptake. In addition, because of partial surface hydrophobicity, these particles may also have a higher uptake by brain endothelial cells, which enables them to cross the BBB. The study provides evidence that the BBB is not inert but rather a highly dynamic system, and that the brain endothelial cells are capable of endocytic uptake of nanoparticles. These results indicate that surface modification of chitosan nanoparticles by coating with polysorbate-80 is effective in drug delivery through BBB.

SUGGESTED READING

1. Banerjee, T., Mitra, S., Singh, A. K., Sharma, R. K. and Maitra, A. (2002) Preparation, characterization and biodistribution of ultrafine chitosan nanoparticles. *Intl. J. Pharm.* **243**, 93-105.
2. Banerjee, T., Singh, A. K., Mitra, S., Sharma, R. K. and Maitra, A. (2002). Comparative evaluation of Stannous chloride and Sodium Borohydride as reducing agents for preparation of Technetium-99m labeled Chitosan nanoparticles. *Indian J. Nucl. Med.* **17**, 11.

3. Sharma, R. K. (2003). Nuclear Medicine Imaging: A development tool in pharmaceutical research. **(Symposium 2: Drug Discovery), 55th Indian Pharmaceutical Congress**, Chennai, 19th December 2003.
4. Reddy, L. H., Sharma, R. K. and Murthy, R. S. R. (2004) Tumor retention and biodistribution studies of etoposide loaded tripalmitin nanoparticles in Dalton's Lymphoma bearing mice. *Alasbimn J.* **6**, July 2004/www2.alasbimnjornal.cl/alasbimn/CDA/sec_b/0,1206,SCID%253D 11063,00.html).
5. Soni, S., Babbar, A. K., Sharma, R. K., Banerjee, T. and Maitra, A. (2004). Pharmacoscintigraphic evaluation of Polysorbate80 coated chitosan nanoparticles for brain targeting. *Indian J. Nucl. Med.* **19**, 144.
6. Reddy, L. H., Sharma, R. K., Mishra, A. K., Chuttani K. and Murthy, R. S. R. (2004) Etoposide incorporated tripalmatin nanoparticles with different surface charge: Formulation, characterization and biodistribution studies. AAPS J. 6, e23 (http://www.aapsj.org).
7. Reddy, L. H., Sharma, R. K. and Murthy, R. S. R. (2004) Enhanced tumour uptake of doxorubicin loaded poly(butyl cyanoacrylate) nanoparticles (DPBC) in mice bearing Dalton's lymphoma tumor. *J. Drug Target* **12**, 443-451.
8. Sharma, R. K. (2005) Nano-particulate carriers for drug delivery. **Symposia Lecture-92nd Indian Science Congress**, Ahmedabad 3-7th January 2005.
9. Sharma, R. K. and Mishra. A. K. (2005) *Magnetopharmaceuticals for the enhancement of the MRI contrast: Current status and Perspectives*. In: Recent advances in MR Imaging and Spectroscopy. Ed. Jaganathan, N. R., Jaypee Brothers Medical Publishers (P) Ltd., Delhi, pp 203-237.
10. Sharma, R. K. (2005) In: Health Technology as Fulcrum of Development for the Nation: The Road Ahead. NISCAIR, New Delhi, pp 214-218.
11. Banerjee, T., Singh, A. K., Sharma, R. K. and Maitra, A. N. (2005) Labeling efficiency and biodistribution of Technetium-99m labeled nanoparticles: Interference by colloidal tin oxide particles. *Intl. J. Pharm.* **289**, 189-195.
12. Sharma, R. K. (2005) Application of nuclear medicine in pharmaceutical research. **Invited Talk. IANCAS workshop on Radiochemistry and application of Radioisotopes**. Panjab University, Chandigarh, May 2-7, 2005.
13. Sharma, R. K. (2005) Nanotechnology for Drug Delivery, CEP on Nanotechnology, SSPL, Delhi, 11th August 2005.
14. Soni, S., Babbar, A. K., Sharma, R. K., Banerjee, T. and Maitra, A. (2005) Pharmacoscintigraphic evaluation of Polysorbate80 coated chitosan nanoparticles for brain targeting. #653, **Proceed. 32nd Annual Meeting Controlled Release Society, USA.**
15. Soni, S., Babbar, A. K., Sharma, R. K., Banerjee, T. and Maitra, A. (2005). Pharmacoscintigraphic evaluation of Polysorbate 80 coated chitosan nanoparticles for brain targeting. *Amer. J. Drug Delivery* **3**, 205-212.
16. Reddy, L. H., Sharma, R. K., Mishra, A. K., Chuttani, K. and Murthy, R. S. R. (2005) Influence of administration route on the uptake and biodistribution of etoposide loaded tripalmitin nanoparticles in Dalton's Lymphoma tumor bearing mice. *J. Controlled Release* **105**, 185-198.
17. Soni, S., Babbar, A. K., Maitra, A. and Sharma, R. K. (2005) Polysorbate80 assisted delivery of N-hexylcabamoyl-5flurouracil loaded nanogels across the blood–brain barrier (Abstract No. H9-Biotechnology and Biotherapeutic section. **57th IPC Hyderabad**, pp 89-90, 2-4 December 2005).
18. Soni, S., Babbar, A. K., Maitra, A. and Sharma, R. K. (2006). Delivery of Hydrophobised 5-Fluorouracil Derivative to Brain Tissue through Intravenous Route using surface modified nanogel. *J. Drug Target* **14**, 87-95.
19. Jain, S., Sharma, R. K. and Vyas, S. P. (2006) Chitosan nanoparticles encapsulated vesicular systems for oral immunization: preparation, in vitro and in vivo characterization. *J. Pharm. Pharmacol.* 58, 303-310.

4

NANO MATERIALS OF CARBON – FULLERENES AND NANOTUBES

V. K. Jindal

ABSTRACT

The purpose of this paper is to highlight the significance of nonmaterials and nanotechnology as a general awareness. I also aim to pay special attention to fullerenes and carbon nanotubes, whose availability and discovery after 1985 laid the foundation for rapid growth of interest and progress in the field of nanotechnology. After introducing these materials in reasonable detail, I also bring out some methods of production of fullerenes and carbon nanotubes used by students at our University. Further, a theoretical methodology employed by our research team to understand the special properties of fullerenes and carbon nanotubes is briefly given with some references for those who wish to look into details. Still, the primary aim of this article remains to introduce fullerenes and carbon nanotubes.

1. INTRODUCTION

Nanomaterials are expected to create huge impact on our society as these are considered to bring the next technological revolution in solid-state electronics. In addition, nanomaterials have been viewed to emerge as new structural materials, to serve as systems for controlled drug delivery and to have a considerable impact in practically all domains of science. Nanostructure based devices taking advantage of their unique functional properties (chemical, optical, magnetic, mechanical, optoelectronics, etc.) are likely to replace existing bulk material based devices.

The existence of a characteristic length scale less than ~100 nm announces the domain of nanomaterials. This length scale can be for instance a particle diameter, grain size, layer thickness, tube diameter or length. The synthesis of materials exhibiting such type of length scales require special fancy techniques which include membrane-template methods, mechanical attrition, sol-gel process, chemical vapor deposition, flame or arc-discharge methods, etc. Reducing this length scale results in systems whose dimensionally can be considered as zero, one or two depending on the number of nano-lengths involved. Reduction to nano-dimensions brings in new properties of materials which are different from those of bulk materials. This primarily results from an increased surface to volume ratio and domination of quantum effects at these length scales. Technique of manipulating individual atoms or molecules to form nanostructures of desired properties is called nanotechnology, and an understanding of this technology leads to nanoscience and development of new computational techniques. Nature knows this technology best – as we see from biological systems – the same atoms of carbon, nitrogen or oxygen are arranged to fabricate various specialized organs or components of living bodies or plants.

Nanotechnology started getting huge attention after new carbon based materials appeared on the horizon; before that carbon existed in its standard forms for centuries. It has been one of the most abundant elements on earth, known to exist in a number of different solid forms. The brilliant diamond and the gray graphite are its well known allotropes in crystalline forms; while soot, an amorphous form of carbon, is formed by deposition of unburnt carbon from fuels, coal being another amorphous form albeit impure. Some twenty years ago, there appeared on the scene new molecules of pure carbon and subsequently (in 1990) their solids. These new forms generated tremendous interest in scientific community of chemists, physicists and material scientists. These new materials, such as fullerene solids, carbon nanotubes or other complex formations show properties completely different from the earlier known forms, diamond and graphite.

2. CARBON AS GRAPHITE AND DIAMOND

Before we introduce the fullerene solids, it would be useful to revise the structure of graphite and diamond. Everyone knows that diamonds are sparkling gemstones, and that graphite is useful as either an electrode of a typical dry cell, or is just the lead in a pencil. What may be a little less known is that these vastly different materials are actually made up of the same atomic building blocks which are carbon atoms, thereby known as allotropes of carbon. The allotropes differ only in structural arrangement of carbon atoms to form a repetitive "unit cell" which is a building block of the bulk graphite or bulk diamond. In graphite, carbon atoms form planar sheets of hexagons (Fig. 1a) where C-atoms occupy vertices of the hexagons. There is weak interaction between these sheets in the perpendicular direction. As a result, the sheets which are compact, can however, slide relatively easily with respect to each other. In diamond, a famous tetrahedral bonding exists, with each carbon atom having four of its neighbors at the vertices of a tetrahedron (Fig. 1b). These two structures are responsible for some of the drastically different properties of graphite and diamond. For example, based upon this structural analysis one explains why diamond is hard while graphite is soft. Further analysis along these lines also explains why diamond is an insulator and crystal clear, while graphite is a conductor and opaque. Diamond and graphite are most important industrial materials. Because of their hardness, diamonds can be mounted on to drill bits to extend the working life of the tip and allow them to cut into extremely hard materials. Graphite is used extensively in the manufacture of motors, heating equipment and electronic components. It is apparent that diamond and graphite are important materials. Currently the study of many other forms of carbon is gaining momentum, one such

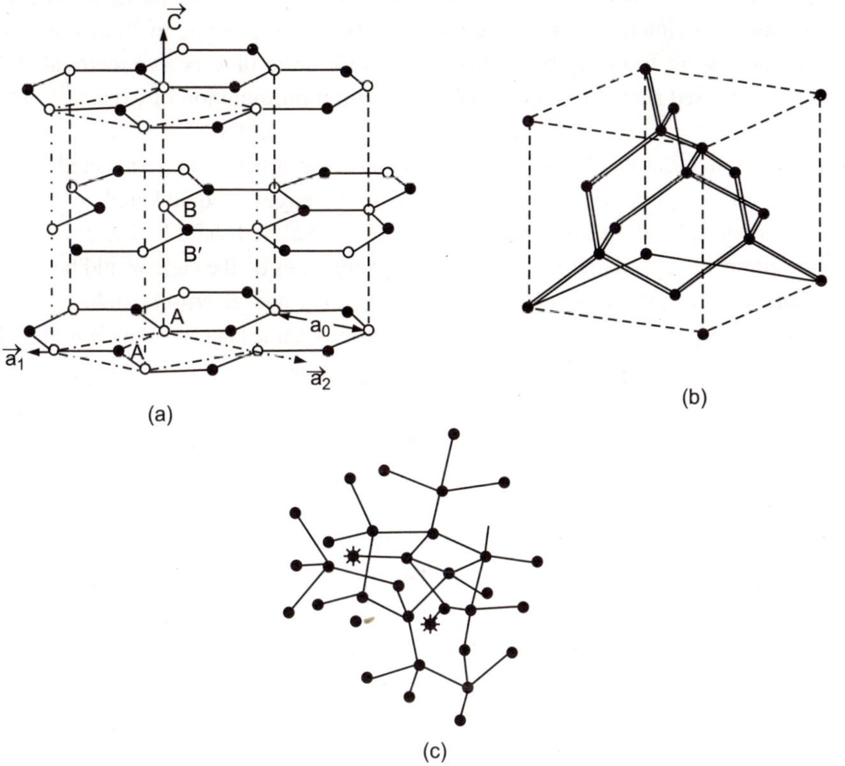

Fig. 1 *Structure of (a) graphite, (b) diamond and (c) amorphous carbon (note the dangling bonds marked*)*

form being amorphous carbon (Fig. 1c). In fact the study of carbon materials is an extremely important area of research.

3. FULLERENES

New forms of the element carbon - called fullerenes - in which the atoms are arranged in closed shells were accidentally discovered in 1985 by Robert F. Curl, Harold W. Kroto and Richard E. Smalley [1-3], which lead to 1996 Nobel Prize award to them in Chemistry. Kroto was particularly interested in carbon-rich giant stars. During his investigation of the spectrum lines of these stars, he had discovered some lines which could be ascribed to a kind of long-chained molecule of only carbon and nitrogen, termed cyanopolyynes. The same sort of molecules is also found in interstellar gas clouds. Kroto's idea was that these carbon compounds had been formed in stellar atmospheres, not in clouds. He collaborated with Richard E. Smalley and Curl, whose research was in cluster chemistry, to vaporize carbon to produce carbon clusters. Interestingly, the clusters could be produced for several carbon atoms, the beam current could be optimized to regulate production of clusters where, predominantly, 60, and also 70 carbon atoms were involved. This was revealed with mass spectrometry, where the clusters now formed showed peaks at mass numbers corresponding to 60 carbon atoms, and a less intense peak also at mass number corresponding to 70 carbon atoms. The clusters were found to be highly stable, indicative of very symmetrical shape, possibly in closed shell structures. Therefore, initial presumption of existence of long chain carbon atom clusters had to be discarded. Instead, the idea arose that the C_{60} (name given to 60-atom cluster) cluster could have the structure of a truncated (cut off) icosahedron. A regular truncated icosahedron has 90 edges of equal length, 60 vertices, 20 hexagonal faces and 12 pentagonal faces to form a closed shell. In this structure, the carbon atoms have no dangling bonds, and the pentagons on the cage would be isolated from one another, thereby creating greater chemical and electronic stability. Direct evidence that C_{60} molecule has icosahedral symmetry came later from nuclear magnetic resonance experiments (NMR), where a single resonance line was observed, consistent with identification of one chemical site from all 60 carbon sites. Similarly, NMR spectra of C_{70} too confirmed a cage like structure, with a few extra lines only from 10 extra atoms. The pattern of a European football, except for curvature, has also a similar structure as that proposed for C_{60} molecule. Further the geodetic dome designed by the American architect R. Buckminster Fuller for the 1967 Montreal World Exhibition had a similar structure, may be because that was conceived by him as an ideal way to form spherical structure using sticks. In his honor, C_{60} was also named buckminsterfullerene, fullerene or simply as bucky ball. In fact, fullerene was also a general name given to various carbon-cluster caged molecules, including C_{60} and C_{70}. These amazing dome shaped buildings can cover large areas of land and the light weight and durable construction makes them ideal structures especially for buildings that have to withstand large changes in temperature and high winds such as are found in the Arctic.

A typical structure of C_{60} molecules has been shown in Fig.2. For chemists, the proposed structure was uniquely beautiful and satisfying. It corresponds to an aromatic, three-dimensional system in which single and double bonds alternated, and was thus of great theoretical significance. Useful and detailed information about the experimental data for fullerene molecules can be found in some of the reviews which have been published from time to time [4-6]. Indeed, these articles also review several other carbon cage clusters in various forms, their solids and several theoretical models, some of which we talk about later in the present text.

The average nearest-neighbor carbon-carbon distance in C_{60} is 1.44 Å, almost identical to that in graphite (1.42 Å nm). Further, as in graphite, each C atom in C_{60} is trigonally bonded to three other C atoms in a sp^2 derived bonding configuration. In C_{60} molecule, the measured values of bond lengths of hexagon-pentagon edge are single bonds of magnitude around 1.46 Å, whereas the hexagon-hexagon carbon bonds are double bonds of magnitude around 1.40 Å. Therefore, there is this slight deviation from C_{60} molecule being a regular truncated icosahedron. Further investigations of C_{60} by attempting to make it react with other compounds, such as hydrogen, nitrous oxide, carbon monoxide, sulphur dioxide, oxygen or ammonia failed to affect the C_{60} peak recorded in the mass spectrometer. This showed that C_{60} was a slow-reacting compound.

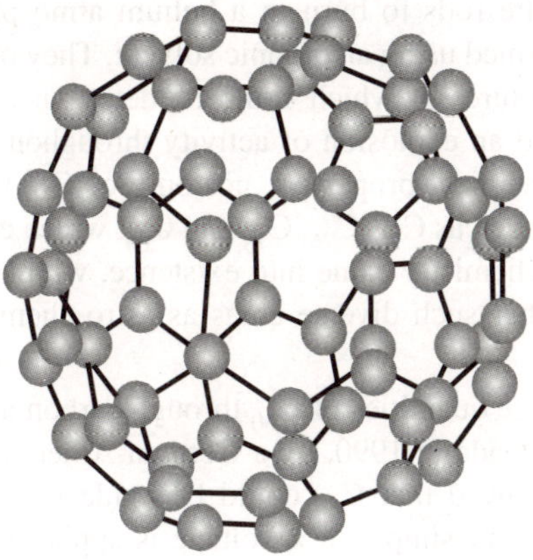

Fig. 2 *Atomic positions of Carbon in C_{60} molecule*

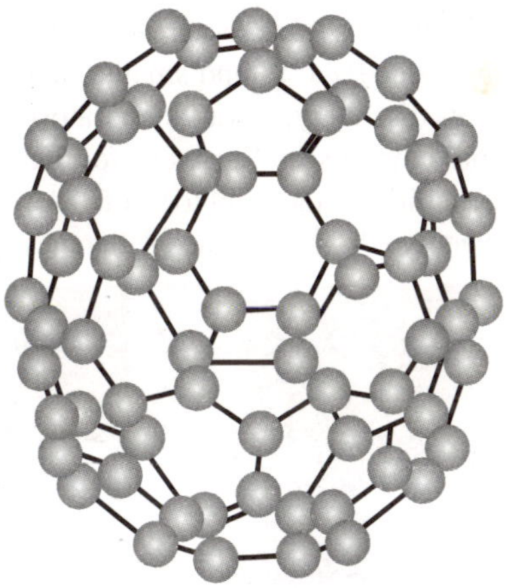

Fig. 3 *The C_{70} molecule with spheres representing the C-atoms*

While the C_{60} molecule is nearly spherical in shape, like a football, the molecule of C_{70} is more like a rugby ball (Fig. 3). The C_{70} molecule can be envisioned by adding a ring of 10 carbon atoms or, equivalently, adding a belt of five hexagons around the equatorial plane of C_{60} molecule which is normal to one of the five fold axes and rotating the two hemisphere of C_{60} so that they fit continuously onto the belt hexagons [6]. In C_{70} molecule there are five inequivalent carbon sites and eight distinct bond lengths. These eight bond lengths were found experimentally by neutron inelastic scattering (NIS) [7]. These bond lengths have been found to be ranging from 1.356Å to 1.475Å.

3.1 Solid Fullerenes - C_{60} and C_{70} Solids

The discovery of the fullerene molecules, predominantly in C_{60} form, was only a first step towards enormous research activity in the fullerene field. Immediately, attention grew in obtaining bulk material of fullerenes. In 1990, physicists W. Krätschmer and D.R. Huffman [8], for the first time produced small quantities of C_{60} solid by causing an arc between two graphite rods to burn in a helium atmosphere and extracting the carbon condensate so formed using an organic solvent. They obtained a mixture of C_{60} and C_{70} solids, the structures of which could be determined. The production of the fullerene material caused an explosion of activity throughout science. This opened a way for studying the chemical properties, not only of C_{60} but also of other carbon clusters and their solids such as C_{70}, C_{76}, C_{78} and C_{84}, which could also be formed. An entirely new branch of chemistry came into existence, with consequences in physics, material science and other such diverse areas as astro-chemistry, superconductivity and polymer science.

The synthesis of large quantities of C_{60} through carbon arc vaporization between two graphite electrodes made in 1990, by a German-American group and the Sussex group independently showed that C_{60} could be made in gram quantities [9]. The technique is essentially very simple; if a voltage is applied to two carbon rods (just touching), an arc will develop between them. If the arc is maintained in helium or argon (instead of in air) clouds of black smoky carbon soot are produced. It turns out that at the correct arc temperature and gas pressure up to 10 % of the black soot is C_{60}. Also present in the soot are about 1% C_{70}, and smaller quantities of larger fullerenes. For the first few years after its discovery, it was a laboratory curiosity, but when it was discovered that macroscopic amounts of C_{60} were formed by heating graphite rods under carefully controlled conditions, the field took off. A simple technique [10] used by students at Panjab University (PU) to produce fullerenes is shown in Fig. 4.

Fig. 4 *Simple setup for fullerene production used by students at PU.,* (For color illustration refer p xvi)

A large bell jar, is connected to a vacuum pump, pressure gauge and argon cylinder. The bell jar was evacuated with the vacuum pump and filled it with inert gas three or four times to a final pressure of 100 torr. The general idea is that when graphite is vaporized, a sample of soot is made which contains fullerenes. Since fullerenes are best produced in 100 torr of an inert gas, the setup is designed to resistively vaporize graphite rods in that environment. The bell jar allows for a vacuum, while the welding transformer supplies the high dc current of about 200 amp that vaporizes the graphite. The current is passed intermittently for a period lasting about 15 seconds during each step. This avoids excessive heating. Further, heating was controlled by passing cold water around a cylindrical enclosure around arc. The graphite rods resist the high current as they are not very good conductors, producing an arc between the rods. The soot contains fullerene molecules. Benzene or toluene dissolves the fullerene component of the soot to form a colored solution. The color of the solution will depend on the concentration of fullerenes and on what type of fullerenes you have made. A sample of concentrated bucky balls will be a dark red and C_{70} will be magenta. So the color of a fullerene/toluene solution can be anywhere from pink to dark red to yellow.

3.2 Doped C_{60} Solids

Since the C_{60} structure is hollow, with room for one or more other atoms, attempts were made to enclose within it a metal atom. A graphite sheet was soaked with a solution of a metal salt (lanthanum chloride, $LaCl_3$) and subjected to vaporization-condensation experiments. Mass spectroscopic analysis of the clusters formed showed the presence of molecules with formula $C_{60}La$. These proved to be photo resistant, i.e. irradiation with intense laser light did ot remove the metal atoms. This reinforced the idea that metal atoms were captured inside the cage structure.

Another kind of doping in the solid C_{60} is also possible, where interstitial sites could be occupied by foreign atoms. The large size (7.1Å diameter) and high electron affinity of C_{60} molecules, in an FCC arrangement (lattice parameter a=14.2Å) provides enough empty intermolecular space and possibility of trapping of metal ions. Because of this large size of fullerene molecules, as compared to alkali metal dopant atoms or ions, the interstitial cavities in a C_{60} lattice can accommodate various guest species.

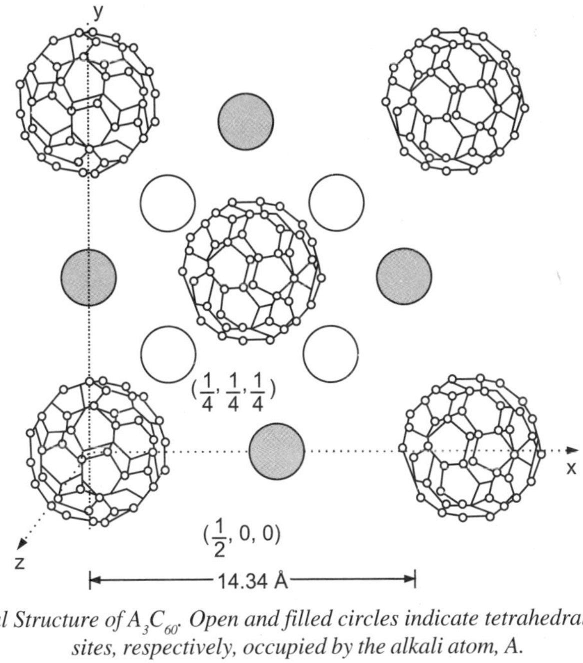

Fig. 5 *Crystal Structure of A_3C_{60}. Open and filled circles indicate tetrahedral and octahedral sites, respectively, occupied by the alkali atom, A.*

Since the nearest distance between two C_{60} molecules is about 10Å, two large tetrahedral void sites (0.25,0.25,0.25) and one large octahedral void site (0.5,0,0) per C_{60} molecule are available for doping it by metal ions. This opens up the possibility of putting other atoms into the structure. See, for example, the structure of A_3C_{60} (Fig. 5) in which tetrahedral as well octahedral sites are occupied by metal cations. Solid C_{60} differs from simple metals in another crucial property: it is a semiconductor. The doped material however is a metal. Simply speaking, an electron is donated from the Rb atom to the buckyball, and this electron can easily move around by jumping from one ball to the next. It is, therefore, expected that A_xC_{60} solids with x ions of metal A can be formed and treated as ionic molecular crystals [11], keeping in view the electron affinity of the C_{60} molecules which tend to take up electrons from doped metal atoms. The free C_{60} molecule in its ground state electronic configuration has a completely filled HOMO level. The LUMO is split into two three-fold degenerate levels which get filled as the bucky-ball is charged with more and more electrons, indicating that the number x can go up to 6 presumably filling the above described levels. That explains why A_3C_{60} could be a metal, while A_6C_{60} an insulator, because of complete filling of band in latter case.

These materials exhibit superconducting behavior on alkali metal doping. Interestingly interstitial doping leads to an increase in lattice parameter and so also an increase in its superconducting transition temperature, and using such arguments, one could possibly fix an upper limit on T_c thus obtained [12].

3.3 Polymerized and Dimerised C_{60} Solids

Recently, it was discovered that fullerenes have a strong tendency to undergo solid state chemical reactions [13-16], provided these are catalyzed by conditions like exposure to light, high pressure or elevated temperature or doping with appropriate amount of alkali ions (A_1C_{60}, A being K, Rb, Cs). In this way, a variety of polymer topologies could be formed and have been investigated by neutron inelastic scattering experiments. Indeed, by quenching the Rb_1C_{60} powder from 450 to 77 K and subsequently heating to 200 K, a new phase forms consisting of $(C_{60})_2$ dimers. These dimers can be single bonded or [2+2] cycloadded double bonded. The latter case most likely leads to a polymer phase, whereas the former leads to a dimer solid phase. The polymer phase can be linear, forming a conducting phase. The dimer phase is an

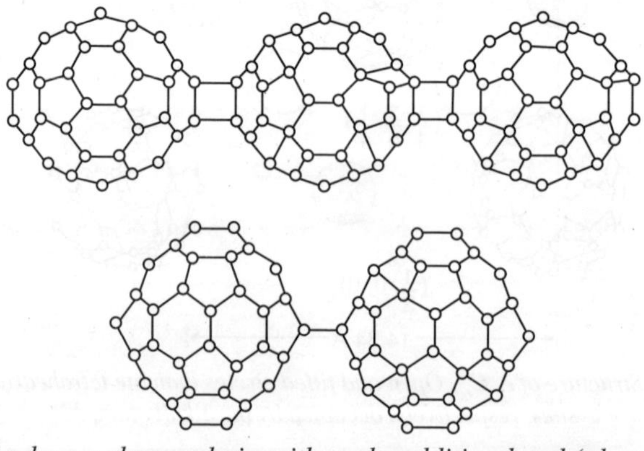

Fig. 6 *Part of a long polymer chain with cyclo-addition bond (above) and a single bond forming a dimer (below) between successive bucky balls.*

insulating phase forming a meta-stable monoclinic structure with space group $P2_1/a$. A dimer molecule consists of two C_{60} molecules bonded by a single bond as shown in Fig. 6. The length of the single bond linking the two bucky-balls, the closest distance between C-atoms on these two C_{60} molecules, is 1.56 Å. The solid of dimer molecules is obviously non-cubic and a monoclinic structure is most likely as has been ascertained by X-ray diffraction experiments. Part of a long chain polymer is also shown in Fig. 6. Here the linkage between the C_{60} units is different.

Passage of high-energy heavy ion beams has also been used to target thin films of C_{60} to cause polymeric phase transformations [17]. Efforts are still on to understand the mechanism of the process and so also for understanding various bulk, structural and thermodynamic properties of dimer and polymer solids [18].

4. CARBON NANOTUBES

The clusters of carbon atoms, in addition to round or nearly round shapes, have also been synthesized in cylindrical or nearly cylindrical forms, called carbon nanotubes (CNT). These nanotubes are in some respects very similar to one-dimensional crystals, as the ratio of length to diameter can be grown to be $\sim 10^6$. Carbon nanotubes were discovered in 1991 by S. Iijima [19]. These are large macromolecules that are unique for their size, shape, and remarkable physical properties. They can be thought of as a sheet of graphite (a hexagonal lattice of carbon) rolled into a cylinder to form a macro-molecule of carbon. These intriguing structures have sparked much excitement in recent years as the inside of the nanotube is extremely fascinating, being hollow, a candidate for trap for various sized atoms and for study of quantum phenomena prevalent at dimensions of its diameter. Indeed, practical applications, including hydrogen uptake, has also generated large amount of research interest dedicated to their understanding. Currently, the physical properties are still being discovered and disputed. Interestingly, nanotubes have a very broad range of electronic, thermal, and structural properties that change depending on the different kinds of nanotube (defined by its diameter, length, and chirality, or twist). Besides having a single cylindrical wall or Single Wall Nanotubes (SWNTs), nanotubes can have multiple walls or Multiwall Nanotubes (MWNTs)—cylinders inside the other cylinders. Nanotubes have been known to be up to one hundred times as strong as steel and almost two millimeters long. The nanotubes have a hemispherical "cap" at each. They may also have, sometimes, caps of other shapes such as conical. They are light, flexible, thermally stable, and are chemically inert. They have the ability to be either metallic or semi-conducting depending on the "twist" of the tube. Infact, they are identified as "armchair" nanotube, "zigzag" type or "chiral" type depending upon folding of graphite sheet, indicated in literature by two integers, (n,m), dictating which atom to join with on folding the graphite sheet. Typical (n,m) tubes obtained on folding a graphite sheet in various twisted or chiral forms are shown in Fig. 7. The chirality in turn affects the conductance of the nanotube, its density, lattice structure, and other properties. A nice description about fullerene nanotubes has been given by Yakobson and Smalley [20, 21]. It turns out that the average diameter of abundant SWNT is around 12 Å. However, nanotubes can vary in size, and they are not always perfectly cylindrical.

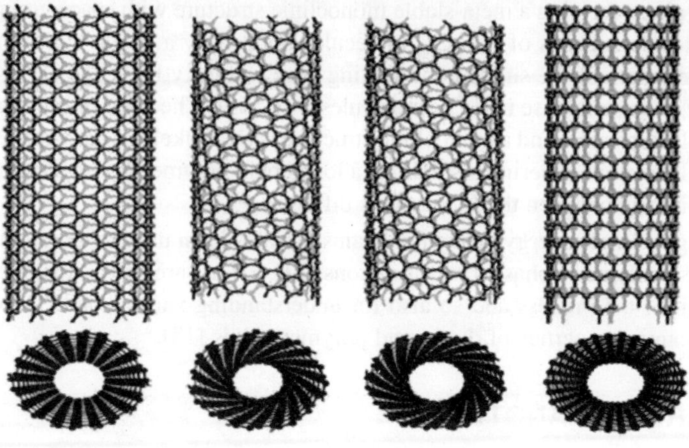

Fig. 7 *Carbon nanotubes of various chiralities without caps*

A chemist might think of a carbon nanotube as a monoelemental polymer, different from usual multi-elemental polymers. If the buckyball is designated by C_{60}, a carbon nanotube might be designated as $C_{1,000,000}$. In physics terms it can be described as a single crystal in one direction, with a unit cell that keeps on propagating and repeating.

There are various methods of production of carbon nanotubes, chemical vapour deposition (CVD) and laser ablation of graphite being more common. We demonstrate a method employed by students using arc ignition of graphite in de-ionsed water under magnetic field [22] in Fig. 8. Also shown are the SEM pictures of the samples so produced.

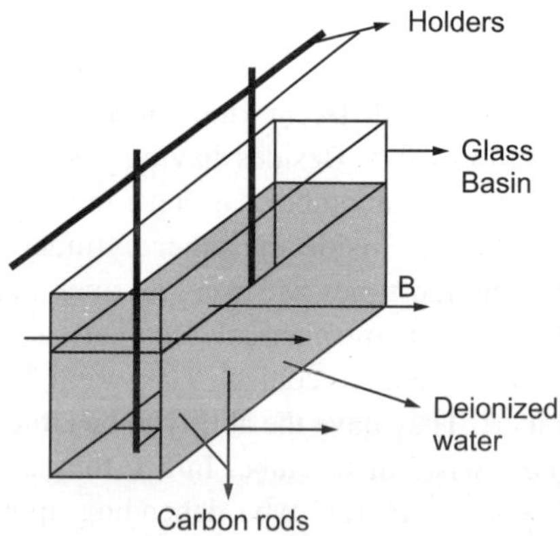

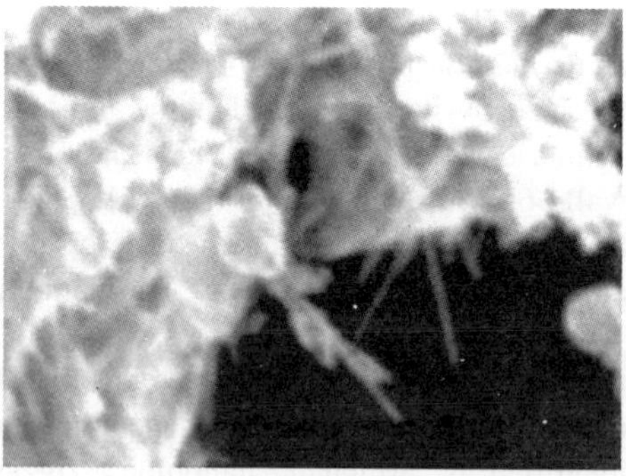

Fig. 8 *Schematics of CNT production (left) and a typical SEM (right) of Carbon soot by Arc ignition under magnetic field using apparatus shown.*

4.1 Nanotube Ropes

Ropes are bundles of tubes packed together in an orderly manner. In 1996, Thess et al. [23] found that the individual SWNTs packed into a close-packed triangular lattice with a lattice constant of about 17 Å, though the actual value depends on whether the material consists of armchair, zigzag or other chirality. This was later confirmed by Gao et al. in 1997 [24]. Armchair tubes had a lattice parameter of 16.78 Å and had a density of 1.33 g/cm^3. For MWNT materials, the space between the tubes was also dependent on chirality. Armchair tubes had a spacing of 3.38 Å, zigzag tubes had a spacing of 3.41 Å, and (2n, n) chiral tubes had an interlayer spacing value of 3.39 Å. These values of spacing compare with the spacing between the layers of graphite sheets, both are about 3.4 Å.

Tersoff and Ruoff [25] provided useful information on the compressibility of nanoropes, where nanotubes under high pressures are assumed to undergo reversible deformations.

5. THEORETICAL MODELS

Theoretical models to explain the structure, dynamics and related properties of C_{60}, C_{70} and nanotube material solids are generally based on model potentials for C-C interaction which are then summed over to obtain molecule-molecule potentials. There are molecular dynamics based methods which are also employed to extract properties of these clusters through so called computer experiments. For the sake of completeness, we would report here some results obtained by us for C_{60}, C_{70}, dimer C_{60} and SWNT and MWNT materials using simple models described below [26-28]. A calculation [17] of phonons in dimer C_{60} is given in Fig. 9.

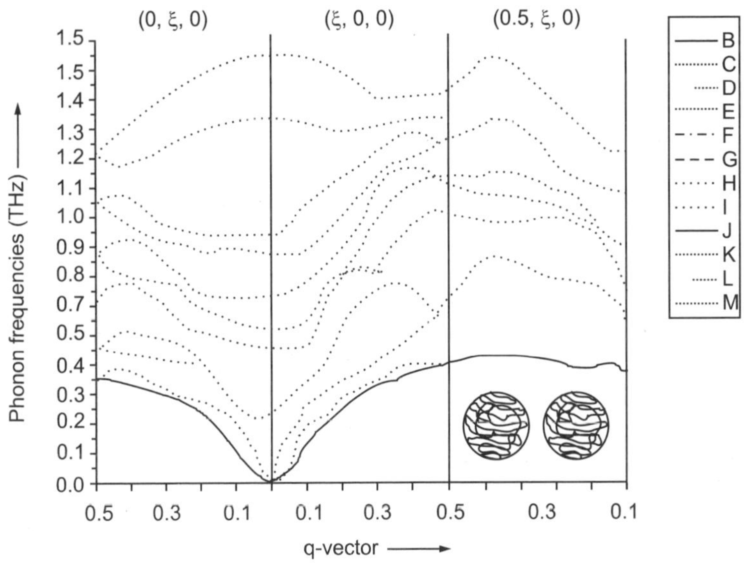

Fig. 9 *Phonon dispersion curves in various symmetry directions for a dimer solid as shown in inset [..] . B,C,D refer to acoustic modes, E,F,G,H,I,K refer to Librational modes and J,L,M refer to Optic modes.*

We also present some structural results [27] for SWNT and MWNT materials in Tables 1 and 2. It is pertinent to mention that Tersoff and Ruoff [25] have also studied structural properties of bundles of carbon nanotubes, and presented detailed calculated results assuming van-der Waals interactions between cylindrical and deformed cylinder of various diameters, for lattice constant and cohesive energy.

The temperature and pressure effects play a significant role in system properties. These macroscopic variables are important parameters as they can be easily altered and response of a system can be studied, giving very useful information about the nature of interactions in the system. The variation of temperature and pressure results in changes of the ordered environment of the system which can undergo phase transitions. In addition, pressure and temperature dependence of structural and dynamical properties becomes available [28]. All these are very useful for interpreting and understanding theoretical models for the system under study.

Table 1 *Potential Energy and Structure of SWNT Crystals*

Type of the tube	*Diameter*	*Lattice parameter (in nm)*	*(in nm)*	*Total potential energy/length*
		Ref. 27	*Expt.*	*(kcal/mol/nm) [27]*
Armchair (10,10)	1.356	1.667	1.678	72.9
Zigzag (17,0)	1.332	1.642	1.652	72.7

Table 2 *Energy and Structure of layered nanotubes (MWNT)*

MWNT Layer	*Diameter (nm)*	*Interwall separation (in nm)*		*Energy/atom (ev/atom*
		Ref. 27	*Expt.*	*Ref. 27*
1st tube	1.35	–	–	–
2nd tube	2.03	0.339	0.338	0.023
3rd tube	2.71	0.339	0.338	0.035

6. APPLICATIONS OF CARBON NANOTUBES

Due to their extremely interesting properties, as calculable from theoretical models as well as indicated by experimental investigations, carbon nanotubes seem to be materials with a great potential for technological applications. Their high electrical conductivity, very high tensile strength at the same time highly flexible behavior - can be bent considerably without damage, very elastic ~18% elongation to failure, high thermal conductivity, low thermal expansion coefficient, good field emission of electrons, high aspect ratio (length = ~1000 x diameter) etc., all make it versatile material of the future. These properties are being exploited for display devices, conductive plastics, energy storage, conductive adhesives and connectors. They open up a new dimension through molecular electronics. These also find applications in making structural composites of high strength, strong and light fibers and fabrics.

The exploration of CNTs in biomedical applications is just underway, but has significant potential. Since a large part of the human body consists of carbon, it is generally though of as a very biocompatible material. Cells have been shown to grow on CNTs, so they appear to have no toxic effect. The cells also do not adhere to the

CNTs, potentially giving rise to applications such as coatings for prosthetics, as well as anti-fouling coatings for ships.

The ability to functionalize (chemically modify) the sidewalls of CNTs also leads to biomedical applications such as vascular stents, and neuron growth and regeneration. It has also been shown that a single strand of DNA can be bonded to a nanotube, which can then be successfully inserted into a cell.

Many researchers and corporations have already developed CNT based air and water filtration devices. It has been reported that these filters can not only block the smallest particles but also kill most bacteria. This is another area where CNTs have already been commercialized and products are on the market now.

A ceramic material reinforced with carbon nanotubes has been made by materials scientists at University of California at Davis (USA). The new material is far tougher than conventional ceramics, conducts electricity and can both conduct heat and act as a thermal barrier, depending on the orientation of the nanotubes.

Ceramic materials are very hard and resistant to heat and chemical attack, making them useful for applications such as coating turbine blades, but they are also very brittle. The researchers mixed powdered alumina (aluminum oxide) with 5 to 10 percent carbon nanotubes and a further 5 percent finely milled niobium. The researchers treated the mixture with an electrical pulse in a process called spark-plasma sintering. This process consolidates ceramic powders more quickly and at lower temperatures than conventional processes.

The new material has up to five times the fracture toughness—resistance to cracking under stress—of conventional alumina. The material shows electrical conductivity seven times that of previous ceramics made with nanotubes. It also has interesting thermal properties, conducting heat in one direction, along the alignment of the nanotubes, but reflecting heat at right angles to the nanotubes, making it an attractive material for thermal barrier coatings.

There is a wealth of other potential applications for CNTs, such as solar collection; nonporous filters; catalyst supports; and coatings of all sorts. There are almost certainly many unanticipated applications for this remarkable material that will come to light in the years ahead, and which may prove to be the most important and valuable ones of all. Many researchers are looking into conductive and or water proof paper made with CNTs. CNTs have also been shown to absorb Infrared light and may have applications in the I/R Optics Industry.

Summary

In this article, I have attempted to present an overall picture summarizing some of the facts gathered about new carbon cluster solids, including carbon nanotubes. Extensive efforts, by many researchers, have gone into molecular dynamics (MD) simulations and other electronic structure calculations. I present a few results based on a model calculation for various carbon cluster materials by us and some production methods. These results indicate that various properties are reproduced broadly by the simple model, which is better than a mere qualitative agreement.

The nanoscale fullerene based structures (fullerenes, tubes and their multi-layered closed structures, like onions) are destined to play a role in the technology of the 21st century. In fact, the carbon cluster solids and carbon nanotubes, because of their size, present very interesting quantum materials, where a trapped particle will have to face all interesting phenomena occurring at such small dimensions. This leads to their novel mechanical and electronic properties and their tremendous potential for future technological applications. Some of these have been mentioned here. These materials present tremendous potential as components for use in nano-electronic and nano-mechanical device applications or as structural elements in various devices.

REFERENCES

1. Kroto, H.W., Heath, J.R., O'Brien, S.C., Curl, R.F. and Smalley, R.E. (1985): Buckminster fullerene. *Nature (London)* **318**, 162-163 and *ibid* (1988) Space, stars, and soot. *Science* **242**,1139-1145.
2. Kratschmer, W., Lamb, L.D., Fostiropouios, K. and Huffmann, D.R. (1990) Solid C_{60}: a new form from carbon. *Nature (London)* **347**, 354-358. *ibid* (1991) *Chem. Phys. Lett.* **17**, 167.
3. Original information can be obtained at this site: http://www.nobel.se/chemistry/laureates/1996/press.html
4. For an overall review, see Copley, J.R.D, David, W.I.F., Neumann, D.A. (1993) *Sc. Rev. Neutron News* **l4**, 20.
5. For a recent overview and for further references, see Ramirez, A.P. (1994) Condensed Matter News **3**, 9.
6. Dresselhaus, M.S., Dresselhaus, G. and Eklund, P.C. (1996) In*: Science of Fullerenes and Carbon Nanotubes*, Academic Press, Inc., New York.
7. Nilolaev, A.V., Dennis, T.J.S., Prassides, K. and Soper, A.K. (1994) Molecular structure of the C_{60} fullerene. *Chem Phys. Lett.* **223**, 143-148.
8. The site by Sussex group describes production method of fullerenes, nanotubes etc.:http://www.susx.ac.uk/Users/kroto/FullereneCentre/workshop/workshop.html
9. Interesting details for simple method of fullerene production: http://192.211.16.13/individuals/thomasel/bucky/
10. Chopra, I.S. and Jindal, V.K. (2004) Production of solid C_{60} by the arc ignition of graphite. *Pb. Univ. Res. J. (Sci.)* **54**, 65-67.
11. Ranjan, K., Dharamvir, K. and Jindal, V.K. (2006) Comparative study of alkali metal-doped solids. *Physica B* **371**, 232-240.
12. Jindal, V.K. Theoretical model upper limit on the lattice parameter of doped C_{60} solid. http://xxx.lanl.gov/abs/cond-mat/0005104.
13. Oszlanyi , G., Bortel , G., Faigel, G., Granasy, L., Bendele, G.M., Stephens, P.W. and Forro, L. (1996) Single C-C bond in $(C_{60})^{-2}_{2}$. *Phys. Rev. B* **54**, 11849-11852.
14. Schober, H. and Renker**,** B. (1999) *Neutron News* **10**, 28.
15. Dulcos, S.J., Brister, K., Haddon, R.C., Kortan, A.R. and Thiel, F.A. (1991) *Nature* **351**, 380.

16. Kaur, N., Bajwa, N., Dharamvir, K. and Jindal, V.K. (2000) Anisotropic compression in Dimer C_{60} fullerene solid. *Int. J. Mod. Phys. B* **14**, 2597.
17. Bajwa, N., Ingale, A., Avasthi, D.K., Kumar, R., Tripathi, A., Dharamvir, K. and Jindal, V.K. (2003) Swift Heavy Ion Induced Modification Studies of C_{60} Thin Films. J. Appl. Phys. 94, 326.
18. Kaur, N., Dharamvir, K. and Jindal, V. K. (2007) Dimerisation and fusion of two C_{60} bucky balls. *Chem. Phys.* (In press)
19. Iijima, S. (1991) Helical microtubules of graphitic carbon. *Nature*, **354,** 56-58.
20. Yakobson, B.I. and Smalley, R.E. (1997) *Am. Sci.* July- August.
21. For the detailed review of the properties of nanotubes, see home-page of David Tomanek and the site compiled by Thomas A. Adams II, http://www.pa.msu.edu/cmp/csc/ntproperties/
22. Jindal, V.K., Dharamvir, K., Suman, V., Chahil, N.S. and Tsomo, M. (2007) Carbon nanotubes production using arc ignition under magnetic field. *Nano Trends: J. Nanotechnol. Appl.* **2**, 48-59.
23. Thess, A., Lee, R., Nikolaev, P., Dai, H., Petit, P., Robert, J., Xu, C., Lee, Y.H., Kim, S.G., Rinzler, A.G., Colbert, D.T., Scuseria, G., Tománek, D., Fischer, J.E. and Smalley, R.E. (1996) Crystalline Ropes of Metallic Carbon Nanotubes. *Science* **273**, 483-487.
24. Gao, G., Cagin, T. and Goddard, W.A. www.wag.caltech.edu/foresight/foresight_2.html
25. Tersoff, J. and Ruoff, R.S. (1994) Structural properties of a carbon-nanotubes crystal. *Phys. Rev. Lett.* **73**, 676-679.
26. Jindal, V.K., Dharamvir, K. and Singh, S. (2000) Temperature and pressure effects on lattice properties of pure solid. *Int. J. Mod. Phys. B* **14**, 51-69.
27. Gupta, S., Dharamvir, K. and Jindal, V.K. (2005) Elastic moduli of single walled carbon nanotubes and their ropes. *Phys. Rev. B* **72**, 165428-165433.
28. Imtani, A. and Jindal, V.K. (2007) Structure of armchair single-walled nanotubes under hydrostatic pressure. *Phys. Rev. B* (In press)

5

SYNTHESIS AND CHARACTERIZATION OF NANOSIZED MATERIALS FOR LITHIUM RECHARGEABLE BATTERIES

S. Selvasekarapandian, T. Savitha, J. Malathi, R. Mangalam and C. S. Ramya

ABSTRACT

Nanosized cathode materials based on Phospho olivine, $LiNiPO_4$ and $LiCoPO_4$ and nano composite polymer electrolyte, PVAc:DMF:$LiClO_4$:TiO_2 have been synthesized for Lithium rechargeable batteries. Cathode materials have been characterized by Thermogravimetric/Differential Thermal Analysis, X-Ray Diffraction, Laser Granulometry and Electrical conductivity studies and composite polymer electrolytes by XRD, and Electrical conductivity studies.

1. INTRODUCTION

Rechargeable lithium batteries have been utilized in a wide range of applications, such as automotive industries, portable devices and aerospace. Depending on the type of the cathode, anode and the electrolyte used in the battery one can expect a range of energy densities. Phospho-olivine $LiMPO_4$ (M = Mn, Fe, Co, or Ni) have been intensively investigated as cathode materials because of their high stability [1]. In the present study, $LiNiPO_4$ and $LiCoPO_4$ have been prepared by Modified Pechini-type Polymerizable method and characterized by TG/DTA, XRD, Laser Granulometry and Electrical conductivity studies. Among various types of polymer electrolytes, composite polymer electrolytes gain more attention among researchers, in which electrochemically inert particulate fillers are incorporated into polymer matrix [2]. Composite formation was found to enhance both the mechanical strength and ionic conductivity leading to their potential use in lithium batteries. In the present study, composite polymer electrolyte based on Polyvinyl acetate (PVAc): Dimethyl formamide (DMF) : Lithium perchlorate ($LiClO_4$) : Titanium dioxide (TiO_2), have been prepared and characterized by XRD, FTIR and Electrical conductivity studies.

2. EXPERIMENTAL DETAILS

Cathode Materials

The raw materials $LiNO_3$, $CoNO_3.6H_2O$, $NiNO_3.6H_2O$ and $NH_4H_2PO_4$ are dissolved in double distilled deionized water separately to prepare $LiCoPO_4$ and $LiNiPO_4$. Citric acid and Ethylene glycol are added as chelating agent and for gelation, respectively. Pyrolysis of the polymer obtained is performed at 350°C for 2 h in a furnace that yields a black powder precursor, that is subjected to TG/DTA analysis using STA 150 Thermal Analyser in oxygen atmosphere.

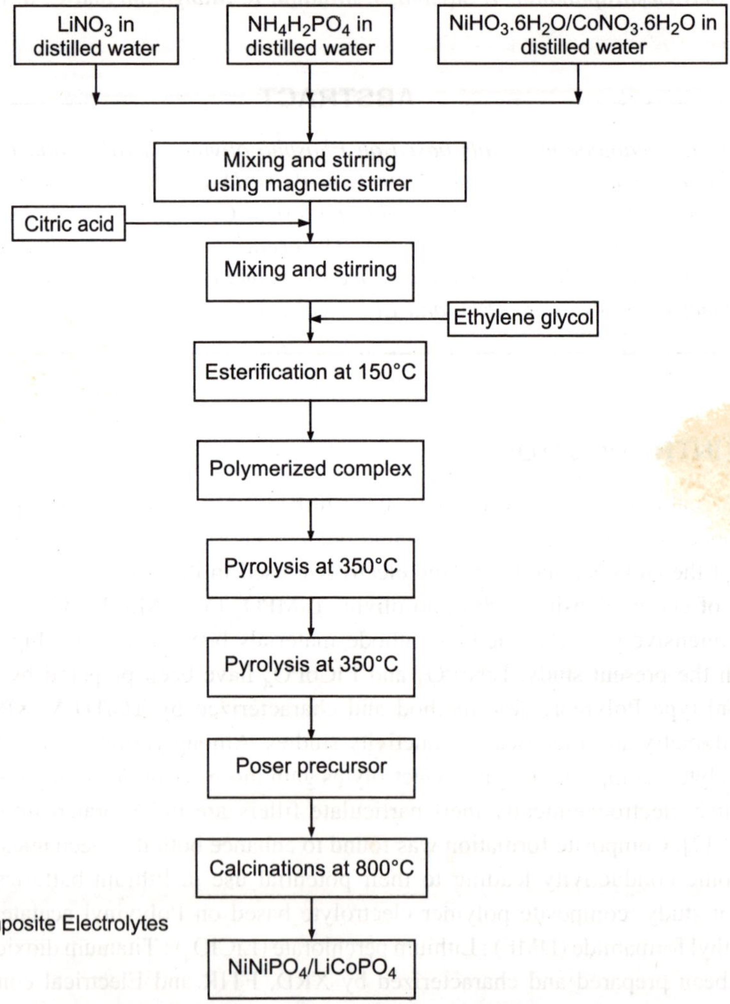

Composite Electrolytes

Polymer electrolyte was prepared using solvent casting technique and filler TiO_2 was prepared by Modified Pechini-type Polymerizable method. The following are the Poly (vinyl acetate) (PVAc) based composite polymer electrolytes that have been investigated in the present study,

Ti : 70mol% PVAc + 20m% DMF + 10m% $LiClO_4$

Ti 1 : 70mol% PVAc + 20m% DMF + 10m% $LiClO_4$ + 5m% TiO_2

Ti 2 : 70mol% PVAc + 20m% DMF + 10m% $LiClO_4$ + 10m% TiO_2

Ti 3 : 70mol% PVAc + 20m% DMF + 10m% $LiClO_4$ + 15m% TiO_2

Ti 4 : 70mol% PVAc + 20m% DMF + 10m% $LiClO_4$ + 20m% TiO_2

Philips X'Pert PRO diffractometer equipped with an X'celerator detector, with CuK_α radiation source was used in the present study. FTIR spectroscopy was performed using FTIR-8400S SHIMADZU spectrophotometer in the wavenumber range 400-4000cm^{-1}. PC based HIOKI 3532 LCZ meter was used to analyse electrical characteristics in the frequency range 42Hz – 5MHz. In the present study for Laser Granulometry studies, LS 13 320, Universal Liquid Module was used to analyse particle size.

3. RESULTS AND DISCUSSION

Electrode Materials

3.1 Thermogravimetric and Differential Thermal Analysis (TG/ DTA)

Figs.1a and 1b show the TG and DTA curves of $LiCoPO_4$ and $LiNiPO_4$ in the temperature range 28°C to 900°C at 5°C/min, respectively. Thermogravimetric analysis of both the samples consists of three regions: from room temperature to 150°C, 300-600°C and 600 – 900°C. For both the samples, upto 150°C the first weight loss (~4%) observed in TGA associated with an endothermic peak in DTA corresponds to the solvent and adsorbed water molecules evaporation [3]. The second weight loss (~3%) can be attributed to the combustion reaction of the raw materials with the evolution of brown gases carbondioxide and nitrogen dioxide.

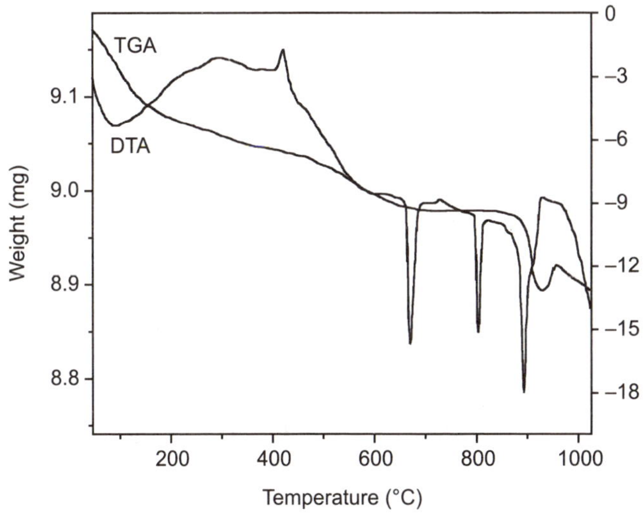

(a)

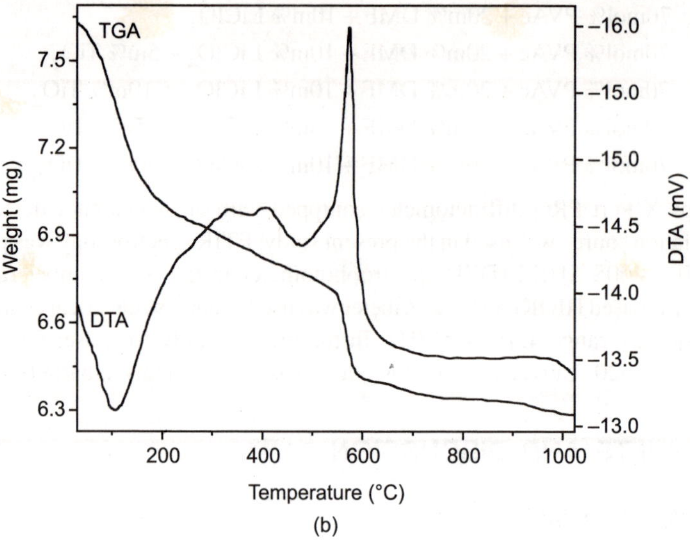

Fig. 1 *TG and DTA curves (a) $LiCoPO_4$ (b) $LiNiPO_4$*

The citric acid added provides the local heat for formation of the compound during the decomposition process because of self-igniting property of its carboxylic group [4]. The exothermic peak in Differential Thermal Analysis in the range 300-600^{0}C corresponds to degradation of polymer and due to conversion of organic compounds into CO_2 & NO_2. At higher temperatures chemical reactions occur with the exothermic heat energy supplied by the combustion of citric acid which facilitates the process of crystallization of the compound.

3.2 XRD Analysis

Figs. 2a and 2b show the XRD pattern of Li_1CoPO_4 and Li_1NiPO_4 prepared by Modified Pechini-type Polymerizable method, respectively. The diffraction patterns for all the samples can be indexed to a single-phase material having an ordered olivine structure.

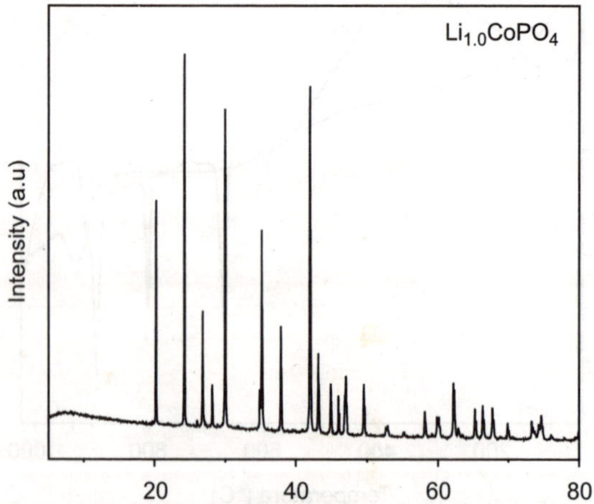

Fig. 2a *XRD pattern of $Li_{1.0}CoPO_4$*

The patterns are isomorphic to the related olivine structure with a Hexagonal Closed Packing (HCP) oxygen array and tetrahedral sites are occupied by P^{5+} forming PO_4 oxo-anions [2]. Olivine structure is preserved in all the samples because of the similarities of the size for small M^{2+} cations (Ni^{2+} – 0.69 Å, Co^{2+} – 0.72Å) when compared to Li^+ (0.69 Å). The olivine structure has orthorhombic symmetry with *Pnmb* space group with alternate a-c planes of Li and M occupying the octahedral sites, which makes it possible for the two-dimensional Li diffusion paths between the HCP oxygen layers to be present [5].

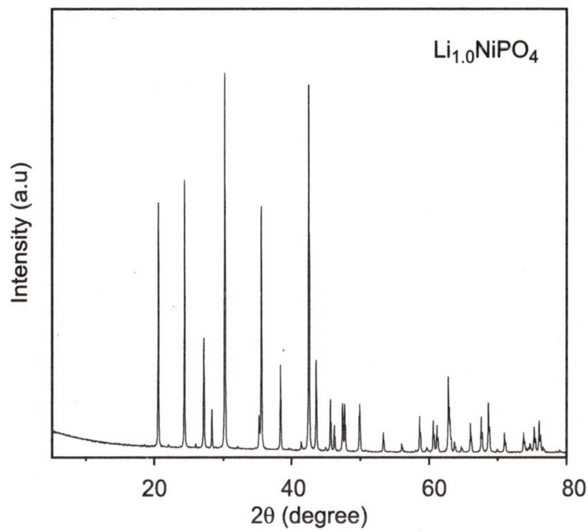

Fig. 2b *XRD pattern of $Li_{1.0}NiPO_4$*

Table 1(a) *Comparison of 2θ values of observed and PCPDF file for $LiCoPO_4$ & $LiNiPO_4$*

S. No.	2θ (degress)		hkl
	PCPDF	$Li_{1.0}CoPO_4$	
1	20.8	20.2	011
2	24.1	24.2	101
3	29.9	29.9	121
4	35.9	35.1	131
5	44.9	44.9	231
6	45.1	45.9	122
7	47.0	47.0	240
8	49.5	49.5	202
9	52.8	52.9	222
10	62.6	62.2	341
11	66.8	66.3	441

Table 1(b) *Comparison of 2θ values of observed and PCPDF file for $LiCoPO_4$ & $LiNiPO_4$*

S. No.	*2θ (degress)*		*hkl*
	PCPDF	*$Li_{1.0}CoPO_4$*	
1	20.8	20.5	011
2	24.3	24.3	101
3	30.5	30.1	200
4	35.5	35.5	220
5	38.9	38.3	140
6	42.5	42.4	122
7	45.4	45.6	122
8	47.6	47.6	240
9	49.8	49.9	202
10	53.3	53.3	042
11	58.0	58.5	331

Tables 1a and 1b show the comparison of 2θ values for maximum intensity peaks obtained in the present systems with that of previous reports. It has been observed that the 2θ values obtained are in good agreement with PCPDF 32-0578 [6] and PCPDF 32-0552 [7] for $LiNiPO_4$ and $LiCoPO_4$, respectively. Average particle size has been calculated using Debye-Scherrer method,

$$\Delta = \frac{0.94\lambda}{\beta_{20} \cos\theta} \quad ...(3.1)$$

where Δ is the average particle size, λ is the wavelength of CuK_{α}, β is the full width half maximum of the reflection peak in radians and θ is the Bragg angle. The particle size calculated using the above formula has been found to be in the range 180 nm and 800 nm for Li_xCoPO_4 and Li_xNiPO_4 respectively.

3.3 Laser Granulometry Analysis

Particle size determination by laser diffraction is relatively rapid and inexpensive, applicable to small and large samples and can be essentially non-destructive. Figs.3a & 3b show the particle size distribution of Li_1CoPO_4 and Li_1NiPO_4 prepared by Modified Pechini-type Polymerizable method using Laser Granulometry Analysis respectively. All the powders have been ultrasonically treated in water for 1 min with a power of 20W to aid the disaggregation of the samples. The graph displays the statistical distributions of the particle size where Differential volume (%) indicates the percentage of grains for the considered size and Cumulative percentile values (i.e., Cumulative < volume (%)) indicates the particle size at which a specified percentage of the particles are finer. It has been observed that all the graphs show bimodal mode which reveals that most of the grains are aggregated. Table 2 provides the grain and the aggregate size of all the samples. From the table it can be seen that all the samples have nanosized grains and micron sized aggregates. The analysis clearly shows that the present synthesis route influences the microstructure of the material as well as the grain size.

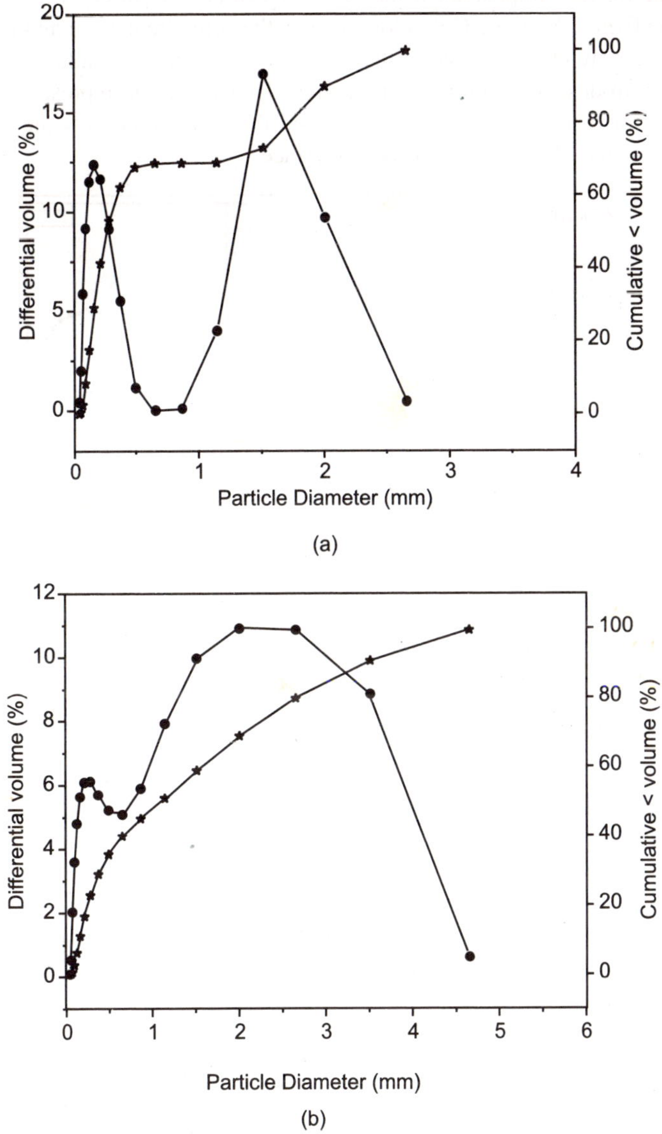

Fig. 3 *Particle size distribution of $Li_{1.0}NiPO_4$*

Table 2 *Grain and Aggregate size calculated from Laser Granulometry Analysis for Li_1CoP_4, Li_1NiPO_4*

Sample Code	*Grain Size (μm)*	*Aggregate size (μm)*
$Li_{1.0}CoPO_4$	0.191	1.669
$Li_{1.0}NiPO_4$	0.062	2.241

3.4 Impedance Analysis

Figs. 4a and 4b show impedance response of Li_1CoPO_4, Li_1NiPO_4 at 673K, 723K and 773K respectively. All Cole-Cole plot contains the semicircular arc and inclined spur which together are characteristic of the behaviour of ionically-conducting solids with blocking electrodes, at different temperatures. The semicircle represents the parallel combination of resistance and capacitance and the spike represents the double layer capacitance at the electrode-electrolyte interface.

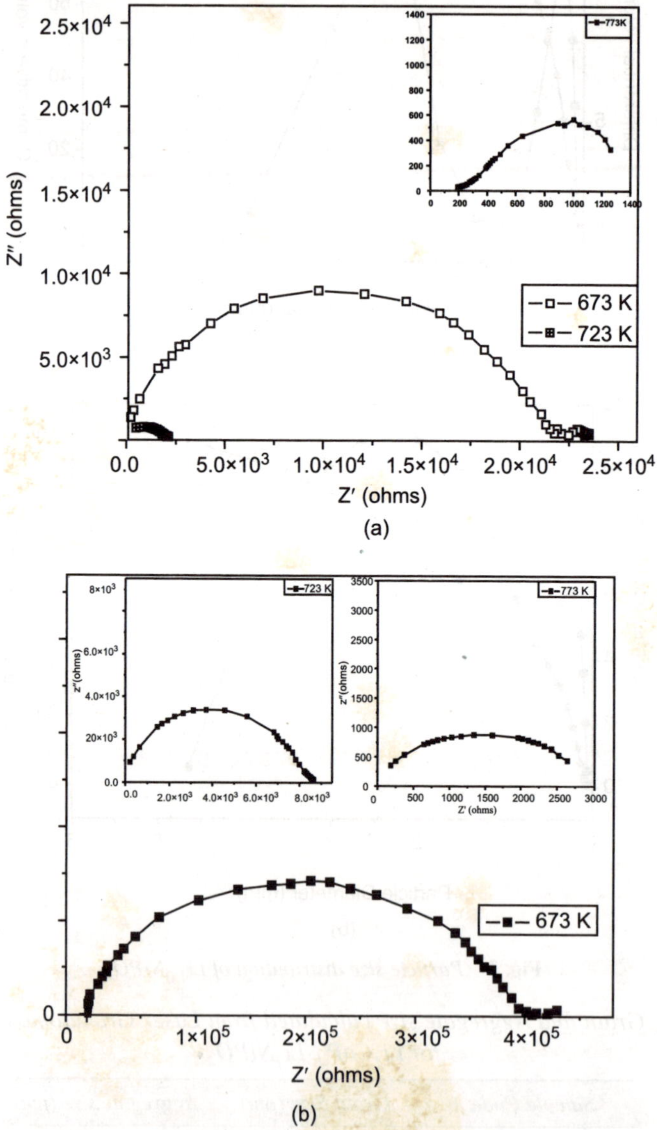

Fig. 4 *Cole-Cole plot of Li_1NiPO_4 at 673K, 723K and 773K.*

The presence of depressed semicircles reveals the non-Debye nature of the sample and the angle by which the semicircle is depressed below the real axis and the amount of inclination in the straight line are related to the width of the relaxation time distribution [8]. At maximum of the semicircle, ωRC=1 condition is satisfied and using this equation capacitance of the semicircle has been calculated and found to be in the order of pF which represents the bulk response of the sample [9]. Bulk resistance R_b, has been calculated from the intercept of the semicircle at the low frequency side of the plot with the real axis. Using R_b, the bulk resistance, A the electrode-electrolyte contact area of the pellet and l thickness of the pellet, bulk conductivity has been calculated by,

$$\sigma_b = \frac{l}{R_b\, A} \quad ...(3.2)$$

Bulk conductivity calculated has been tabulated in the Table 3. It has been observed that when the temperature increases bulk resistance value decreases for all the samples within the frequency range measured which results in the increase in conductivity.

Table 3 *Bulk conductivity values of Li_1CoPO_4 and Li_1NiOP_4 at 673, 723K and 773K*

Sample code	σ_b (Scm^{-1})		
	673 K	*723K*	*773 K*
$Li_{1.0}CoPO_4$	4.7×10^{-7}	55×10^{-6}	4.5×10^{-5}
$Li_{1.0}NiPO_4$	5.2×10^{-7}	23×10^{-5}	66×10^{-5}

Composite Polymer Electrolytes

3.5 XRD Analysis

Fig. 5 (a, b, c, d, e) shows the XRD pattern for Pure PVAc,Ti, Pure TiO_2,Ti 1,Ti 4 respectively. Fig. 5a shows two broad peaks at 2θ = 14.8° and 21.6° ascribed to pure PVAc indicating amorphous nature. The diffraction peak at 2θ =14.8° found in pure PVAc was found to be absent in the plasticized PVAc + DMF + $LiClO_4$ (Fig. 5b). Also the diffraction peak at 2θ = 21.6° has been slightly shifted in the plasticized system. The diffraction peaks of pure $LiClO_4$ at 2θ = 18.392°, 21.90°, 23.2°, 31° and 32° (PCPDF-30-0751) [10] were found to be absent in the polymer complex indicating the complete dissociation of the salt in the polymer and also the complexation. Fig. 5c reveals clearly that the obtained TiO_2 is a pure rutile phase (PCPDF-34 – 0180) as shown by the literature. The diffraction peak at 2θ = 27.5°, 36.5°, 54.5°, 56.8° and 69° are attributed to TiO_2 (PCPDF- 34 – 0180) [11], which indicates the dispersion of the nano filler particles in Fig. 5d-e. The intensity of peaks corresponding to 2θ values of the TiO_2 were found to increase in the Fig. 3.5e indicating a higher concentration of the filler (20 mol% - TiO_2).

3.6 Impedance Analysis

Fig. 6a shows impedance plots for Ti, Ti 1, Ti 2, Ti 3 and Ti 4 respectively at 303K. The plot consists of high frequency depressed semicircle represented by a frequency dependent capacitor C_g parallel to a bulk resistor (R_b) and a low-frequency spike represented by a constant phase element. In this case, the migration of ions may occur through the free volume of matrix polymer, which can be represented by a resistor. The immobile polymer chains become polarized in the alternating field, which can be represented by a capacitor. The ionic migration and bulk polarization are physically in parallel, and therefore the semicircle at high frequency can be observed. The inclined straight line at the low frequency region could be the effect of electrode and electrolyte interface [12]. The ionic conductivity is calculated using the equation

$$\sigma_b = \frac{l}{R_b A} \quad \text{...(3.3)}$$

The highest room temperature conductivity value is found to be 3.23×10^{-4} for the 15m% TiO_2 doped system. The ionic conductivity for all the samples at different temperatures has been tabulated in Table 4. It reveals that, as the temperature increases, the ionic conductivity of all the samples increase.

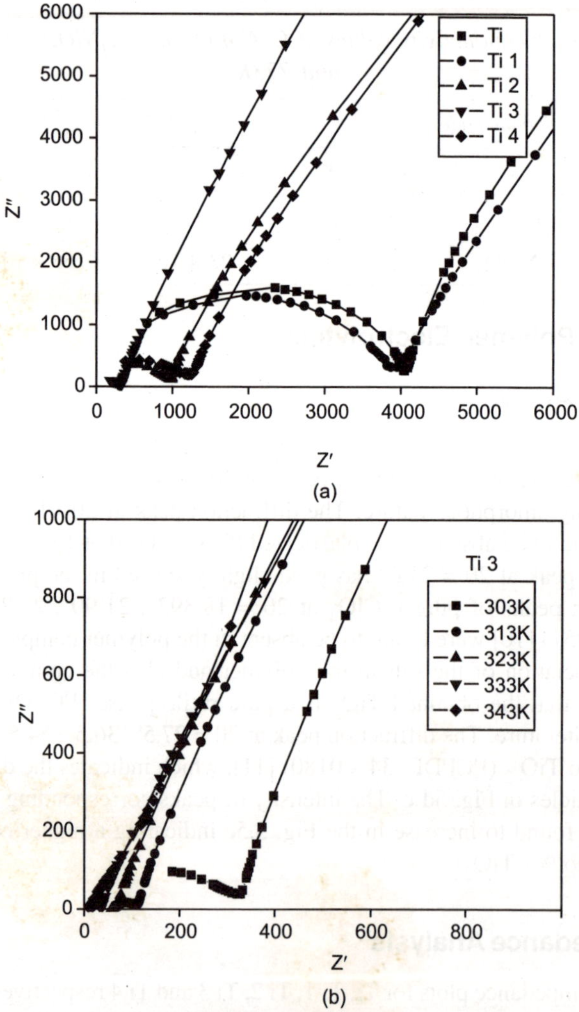

Fig. 5 *(a) Impedence plot for all samples at 303 K (b) Impedence plot for the PE 303K with 15m% of TiO_2 at various temp*

Table 4 *Bulk conductivity values for different concentrations of TiO_2*

Sample code	*Conductivity (Scm^{-1})*		
	303 K	*323 K*	*343 K*
Ti	2.53×10^{-5}	1.2×10^{-4}	9.38×10^{-4}
Ti 1	3.23×10^{5}	1.9×10^{-4}	9.47×10^{-4}
Ti 2	1.01×10^{-4}	1.14×10^{-3}	4.9×10^{-3}
Ti 3	3.23×10^{-4}	1.55×10^{-3}	0.74×10^{-2}
Ti 4	7.10×10^{-5}	1.69×10^{-3}	2.66×10^{-3}

3.7 Temperature Dependent Ionic Conductivity

Figure 7a shows the temperature dependent ionic conductivity for composite polymer electrolytes. The temperature dependence of electrical conductivity of polymer films is non-linear, indicating that the ion transport is due to polymer segmental motion. Ionic conductivity has been found to increase with temperature. The conductivity values do not show any abrupt jump with temperature indicating that these electrolytes exhibit a completely amorphous structure [13]. The conductivity can be expressed as

$$\sigma_b = \sigma_0 \exp(-E_a/kT) \qquad ...(3.4)$$

where σ_0 is the pre-exponential factor and k is the Boltzmann constant. Fig.7b shows the dependence of ionic conductivity on the addition of nano-sized TiO_2 at different temperatures. Addition of TiO_2 to the polymer electrolyte has been found to enhance the ionic conductivity of composite polymer electrolytes. The nanosized ceramic filler, due to its large surface area, readily interacts with the Li^+ ion and enhances conductivity.

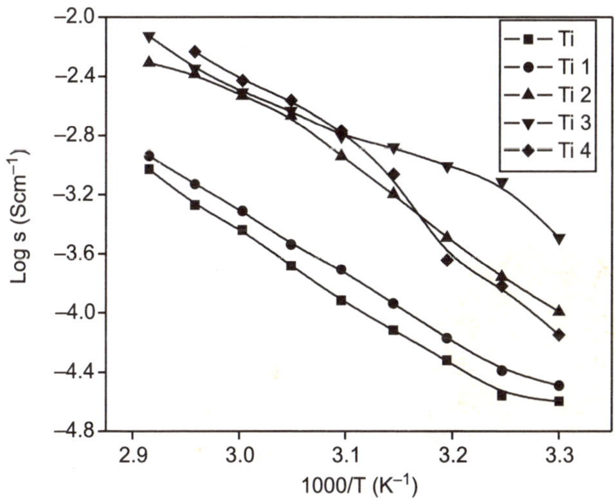

(a)

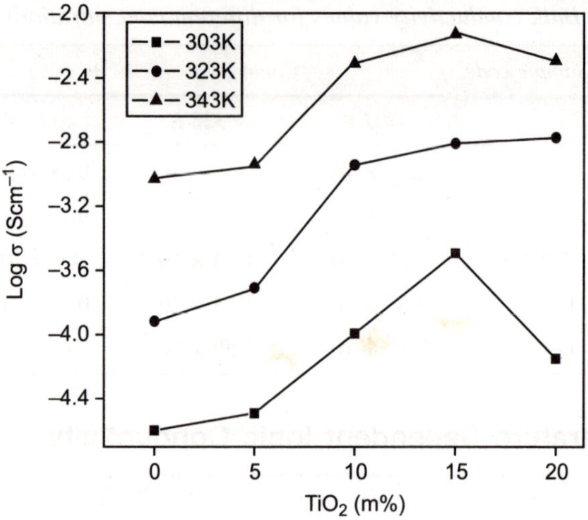

(b)

Fig. 7 *(a) Temp dependent ionic cond for all samples at 303 K (b) Dependence of ionic cond on filler conc at three diff temp.*

Figure 8a shows conductance spectra for Ti, Ti 1, Ti 2, Ti 3 and Ti 4 at 303K. The plot shows three regions, low frequency dispersion ascribed to the space charge polarization at the blocking electrodes, frequency independent plateau region and high frequency conductivity dispersion prominent at low temperature. As the temperature increases, more and more charge accumulation occurs at the interface between the electrode and the electrolyte, which leads to a decrease in the conductivity at low frequencies. The second region corresponds to the frequency-independent plateau region and the extrapolation of the plateau to zero frequency gives the value of DC ionic conductivity at different temperatures. Conductivity has been found to increase with increase of TiO_2 concentration up to 15 m% after which the conductivity ceases. Fig. 8b shows the variation of conductivity as a function of frequency for the sample with 15m% TiO_2. As the temperature increases, the frequency at which the dispersion due to electrode effects becomes prominent and gets shifted to high frequency region and the frequency independent conductivity region decreases.

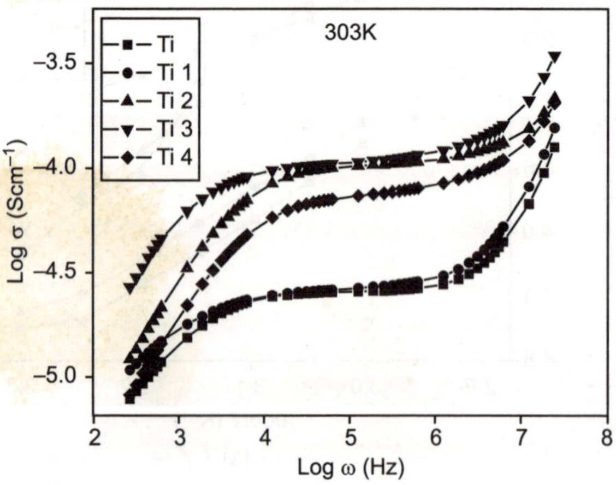

(a)

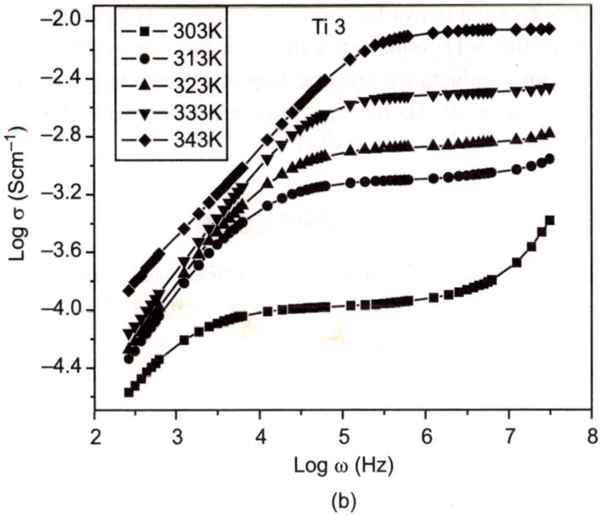

Fig. 8 *(a) Variation of σ with freq. for all samples at 303 K (b) Variation of σ as a function of freq. for the sample with 15m% TiO$_2$*

Table 5 *Hopping frequency and Relaxation time for all the samples at RT*

Sample code	*Hopping frequency (ω_p) in Hz, at 303 K*	*Relaxation Time (τ) in sec, at 303K*
Ti	4.58×10^6	1.18×10^{-5}
Ti 1	6.22×10^6	8.38×10^{-6}
Ti 2	2.48×10^7	2.36×10^{-6}
Ti 3	7.67×10^7	1.69×10^{-6}
Ti 4	1.74×10^7	2.64×10^{-6}

Conclusion

Li_1NiPO_4 and Li_1CoPO_4 have been prepared by Modified Pechini-type Polymerizable method. Thermogravimetric and Differential Thermal Analysis (TG/DTA) confirm the formation of the samples at 800°C. X-ray diffraction (XRD) pattern of all the samples shows the formation of pure phase with ordered olivine structure. Laser Granulometry studies reveal the particle size distribution with nanosized particles and micron sized aggregates. Electrical conductivity studies reveal that conductivity of all the samples increase from ~10^{-7} to ~10^{-5} Scm^{-1} when temperature increases from 673K to 773K.

Well defined peaks for pure TiO_2 in XRD results, are an evidence for the formation of rutile phase TiO_2. Absence of $LiClO_4$ peaks indicates the dissociation of the salt in the polymer matrix. From impedance analysis the ionic conductivity has been found to be high at 3.23×10^{-4} Scm^{-1} at 303K for the composite polymer electrolyte with 15m% TiO_2. This enhancement in conductivity is due to the dispersion of nano TiO_2

filler which creates free pathways for the ions to move. Temperature dependent ionic conductivity obeys the VTF equation which reveals the segmental motion of the polymer chain. From conductance spectra, hopping frequency has been found to be high of the order of 7.67×10^7 Hz for the composite with 15m% of TiO_2 and hence the low value of relaxation time (1.69×10^{-6} s).

REFERENCES

1. Author 1, Tadashi Matsushita, Author 2, Kaoru Dokko and Author 3, Kiyoshi Kanamura (2005) Comparison of Electrochemical Behavior of LiCoO2 Thin Films Prepared by Sol-Gel and Sputtering Processes *Journal of The Electrochemical Society* **Vol. 152 (11),** Pg. no A2229-A2237
2. Author 1, Wu Yang, Author 2, Youli Qi, Author 3, Yongjun Ma and Author 4, Xien Li, (2004) Synthesis of Nd2O3 nanopowders by sol–gel auto-combustion and their catalytic esterification activity *Materials chemistry and physics* **Vol. 84** Pg. no 52-57
3. Author 1,Felix B. Dias, Author 2, Lambertus Plomp and Author 3, Jakobert B. J. Veldhuis (2000)Trends in polymer electrolytes for secondary lithium batteries *Journal of Power Sources* **Vol. 88** Pg.no 169-191
4. Author 1, Gianni Antonioli, Author 2, Pier Paolo Lottici, Author 3, Ivano Manzini and Author 4, Guglielmina Gnappi (1994) An EXAFS study of the local structure around Zr atoms in Y2O3-stabilized ZrO2 by the sol-gel method *Journal of Non-crystalline solids* **Vol. 177** Pg.no179-186
5. Author 1, Shigeto Okada, Author 2,Shoichiro Sawa, Author 3, Minato Egashira and Author 4, Jun-ichi Yamaki (2001) Cathode properties of phospho-olivine LiMPO4 for lithium secondary batteries Journal of Power Sources Vol. 97-98 Pg.no 430-432
6. Author 1,D. Morgan, Author 2, A. Van der Ven and Author 3, G. Ceder (2004) Li Conductivity in LixMPO4 (M = Mn, Fe, Co, Ni) Olivine Materials *Electrochem. Solid-State Lett.*, **Volume 7, Issue 2**, pp. A30-A32
7. PCPDF 32-0578, Krabbenhoft.D, McCarthy.G, North Dakota State University, Fargo, North Dakota, USA ICDD Grant-in-Aid (1980)
8. PCPDF 32-0552, Krabbenhoft.D, McCarthy.G, North Dakota State University, Fargo, North Dakota, USA ICDD Grant-in-Aid (1980)
9. Author 1, O. Bohnke, Author 2, S. Ronchetti and Author 4, D. Mazza (1999) Conductivity measurements on nasicon and nasicon-modified materials *Solid State Ionics*, **Volume 122, Issues 1-4**, Pages 127-136
10. PCPDF 30-0751, Pustelnik, Soloniewicz, Rocz. Chem., 46, (1972) 1961
11. PCPDF 34-0180, Smith, Dept. Geol. Sciences, Penn State Univ, University Park,USA, (1979)
12. J.R. Macdonald (Ed.), Impedance Sepctroscopy, John Wiley & Sons, New york, 1987 p12-23
13. Author 1, M. S. Michael, Author 2, M. M. E. Jacob, Author 3, S. R. S. Prabaharan and Author 4, S. Radhakrishna (1997) Enhanced lithium ion transport in PEO-based solid polymer electrolytes employing a novel class of plasticizers *Solid State Ionics*, **Volume 98, Issues 3-4,** Pages 167-174

6

CLAY NANO-COMPOSITES AS ENVIRONMENTAL FRIENDLY CATALYSTS FOR GREEN CHEMICAL SYNTHESIS

Vasundhara Singh

ABSTRACT

Catalyst is one of the fundamental pillars of 'Green Chemistry', which is defined as the design of chemical products and processes that reduce or eliminate the use and generation of hazardous substances. Heterogeneous catalysis in particular addresses the goals of 'Green Chemistry' by providing the ease of separation of products and catalyst, recyclability, increased yields and selectivity, thereby eliminating the need for separation through distillation or extraction or other unit operations. Another class of liquid–liquid biphasic catalysis using ionic liquids as green solvents due to their non volatility and easy recylability has been exploited in synthesis. The development of new methodologies based on modified forms of clay and clay nano-composite materials give an insight both in the studies towards the behavior and synthetic potential of clays in conjunction with ultrasonic and microwaves and their applications in natural product synthesis and biologically active compounds.

1. INTRODUCTION

Developments of recoverable and recyclable catalysts with high TON and TOF, improved selectivity and reactivity for industrial applications have become important from both environmental and economical point of view and as per the guiding principles of green chemical synthesis [1]. To achieve this objective several catalysts are anchored or immobilized onto polymeric and inorganic supports. Design and synthesis of organic-inorganic composite catalysts is necessary as industry still prefers heterogeneous catalysts for greener processes due to operational simplicity, recyclability and ease of separation. A large number of active homogenous catalysts, particularly organometallic complexes exist and are being developed for catalyzing several important organic reactions. The encapsulation of metal complexes into solid supports has attracted much interest due to the possibility of obtaining stable heterogeneous catalytic systems. Among aluminosilicates, zeolites or clays are often materials of choice as they are naturally occurring materials, which can be modified to serve as hosts to organic moieties due to their characteristic structural properties. Commonly used organophilic clays are modified with alkyl ammonium cations, which hold apart the aluminosilicate layers permanently. The interlayer distance varies with the chain length of the alkyl group. Intercalation of cationic complexes is possible and expansion up to 25Å is possible depending upon the nature of the cation. Thus the composite formed with the active functional group in the interlayer spacing of nanometer scale are good systems for designing nanoreactors.

2. CLAYS

Two broad classes of clays may be identified: cationic clays or clay minerals, widespread in nature, and anionic clays or layered double hydroxides LDHs, more rare in nature but relatively simple and inexpensive to synthesize on laboratory and industrial scales.

The cationic clays have negatively charged alumino-silicate layers, with small cations in the interlayer space to balance the charge, while the anionic clays have positively charged brucite-type metal hydroxide layers with balancing anions and water molecules located interstitially. Cationic clays are mainly prepared starting from the minerals whereas anionic clays are usually synthesized. The physical and chemical properties of these clays with their many industrially important catalytic applications/ reactions such as nitration, alkylation, oxidations, reductions and asymmetric catalysis have been well documented in literature [2]. A general diagrammatic representation of different clays is shown as above in Figure 1.

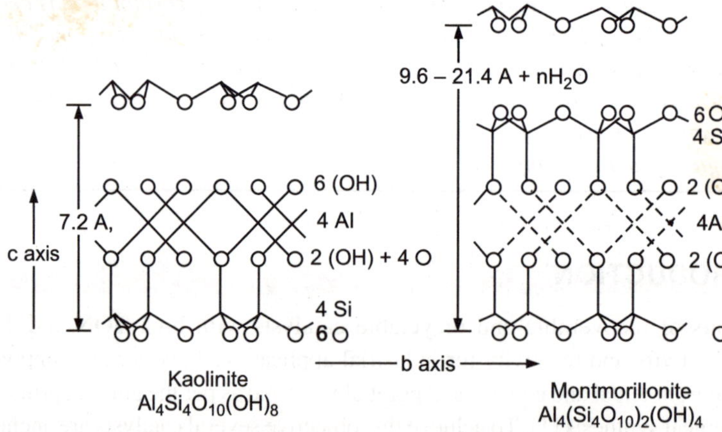

Fig. 1 *Clay layer structuresVertical dimensions: 7-14 Å = 0.7- 1.4 nm.*

2.1 Cation Exchanged Clays

The acidity of Montmorillonite clay is further enhanced by ion exchange with aquometal ions such as Al^{3+}, Cr^{3+}, Cu^{2+}, and Ti^{4+} etc to a level similar to mineral acids. The presence of water influences the acidic properties of clays. For cation-exchanged clays, Bronsted acidity is induced by polarizing power of the cation (e.g., Al^{3+}, Fe^{3+}), which depends on the size and the charge of the ion. Calcination of these cation-exchanged clays results in a marked reduction in Bronsted acidity and assumes much more pronounced Lewis acid character, which is due to the transformation of Bronstead acidity into Lewis acidity [3].

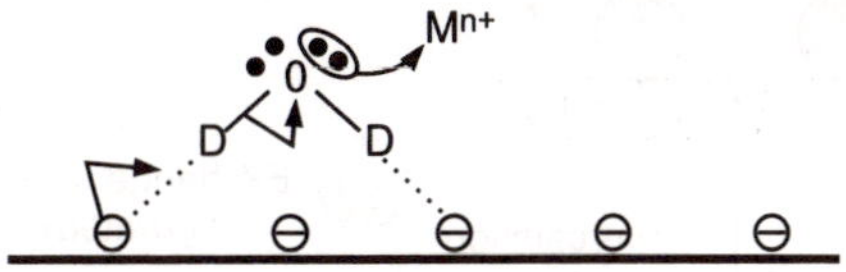

Fig. 2 *Bronsted acidity is induced by polarizing power of the cation*

2.2 Oranometallic Complexes Immobilization in Clay

Immobilization of a metal complex catalyst in smectite galleries demands a net positive charge on the complex. The positive charge ensures electrostatic binding of the polyanionic layers. Cationic analogs of neutral metal complex catalysts can be prepared by replacing neutral ligands with positively charged ligands.. The supported catalysts so obtained, called inorganic-organic hybrid materials, show catalytic activity similar to their homogeneous counterparts and can be recovered easily for recycling. Fernandez and coworkers [4] have prepared highly active recyclable Pd catalysts modified with a rigid C, N, C-tridentate pincer carbene ligand and successfully applied to the coupling of aryl halides with terminal acetylenes.

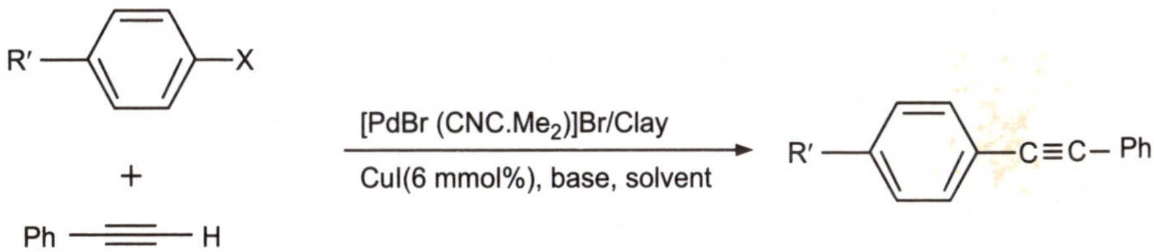

2.3 Enzyme Immobilization

Alcantara et al. [5a] have studied the enzymatic activities resulting from the adsorption of *Rhizomucor miehei* lipase (RML) and *Candida cylindracea* lipase (CCL) onto different phyllosilicates (sepiolite, palygorskite and montmorillonite) as supports for lipase immobilization. Recently, Rahmana et al. [5b] have studied the immobilisation of lipase from *Candida rugosa* on layered double hydroxides of Mg/Al and its nanocomposite as biocatalyst for the synthesis of ester.

2.4 Pillared Clays

Intercalated/pillared clays are a new class of porous materials having many characteristics of 2D zeolite like materials, developed by molecular engineering, in which a homogeneous distribution of micro-pores can be obtained with pore openings varying from 4 to 18Å according to the type of pillars. These are larger than those found in traditional zeolites. These solids are attractive because they can adsorb and convert larger molecules.

After pillaring, the presence of this new porous structure and the incorporation of new active sites present several possible applications of these materials. The pillaring process is summarized in **Figure 3** [6].

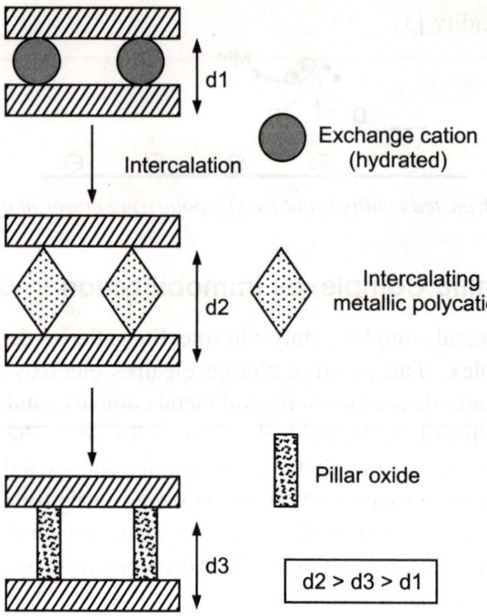

Fig. 3 *Schematic representation of the pillaring process*

Several inorganic hydroxy species have been used for intercalation including those of aluminium, zirconium, chromium, iron, silicon, gallium, tantalum, nickel and titanium [7].

Several important reactions have been accomplished using modified forms of clays [8] and some of them are as given below:

I. Friedel-Craft Reaction

Figueras et al. [9] have used a series of clays obtained by exchanging the original cations of montmorillonite K-10 by Ti^{4+}, Fe^{3+}, Zr^{4+}, Cu^{2+}, Ce^{3+}, Cr^{3+} and Sn^{2+}- cations.

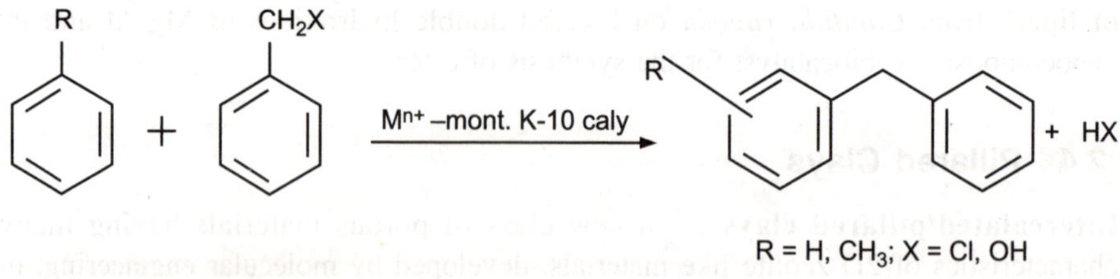

II. Condensation

Bandgar et al. [10] have studied condensation of ethyl diazoacetate with aldehydes under mild conditions using mont K-10.

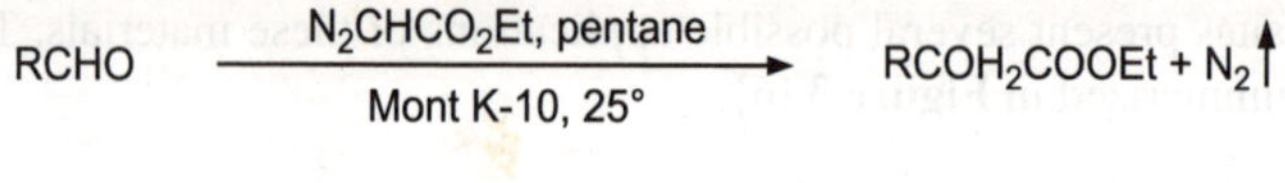

III. Heck reaction

Sudalai et al. [11] have found palladium and copper exchanged mont. K-10 clay as an effective, efficient and recyclable catalyst for vinylation of aryl halides.

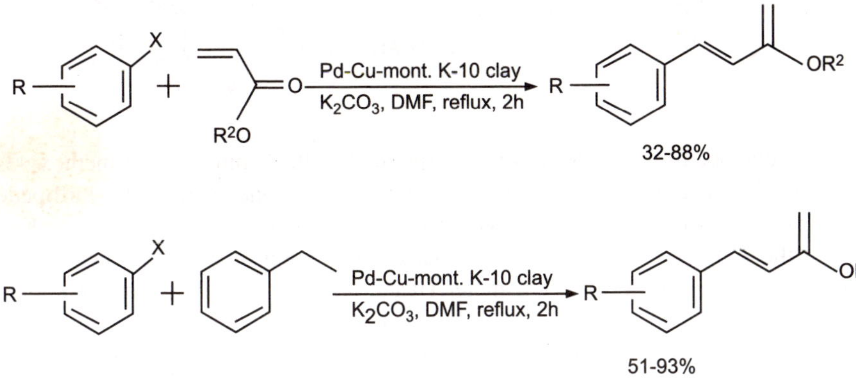

2.5 Oxidation

Choudary and coworkers [12] have developed a simple and convenient method for allylic oxidation of cyclic and acyclic alkenes to the corresponding α,β-unsaturated carbonyl compounds in good yields using catalytic amounts of chromia-pillared montmorillonite and equimolar quantities of *tert*-butyl hydroperoxide.

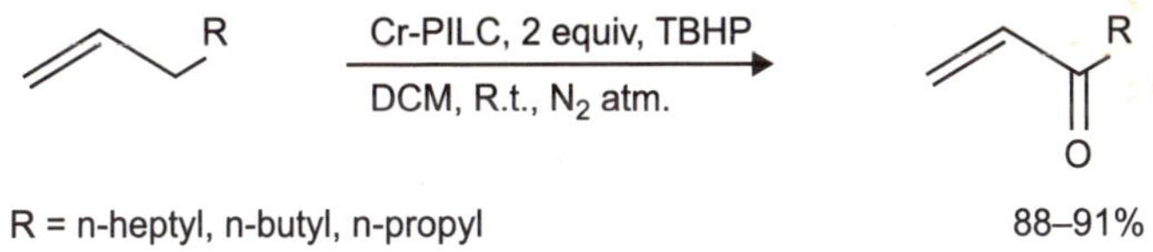

2.6 Characterization of Heterogeneous Catalysts

Various physical techniques are used for structure determination of these naturally occurring and modified solid heterogeneous catalysts are X-ray diffraction analysis **(XRD),** Electron-X-ray dispersive coupled with Scanning electron microscope **(EDX-SEM)** technique, Atomic Absorption Spectroscopy **(AAS), ICP-AES** and X-ray-Flourescent Spectroscopy **(XFS)** is used for determining the metal ion content. Temperature programmed desorption **(TPD, TPT or TP)** techniques are used to determine the surface acidity of the catalyst. Both Bronstead and Lewis acidity can be easily detected using FT-IR TPD experiments to mention a few.

Some of the Work Done in our Research Group is as Follows

Part A: K-10 Clay has been used for the selective bromination of benzylic alcohols using NaBr on clay [13a]. Facile synthesis of Elvirol, Curcuphenol and Sesquichamaenol using Montmorillonite K-10 clay in dry media using the Friedal Craft reaction was achieved [13b].

Part B: Ultrasound-activated oxidation of hydroquinone to quinones catalyzed by ceric ammonium nitrate doped on metal-exchanged montmorillonite-K10 clay [13c]. The general reaction is as below:

General Reaction

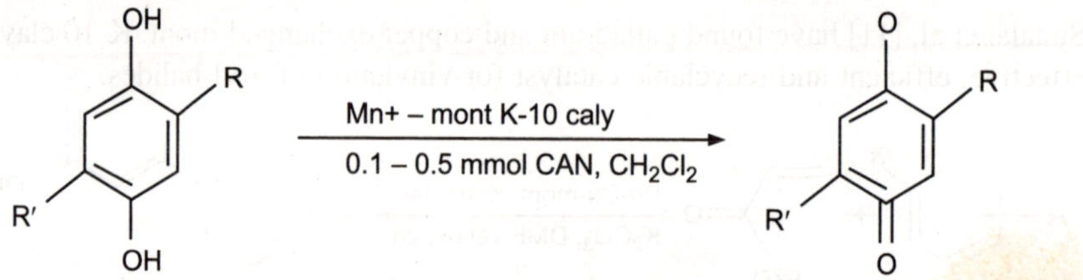

The developed methodology was incorporated in the synthesis of 2-methoxy-3-methyl-1, 4-benzoquinone; a defense and alarm pheromone secreted by millipede *Narceus gordanus* from the glands along its body, when it is attacked by enemies. Key steps in the synthesis are; microwave-assisted preparation of dimethylether of 2,6-dihydroxytoluene followed by Vilsmeier-Haack reaction and ultrasound promoted Dakin in aqueous media. Finally the hydroxylated aromatic compound was subjected to our own developed methodology

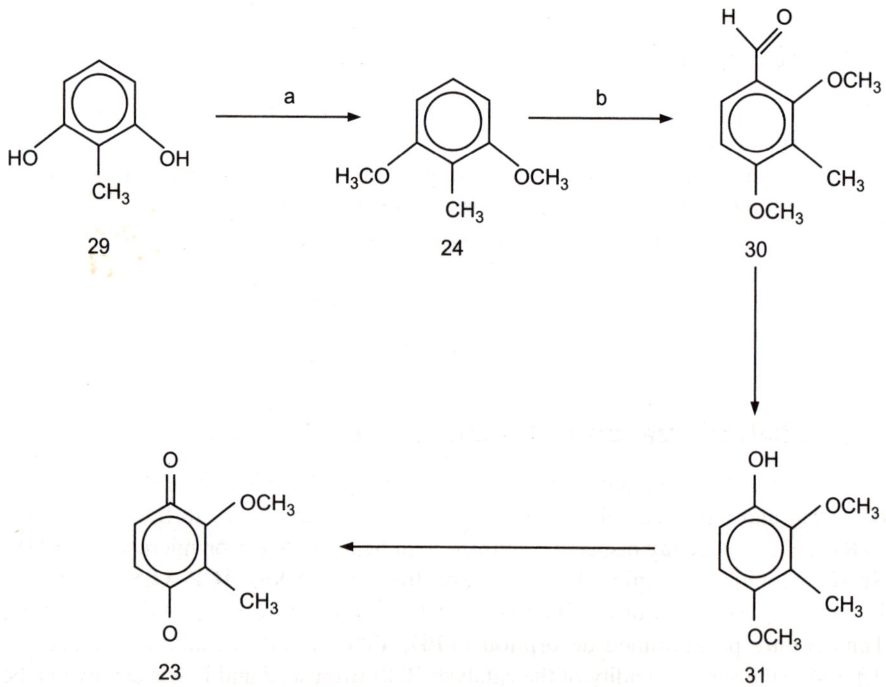

Reagents: (a) aq. NaOH, C_2H_5OH, $(CH_3)_2SO_4$, MW(630 W); (b) $POCl_3$, DMF; (c) Na_2CO_3.1.5 H_2O_2, THF, H_2O, N_2, CH_3COOH; (d) Fe^{3+}-mont K-10, CAN(1.5 eq.), H_2O, 30 min.

Part C embodies the synthesis of chimaphilin and hydroxychimaphilin isolated form *Chimaphilin sp.* and *Pyrola sp.* Chimaphilin and hydroxychimaphilin are yellow naphthoquinones, for long known to have antifungal, antibacterial, antihemorragic properties. The regio- and stereo control depends on the nature of the cation exchanged in the clay. In our efforts to study regio-selectivity in Diels-Alder, we have particularly studied the cation-exchanged clays (La^{3+}, Dy^{3+}, Sm^{3+}, Tb^{3+}, Nd^{3+}, Pr^{3+}, Gd^{3+}, Fe^{3+}, Zn^{2+}) as catalysts, which are mild Lewis acids as shown in **Figure 4.**

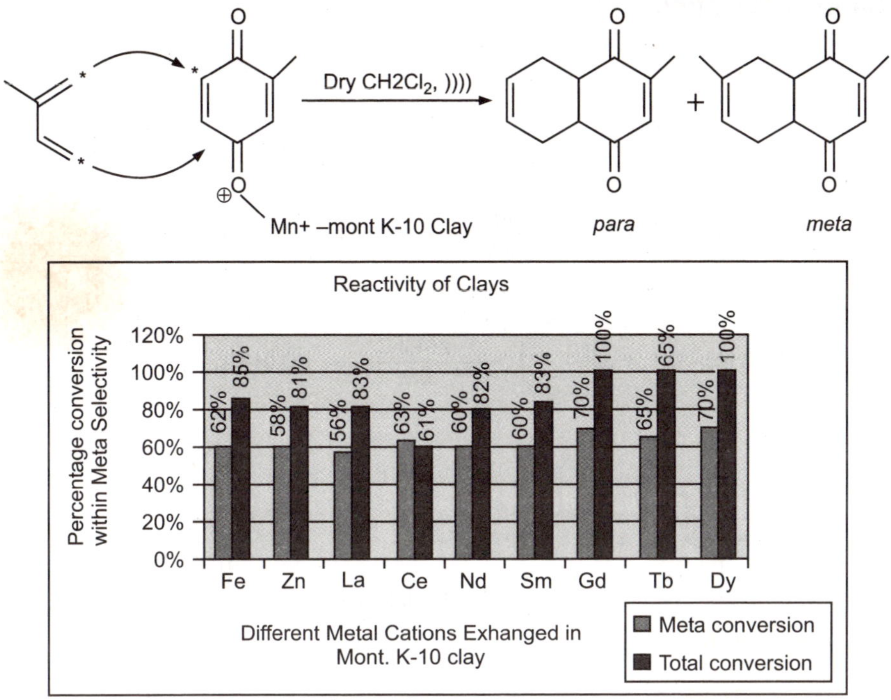

Fig. 4 *Reactivity of various cation exchanged clays,* (For color illustration refer p xvii)

Part D: incorporates the preparation and characterization of ZrO_2-pillared clay [13d] catalyst using microwave in conjunction with ultrasound energy as shown in **Figure 5.**

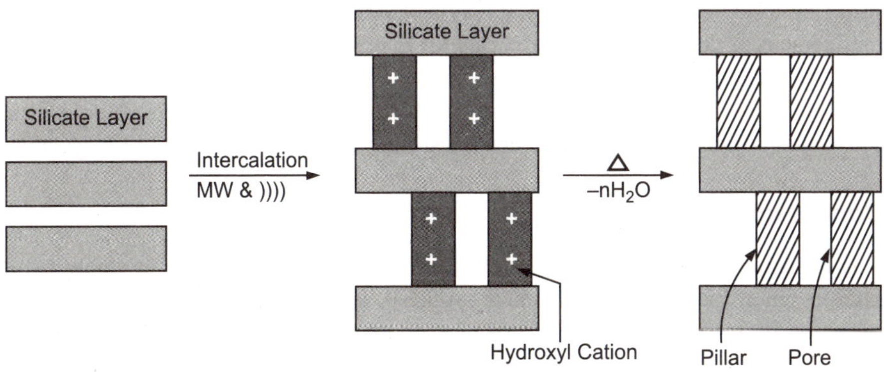

Fig. 5 *ZrO_2-pillared clay prepared by ultrasonic and microwave activation,* (For color illustration refer p xvii)

Applications of ZrO_2-pillared clay for the selective monotetrahydropyranylation and monoacetylation of different 1,n-symmetrical diols [13e] and for one-pot Biginelli condensation reaction including the synthesis of monastrol and nitractin (anticancerous, antiviral and antitrachomal drugs) as the key step [13f] has been achieved as shown below:

General Reaction

$$R\text{–}OH + Ac_2O \xrightarrow[MW]{Zr\text{-}PILC} R\text{–}OAc$$

$$HO(CH_2)_nOH + (AC_2O) \xrightarrow[MW]{Zr\text{-}PILC} OH(CH_2)_nOAc + AcO(CH2)_nOAc$$

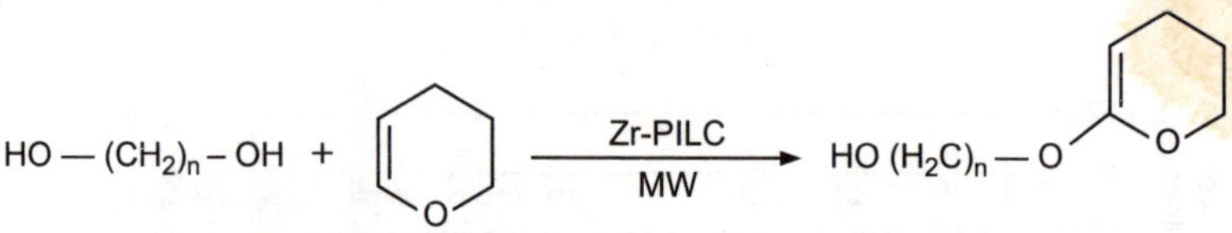

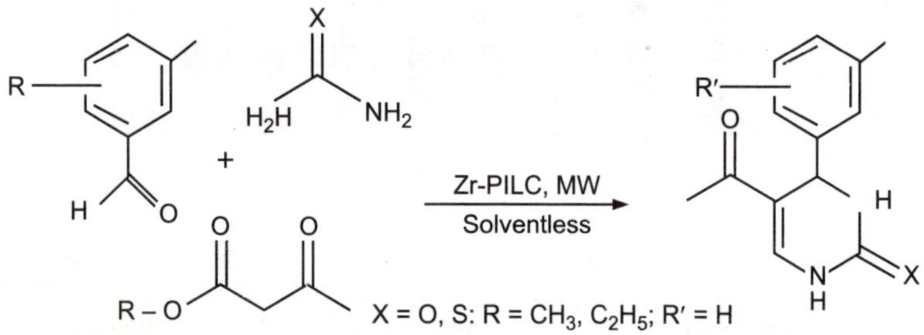

Part E: A ionic supported palladium complex has also been synthesized and immobilized in clay making an organic – inorganic hybrid catalyst in ionic liquid medium which acted as an excellent catalyst in Heck, Suzuki and Sonogashira reaction occurring in the interlayer spacing of clay and surprising results due to the restricted confinement of the reactants and the catalyst is observed. Micro molar concentration of catalyst at lower temperature and shorter reaction time and was active even after seven recycles. The catalytic activity of these catalysts was found to be better in ionic liquid medium than molecular solvents. Also the solvent and catalyst are recyclable.

Conclusion

Clay nanocomposites can be used as efficient recyclable and recoverable catalysts for environmental friendly green chemical synthesis as demonstrated from literature and our work. Further work in making organic – inorganic hybrid catalysts is in progress. Use of ionic liquids as green solvents is also being explored.

Acknowledgement

This work was possible due to the my dedicated co-workers, Dr.Varinder Sapehiyia, Sukhbir Kaur, Vivek Srivastava, Rajni Ratti and financial support from DST, New Delhi, AICTE, New Delhi (for Young Teachers Career award) and Indo French Centre for promotion of advanced research (IFCPAR), New Delhi.

REFERENCES

1. (a) Anatase, P. T. and Warner, J. C. (1988) '*Green Chemistry: Theory and Practical*' Oxford Press, Oxford (b) Anastas, P. T. and Kirchhoff, M. M. (2002) *Acc. Chem. Res.* **35**, 686
2. Vaccari, A. (1999) *Appl. Clay Sci.* **14**,161; (b) Nikalaje, M. D., Phukan, P. and Sudalai, A. (2000) *Org. Prep. Proced. Int.* **32**, 1.
3. Balogh, M., and Laszlo, P. (1993) In '*Organic Chemistry Using Clays*' Springer-Verlag: New York.
4. Poyatos, M., Marquez, F., Peris, E., Claver, C. and Fernandez, E. (2003) *New J. Chem.* 27, 425 and reference cited therein.
5. (a) de Fuentes, I. E., Viseras, C. A. Ubiali, D., Terreni, M. and Alcantara, A. R. (2001) *J. Mol. Catal. B: Enzymatic* **11**, 657.(b) Rahmana, M. B. A., Basri, M., Hussein, M. Z., Idris, M. N. H., Rahmanb, R. N. Z. R. A. and Salleh, A. B. (2004) *Catal. Today* **93–95**, 405.
6. Gil, A., Gandia, L. M. and Vicente, M. A. (2000) *Catal. Rev.-Sci. Eng.* **42**, 145.
7. (a) Brindley, G. W. and Sempels, R. E. (1977) *Clays Clay Minerals* **12**, 229; Chae, H. J., Nam, I.-S., Ham, S. W. and Hong, S. B. (2001) *Catal. Today* **68**, 31; Nagendrappa, G. (2002) "Organic Synthesis using Clay Catalysts" *RESONANCE* 64.
8. (a) Cseri, T., Békássy, S., Figueras, F. and Rizner, S. (1995) *J. Mol. Catal. A: Chemical* **98**, 101.
9. Selvam, P. and Gupta, N. M. (2004) *Appl. Catal. A: General* **276**, 5.
10. Bandgar, B. P., Pandit, S. S. and Sadavarte, V. S. (2001) *Green Chem.* **3**, 247.
11. Sudalai A.; Choudhary V.R.; Wakharkar R.D.; Vinod M.P.; Uphade B.S.; Ramchandani R.K.;(1997) *Chem Commun.,* 2071.
12. Choudary B. M. P., Durga Prasad A. and Vinukonda S. (1992) *Tetrahedron* **48**, 953.
13. (a) Singh, V., Kad, G. L., Kaur, K. and Singh, J. (1997) *Tetrahedron Lett.* **1079**. (b) Singh, V., Sapehiyia, V., Srivastava, V. and Kaur, S. (2002) *J. Chem. Soc. Perkin Trans.* **1**, 1766 (c) Singh, V., Sapehiyia, V. and Kad, G. L. (2003) *Synthesis* 198 (d) Singh, V., Sapehiyia, V. and Kad, G. L. (2004) *J. Mol. Catal. A: Chemical* 119-124 (e) Singh, V., Kaur, S., Sapehiyia, V., Singh, J. and Kad, G. L. (2005) *Catal. Commun.* **6**, 57. (f) Singh, V., Sapehiyia, V., Srivastava, V. and Kaur, S. (2006) *Catal. Commun.* 456.

7

MICRONUTRIENT DISORDERS – CHALLENGE TO SUSTAIN FOOD GRAIN PRODUCTION

Upkar S. Sadana

ABSTRACT

Micronutrient deficiencies have become one of the major constraints in sustaining agricultural production in India. Analysis of more than two lakhs soil samples in different states of the country revealed the predominance of Zn deficiency in 48 per cent samples followed by Fe and Mn deficiency in 12 and 5 per cent samples, respectively. Coarse textured, alkali, flood plain and calcareous soils, which have relatively high pH and/or low organic matter content, are more prone to micronutrient deficiencies. As such it is not advisable to use micronutrient mixtures for combating Zn, Fe or Mn deficiency in crops. Organic manures are not only a good source of micronutrients but have also been found to enhance their availability in soil through chelation and reduction in pH. Judicious and efficient management of costly micronutrient fertilizers holds the key to maintain sustainable agricultural production. Zinc has received the major focus of attention of the micronutrient scientists due to its wide spread deficiency all over the country. Soil application of zinc sulphate is the most commonly used method to correct Zn deficiency in different crops. Severe Fe and Mn deficiency is difficult to manage with soil application due to the oxidation of soil-applied iron and manganese sulphate at high soil pH. Foliar application of Fe and Mn is an immediate effective measure to combat their deficiency; however, Fe and Mn have to be applied repeatedly through foliar application. Other approach could be sowing of nutrient-efficient crops that can grow well on soils low in micronutrients. For nutrients such as Fe, Mn and Zn which generally have an extremely low solubility in soils of arid and semi-arid regions, the response to nutrient deficiency stress is mainly biochemical. The response may involve changes in rhizosphere pH, redox potential, root exudates and microbial activities. Identification and / or evolving of micronutrient efficient crop cultivars having either low micronutrient requirement for potential yields or capable of mining micronutrients from the less available pools should be given priority.

1. INTRODUCTION

Micronutrient deficiencies have become one of the major obstacles in sustaining crop production in the present agriculture in India. These deficiencies appeared much faster in the northern states as compared to other parts of the country, primarily because of fast adoption of new agricultural technologies like cultivation of high yielding crop varieties, increase in cropping intensity, expansion of irrigation facilities, increased use of pure fertilizers and poor quality irrigation water. Though food grain production increased tremendously and made the country self-sufficient in food grain production, it also resulted in the faster depletion of the finite micronutrient reserves of soils. There is need to make a quantum jump in raising food grain production from the present around 200 million tones to meet ever increasing demand of growing population. India with 17% of the world population is presently producing around 10% of the world food demand. Intensive cultivation is the only way to feed the growing population, which would cause deterioration of soil health and may result in newer production problems.

The adoption of rice-wheat system particularly in the non-traditional rice growing north Indian states has resulted in the over-exploitation of the natural soil resource base and this trend has been aggravated by the imbalanced use of inputs. The increased use of poor quality irrigation water to meet the water requirement of this cropping system has further aggravated the problems of micronutrient deficiencies.

The first fall out of such exploitive agriculture practices was the appearance of zinc (Zn) deficiency in many parts of the country. Field scale deficiency of Zn was first noticed in rice on Tarai soils in 1965 and in wheat on sandy soils of Punjab in 1970 and later on in most of the intensively cultivated areas, particularly where rice and wheat were grown. Subsequently, the deficiencies of iron (Fe), manganese (Mn) and copper (Cu) were recorded in particular situations and their severity depended on the soil conditions and the crop grown.

1.1 Extent of Micronutrient Deficiencies

The total micronutrient contents of soils are generally of limited value as far as plant growth and responses to their application are concerned. In most cases total contents are not significantly related to the plant content. In order to match the levels of micronutrients in soil with plant requirement, their available contents are determined. Most researchers used DTPA soil test method for determining the available content of Zn, Cu, Fe and Mn particularly in alkaline calcareous soils. Like total contents, the available micronutrient status of soils is also highly variable. Available Zn, Cu, Fe and Mn content of Indian soils ranged from 0.2 to 6.9, 0.1 to 8.2, 0.8 to 196 and 0.2 to 118 mg/kg soil, respectively [15]. Soil properties exercise a considerable influence on the availability of micronutrients. The extent of micronutrient deficiency, therefore, varies not only in different states but also in different districts of the same state or in various blocks of the same district depending upon the soil characteristics and other management conditions.

Among micronutrients, Zn deficiency is the most common and widespread in the country. Analysis of more than two lakh samples in different states of the country revealed the predominance of Zn deficiency in divergent soils (Table 1). Of these samples 48, 12, 5 and 3 per cent samples were found to be deficient in available Zn, Fe, Mn and Cu, respectively [15]. The proportions of Zn deficient soils in different states vary from 8% in Pondichery to 72.8% in Karnataka. Despite the fact that soil analysis is extremely useful in identifying the nature and extent of deficiencies of a particular micronutrient, facilities for testing micronutrients in soils are not adequate to cope up with the required volume of analysis. Under such situations one has to look for the soils and situations under which the deficiency of a particular micronutrient is likely to show up in crops when adequately fertilized with NPK and properly managed.

Table 1 *Extent of Zn, Cu, Fe and Mn deficiency in India*

Name of State/ Union Territory	*No. of soil samples analyzed*	*Percent samples deficient*			
		Zn	*Cu*	*Fe*	*Mn*
Andhra Pradesh	8158	49.4	<1	3	1
Assam	12165	34.0	<1	2	20
Bihar	19214	54.0	3	6	2
Gujarat	30152	23.8	40	8	4
Haryana	21848	60.5	2	20	4
Karnataka	27860	72.8	5	35	17
Madhya Pradesh	32867	43.9	<1	7	1
Orissa	16040	54.0	–	0	0
Pondichery	4108	8.0	4	2	3
Punjab	16483	48.1	1	14	2
Tamil Nadu	28087	58.4	6	17	6
Uttar Pradesh	26126	45.7	1	6	3
West Bengal	6547	36.0	0	0	3
All India	251660	48.5	3	12	5

Source: Singh [15]

The extent of Fe deficiency is approximately one fourth that of Zn deficiency in the country and is largely prevalent in vast areas of alkaline to calcareous soil tracts. Iron deficiency seems to be only second in importance to Zn deficiency in the states of Karnataka, Haryana and Punjab. Manganese deficiency for field crops in the country is mainly at localized sites where rice-wheat crop rotation is practiced in coarse-textured soils. Copper deficiency is not widespread except in soils of Gujarat. Diagnostic clinical approach (DCA) in solving intricate problems related to micronutrient stresses in soils and plants is a prerequisite [3]. Development of visual deficiency symptoms for micronutrients under solution, sand and field level be correlated and refined for better understanding of symptomology of micronutrient deficiencies in crops.

1.2 Management of Zn Deficiency

Zinc has received the major focus of attention of the scientists working on micronutrients in view of its wide spread deficiency all over the country. Coarse textured soils, alkali soils, flood plain soils and calcareous soils, which have relatively high pH and low organic matter content, are more prone to Zn deficiency [6, 14]. Similarly, fine textured calcareous black soils (vertisols) and highly leached red soils are expected to show greater incidence of Zn deficiency. Awareness amongst farmers on the menace of soil Zn deficiency has resulted in the increased use of zinc sulfate. As a result, the status of available Zn in soil has increased in many areas and the extent of Zn deficiency in many states has decreased depending on the rate and frequency of zinc sulfate application. Quite a large number of multi micronutrient mixtures have been introduced under different trade names in the market with a philosophy of supplying all micronutrients irrespective of their deficiencies. Evaluation of such mixtures, used mainly for foliar application or seed treatment, has proved these to be distinctly inferior carrier of Zn owing to relatively low content of Zn in these mixtures and their inability to meet the Zn requirement of the crops. As such it is not advisable to use these micronutrient mixtures for combating Zn deficiency of crops. Organic manures are not only a good source of micronutrients but have also been found to enhance the availability of the native soil Zn through chelation. Their application to soil builds up available Zn status in all types of soils.

Among the methods of Zn application, the broadcast and mix method has been reported to be the most efficient compared to its drilling or band placement and topdressing at the optimum rate of application. In view of the very low efficiency of soil applied Zn owing to its conversion to unavailable forms, efforts have been made to correct Zn deficiency by seed coating or soaking in Zn solution and dipping or soaking seedling roots. By and large all these methods, except dipping seedling roots in ZnO suspension, proved either inferior to or at par with soil mode of Zn application [6]. Dipping of rice seedlings in ZnO suspension showed some promise in increasing Zn use efficiency, but it could not be popular among farmers because of the practical difficulties in carrying out the practice. Enrichment of rice nursery with Zn by applying 150 kg $ZnSO_4$ ha^{-1} significantly increased the paddy grain yield it could not match $ZnSO_4$ applied to soil @ 60 kg ha^{-1} at the time of transplanting rice. Cultivation of Zn-efficient crops and their cultivars under Zn stress conditions is another alternative to combat Zn deficiency. As a result of the work carried out in this direction, a number of cultivars of different crops have been identified as tolerant to Zn-stress. However, due to the availability of the easier and efficient techniques for correcting Zn deficiency with Zn fertilizers, the cultivation of these varieties has not become popular among the farmers.

1.3 Management of Fe Deficiency

Iron deficiency is a widespread mineral disorder in higher plants grown in calcareous soils. Poor availability of Fe in the soil, insufficient uptake and iron inactivation within the plants are reported to be the main causes of iron chlorosis in crops grown on such soils. In India, Fe is considered to be the second limiting micronutrient in crop production after Zn. Of the various inorganic carriers of Fe, ferrous sulphate ($FeSO_4.7H_2O$) has been rated as the most efficient and most commonly used inorganic carrier for correcting Fe deficiency in rice-wheat system. Use of organic manures alone or in combination with $FeSO_4.7H_2O$ has been found to be effective in correcting Fe deficiency. In view of the high sensitivity of Fe to electro-chemical changes in soils, addition of easily decomposable organic carbon materials is expected to increase the availability of the native soil Fe. Sadana and Nayyar [9] reported that the adoption of irrigated upland rice-wheat cropping system on coarse textured soils caused widespread Fe deficiency that can be rectified by pre-transplanting incorporation of *Sesbania aculeata* green manure with 3 sprays of 1 % $FeSO_4 7H_2O$. However, the deficiency of Fe cannot be remedied through the use of fertilizers as the problem is not of abundance, but rather of solubility. Hence selection and screening of Fe efficient crops should get prime importance to increase the food grain production.

1.4 Management of Mn Deficiency

Severe Mn deficiency is difficult to manage with soil application due to oxidation of soil-applied Mn at high soil pH. Several sources of Mn viz. manganese sulphate ($MnSO_4$), manganese oxide (MnO) and manganese-frits as well as some micronutrient mixtures have been evaluated for their efficiency to correct Mn deficiency. While $MnSO_4$ is the most efficient and economical source of Mn. Manganese oxide proved to be extremely inefficient as the plants in pots treated with it to supply 40 mg Mn kg^{-1}, could not even reach the reproductive stage under greenhouse conditions. Comparing a number of multi micronutrient mixtures with $MnSO_4$ as carriers of Mn under field condition, Sadana et al. [7] have shown them to be distinctly inferior to $MnSO_4$. Foliar application of Mn is an immediate effective measure to combat Mn deficiency in wheat [5, 8, 16] though it has to be applied every year. The other preferable approach could be sowing of Mn-efficient crops and their cultivars that can grow well on low Mn soils [10, 11, 12, 13]. Bansal and Nayyar [1] screened wheat cultivars under field conditions and found that wheat cultivars, PBW 34, PDW 233, and PDW 215 were Mn inefficient and WH 542, PBW 343 and HD 2329 were Mn efficient. Under field conditions, wheat cultivar PBW 343 produced 1.5 times higher grain yield than PDW 233, when grown on low-Mn soil. D*urum* wheat cultivars are usually more sensitive to Mn deficiency as compared to *aestivum* cultivars [4, 10]. Wheat cultivars (*Triticum aestivum* L. cv. PBW 373, WH 542, PBW 343 and PBW 502 and *Triticum durum* L. cv. PDW 274) were grown on cultivator's field on Mn deficient soil. All the 5 wheat cultivars showed Mn deficiency symptoms when grown on Mn deficient soil having DTPA-extractable Mn 2.81 mg/kg soil. However, Mn deficiency was more severe in *durum* cv. PDW 274 than *aestivum* wheat cultivars. *Durum* cv. PDW 274 produced only 65% of its maximum grain yield under Mn deficiency conditions; where as *aestivum* wheat cv. PBW 373 produced 93% of its maximum grain yield followed by 83-85% grain yield by other 3 *aestivum* wheat cultivars. Wheat grain yield under field conditions was directly related to total Mn uptake. Further, restriction in root growth was least in cv. PBW 373 but more in durum wheat cultivars. Therefore, the higher Mn efficiency of *aestivum* wheat cultivars could be related to higher Mn uptake and less restriction in root growth under Mn deficiency conditions.

The uptake efficiency of a plant depends on the size of the root system and the efficiency of uptake of each root segment i.e. the influx [2] that depends on the concentration gradient in the rhizosphere. The possible mechanisms by which plants can increase the concentration gradient in the rhizosphere are: a) by decreasing the Mn concentration at the root surface to a lower value b) by increasing the concentration in soil solution through root exudates that increase Mn solubility in soil c) or by both.

Several authors {10, 11,12,13} have used mathematical models to improve the understanding of basic principles involved in the process of nutrient uptake by plant roots growing in soil. These models are based on ion transport from soil to roots by means of mass flow and diffusion and on nutrient uptake kinetics, mostly following Michaelis-Menten kinetics. Uptake kinetics are usually described by a modified Michaelis-Menten equation with the parameters, maximum influx (I_{max}), the Michaelis constant (K_m) and the minimum concentration in solution at which net influx is zero (C_{Lmin}). The roots of nutrient efficient crops should have the capacity to decrease nutrient concentration at root surface to a lower value either by increasing I_{max} or decreasing K_m i.e. a more efficient uptake kinetics.

Sadana et al. [12] using nutrient uptake model calculations showed that even in a soil of low Mn supply, initial Mn concentration of 0.19 mM was decreased at the root surface to only 0.17 mM after 30 days of uptake (Figure 1).

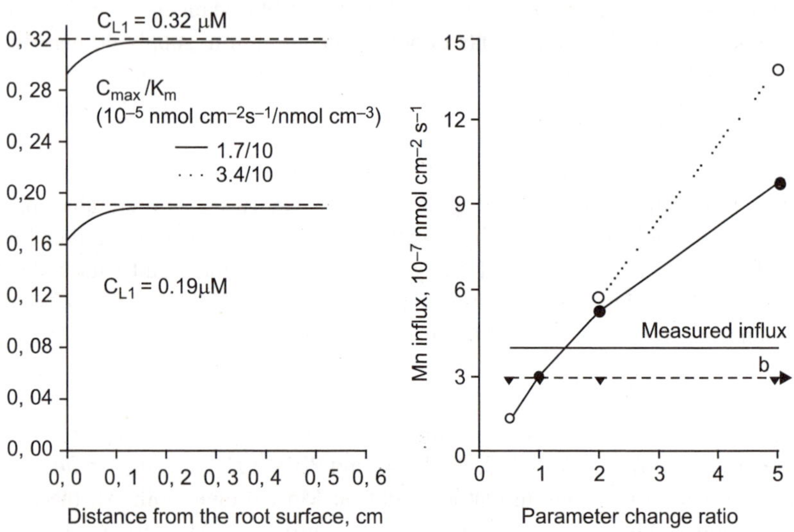

Fig. 1 *Calculated concentration profile of Mn concentration in the rhizosphere of wheat cv. PBW 138 for varying I_{max} /K_m (a), Sensitivity analysis of Mn influx into roots of wheat cv. PBW 138 at initial soil solution Mn concentration, C_{Li} 0.19 mM (b).*

However, increasing the slope of the uptake isotherm (I_{max} /K_m) by two times further decreased Mn concentration at the root surface to 0.15 mM and calculated Mn influx increased from 3.0´10^{-7} nmol cm^{-2} s^{-1} to 5.30´10^{-7} nmol cm^{-2} s^{-1}. These results show that wheat cultivars with a higher slope of the uptake isotherm (I_{max} /K_m) would decrease Mn concentration to a lower value, which would cause a higher concentration gradient and result in greater transport of Mn to the root surface and hence higher Mn influx. Manganese transport to the root surface and thereby the Mn influx would have been much higher had the plant been able to decrease the Mn concentration to a lower value. Crops grown under Mn deficiency conditions can obtain higher Mn influx if there is stronger decrease in Mn concentration at the root surface because of more efficient uptake kinetics.

Sensitivity analysis is a useful tool for evaluating the significance of individual soil and plant parameters regulating Mn influx. In such an analysis, several model calculations are performed and only one input parameter is changed while keeping all other input data constant. The results of the sensitivity analysis for cv. PBW 138 at low soil Mn is depicted on the right side of Figure 1. It can be seen that the initial Mn soil solution concentration (C_{Li}) was the most sensitive parameter controlling Mn influx followed by the slope of the uptake isotherm (I_{max} /K_m). Under low Mn supply conditions, increasing C_{Li} would be helpful in ameliorating Mn-deficiency. The other possibility could be selecting cultivars having more Mn efficient uptake kinetics i.e., the roots should have the capacity to decrease Mn concentration at root surface to a lower value either by increasing I_{max} or decreasing K_m. Manganese availability to wheat plants could also be increased if roots are able to decrease Mn concentration at the root surface to lower value i.e., more depletion on Mn in the rhizosphere causes higher Mn influx.

It is disturbing to note that micronutrient efficiency of present day high yielding cereal cultivars is quite low. It has been hypothesized that plant breeding is carried out under fertile soil conditions. Thus the selection pressure for nutrient efficiency factor is low or even zero. The susceptibility to micronutrient deficiencies is discovered only at the farmer's fields. It would appear that certain plant genotypes are more efficient than others in the mobilization and uptake of nutrients from soils. In most cases this genotypic variation is caused by differences in nutrient acquisition and is associated with morphological and physiological/biochemical properties of the plant roots. For nutrients such as Fe, Mn and Zn which generally have an extremely low solubility in soils of arid and semi-arid regions, the response to nutrient deficiency stress is mainly biochemical. The response may involve changes in rhizosphere pH, redox potential, root exudates and microbial activities.

Conclusion

Incidences of micronutrient deficiencies are increasing at an alarming rate that is a serious threat to sustain agricultural production for food security in India. Integrated use of organic materials along with costly micronutrient fertilizers improves soil health and ameliorates micronutrient deficiency. Other approach could be sowing of nutrient-efficient crops that can grow well on soils low in micronutrients. Selection and screening of micronutrient efficient crops should be carried out on a priority basis.

REFERENCES

1. Bansal, R. L. and Nayyar, V. K. (1998) Screening of wheat (*Triticum aestivum*) varieties tolerant to manganese deficiency stress. *Ind. J. Agric. Sci.* **68**, 66-69.
2. Jungk, A. O. and Claassen, N. (1997) Ion diffusion in the soil-root system. *Adv. Agron.* **61**, 53-110.
3. Malewar, G. U. (2005) Micronutrient stress in soils and crops: Serious sickness and clinical approach for sustainable agriculture. *J. Ind. Soc. Soil Sci.* **53**, 484-499.
4. Nayyar, V. K. (1999) Micronutrient management for sustainable intensive agriculture. *J. Ind. Soc. Soil Sci.* **47**, 666-680.

5. Nayyar, V. K., Sadana, U. S. and Takkar, P. N. (1985) Methods and rates of application of Mn and its critical levels for wheat following rice on coarse textured soils. *Fertil. Res.* **8**, 173-178.

6. Sadana, U. S. and Takkar, P. N. (1983) Methods of zinc application to rice on sodic soil. *Int. Rice Res. Newsl.* **8**, 21-22.

7. Sadana, U. S., Nayyar, V. K. and Kaur, N. P. (1989) Response of wheat grown on manganese deficient soil to foliar application of different micronutrients formulations. Micro. News **II**, 1-2.

8. Sadana, U. S., Nayyar, V. K. and Takkar, P. N. (1991) Response of wheat grown on manganese-deficient soil to methods and rates of manganese sulphate application. *Fert. News* **36**, 55-57.

9. Sadana, U. S. and Nayyar, V. K. (2000) Amelioration of iron deficiency in rice and transformations of soil iron in coarse textured soils of Punjab, India. *J. Plant Nutr.* **23**, 2061-2069.

10. Sadana, U. S., Kusum-Lata and Claassen, N. (2002) Manganese efficiency of wheat cultivars as related to root growth and internal manganese requirement. *J. Plant Nutr.* **25**, 2677-2688.

11. Sadana, U. S., Samal, D. and Claassen, N. (2003) Differences in manganese efficiency of wheat (T*riticum aestivum* L.) and raya (B*rassica juncea* L.) as related to root-shoot relations and manganese influx. *J. Plant Nutr. Soil Sci.* **166**, 385-389.

12. Sadana, U. S., Sharma, P. Castañeda O. N., Samal, D. and Claassen, N. (2005) Manganese uptake and Mn efficiency of wheat cultivars explained by root growth and Mn uptake kinetics. *J. Plant Nutr. Soil Sci.* **168**, 581-589.

13. Samal, D, Sadana, U. S. and Gill, A. A. S. (2003) Mechanistic approach to study manganese influx and its depletion in the rhizosphere of wheat and raya. *Comm. Soil Sci. Plant Analysis* **34**, 3039-3050.

14. Sharma, B. D., Takkar, P. N. and Sadana, U. S. (1982) Evaluation of levels and methods of zinc application to rice in sodic soils. *Fertil. Res.* **3**, 161-167.

15. Singh, M. V. (1999) *Micronutrient deficiency delineation and soil fertility mapping.* In: National Symposium on "Zinc Fertilizer Industry – Whither To", Ed. Singh, R. and Kumar, A.

16. Takkar, P. N., Nayyar,V. K. and Sadana, U. S. (1986) Response of wheat on coarse textured soils to mode and time of manganese application. *Expl. agric.* **22**, 149-152.

8

USING PLANTS FOR REMEDIATION OF METAL POLLUTED SOILS

G. K. Psaras

ABSTRACT

Organic compounds and inorganic elements constitute the key pollutants of soil. Efforts to clean up the soil using a 'green' technology are on. It has been shown that some plants have the capability of eliminating toxic metals from soil. Metal phytoextraction, either for phytomining, or for the so-called phytoremediation of contaminated soil involves the growth of hyperaccumulating plants to concentrate the metal. The metal is extracted from the harvested crop. There are three points that need to be kept in kind. The first is the discovery of plants that can act as hyperaccumulators. Second is the identification of mechanisms by which the plant handles the toxic metals. The third point is to deal with ecological repercussions of hyperaccumulation. We have used scanning electron microscopy, X-ray microanalysis and atomic absorption spectroscopy to study metal hyperaccumulation.

The major soil pollutants are either organic compounds or inorganic elements. While organic compounds could be biologically degraded, inorganic elements and particularly heavy metals should be removed by another way. Physicochemical methods are costly and impose negative impacts on the ecosystem. A great deal of research has shown that plants possess the genetic potential to remove toxic metals from the soil. Indeed, some plant species are able to accumulate metal ions that are toxic for all other organisms. The question is whether this feature could be used to clean up contaminated soils.

As early as in 19th century, Baumann [1] identified plants capable to accumulate high Zn levels. In 1935, Byers [2] reported the accumulation of selenium in *Astragalus* sp. In 1948, Minguzzi and Vergnano [3] identified plants that concentrate high amount of nickel in their above ground parts. Plants that accumulate high concentration of metal are known as metal hyperaccumulators [4]. For most accumulated elements, the threshold concentration to characterize a plant as hyperaccumulator is 1000 μm/g of dry mass (0.1%).

The idea to use plants to extract metals from the soil is relatively new [5]. Metal phytoextraction, either for phytomining, or for the so-called phytoremediation of contaminated sites [6, 7] involves the growing of hyperaccumulating plants to concentrate the metal. Subsequently the crop is harvested and the metal is extracted. Though phytomining has not yet been tested on a commercial scale, it seems to be a promising "green" and environmentally-friend tool.

The very first part of the research concerning metal accumulators was focused on identifying the species which are hyperaccumulators [8]. More than 400 plant species that hyperaccumulate metals have been identified worldwide. The most frequent metal accumulating species (about 300) are also reported to be Ni-hyperaccumulators [9].

The next phase of research on metal hyperaccumulating plants was to understand the physiological mechanisms underlying the normally toxic, metal manipulation by the plant [10, 11, 12]. The "unusual appetite" of plants for metal uptake has been an interesting subject of scientific research [13, 14]. With reference to nickel hyperaccumulation, the physiological point of view raises interesting questions concerning the mechanism of metal uptake, sequestering, and toxicity avoidance [5, 12, 15, 16].The distribution of the metal in the plant body received also much attention. Leaves seem to be the main sinks for nickel [5, 17]. According to the present knowledge, nickel is concentrated in leaf epidermis [18, 19]. The metal is found in large epidermal cells away of stomata [19], in the vacuoles of epidermal cells [20] or in outside cell walls [21].

A third part of research is dealing with the ecology of the phenomenon of hyperaccumulation [14]. It has been hypothesized that the role of hyperaccumulation is multiple [14, 22, 23]. The most attractive hypothesis is the defense of the metal accumulating plant against herbivores or pathogens.

Ideas concerning metal phytoextraction, either for phytomining, i.e. the use of plants to extract metals from soils with high metal concentration [6, 24] or the so called phytoremediation of contaminated sites [7] gained also attention. Besides the discovery of fast growing hyperaccumulating species producing considerable biomass, soil fertilization could enhance the biomass production and /or metal uptake. Nowadays, when mechanical genetics is very modern, there is also a transgenic approach: the goal is to transfer the regulatory genes involved in metal accumulation to fast growing species.

Almost one-quarter of the Ni-hyperaccumulators belong to the family of Brassicaceae [9]. Though certain members of Asteraceae have been characterized as Ni-hyperaccumulators [25], *Centaurea* species have not been referred to as hyperaccumulators until recently. Reeves and Adigüzel [26] found 13 plants within the genus *Centaurea* from the Turkish serpentine flora that hyperaccumulate nickel. Some of them appear to be serpentine endemics some others grow on a variety of substrates. *Centaurea* species from the Greek serpentine flora were not been so far investigated from the viewpoint of metal hyperaccumulation. A species useful in the direction of phytomining should be fast-growing, producing considerable biomass.

We started recently working with metal hyperaccumulators using scanning electron microscopy and x-ray microanalysis [19] and atomic absorption spectroscopy. During our screening of the serpentine endemics of Greece, among other hyperaccumulators found, a member of *Centaurea* was also identified as hyperaccumulator (Constantinidis and Psaras, unpublished). According to our data *Centaurea thracica* is a Ni-hyperaccumulator (a content of 3800 ±300μg/g dry mass was detected) which is a relatively high biomass producing species. It could be one of the species capable for remediation of Ni-polluted soils.

REFERENCES

1. Baumann, A. (1885) Das Verhalten von Zinksalzen gegen Pflanzen und in Boden. *Landwirtschafltliche Versuchstation* **31**, 1-53.
2. Byers, H. G. (1935) Selenium occurrence in certain soils in the United States, with a discussion of the related topics. *U.S. Department of Agricultural Technology Bulletin* **482**,1–47.
3. Minguzzi, C. and Vergnano, O. (1948) Il contenuto di nichel nelle ceneri di Alyssum bertolonii Desv. *Memorie dagli Società Toscana di Scenze Naturali Serie A* **55**, 49-74.
4. Brooks, R. R., Lee, J., Reeves, R. D. and Jaffre, T. (1977) Detection of nickelliferous rocks by analysis of herbarium specimens of indicator plants. *Journal of Geochemical Exploration* **7**, 49-57.
5. Brooks, R. R. (1998) Phytochemistry of hyperaccumulators. *Plants that hyperaccumulate heavy metals* (ed. Brooks R. R.), p. 15-53, New York, USA: CAB International.
6. Anderson, C. W. N., Brooks, R. R., Chiarucci, A., LaCoste, C. J, Leblanc, M., Robinson, B. H., Simcock, R. and Stewart, R. B. (1999) Phytomining for nickel, thallium and gold. *Journal of Geochemical Exploration* **67**, 407-415.
7. McGrath, S. P. and Zhao, F. J. (2003) Phytoextraction of metals and metalloids from contaminated soils. *Current Opinion in Biotechnology* **14**, 277-282.
8. Reeves, R. D., Baker, A. J. M., Borhidi, A. and Berezain, R. (1996) Nickel-accumulating plants from the ancient serpentine flora of Cuba. *New Phytologist* **133**, 217-224.
9. Freeman, J. l., Persans, M. W., Nieman, K., Albrecht, C., Peer, W., Pickering, I. J. and Salt, D. E. (2004) Increased glutathione biosynthesis plays a role in nickel tolerance in *Thlaspi* nickel hyperaccumulators. *Plant Cell* **16**, 2176-2191.
10. Ernst, W. H. O. (1976) Physiological and biochemical aspects of metal tolerance. In: *Effects of air pollutants on plants* (ed. Mansfield, T. A.), p. 115-133. Cambridge University Press, UK.
11. Salt, D. E. (2001) Nickel hyperaccumulation in *Thlaspi goesingense*: a scientific travelogue. *In Vivo Cellular and Developmental Biology, Plant* **37**, 326-329.
12. Hall, J. L. (2002) Cellular mechanisms for heavy metal detoxification and tolerance. *Journal of Experimental Botany* **53**, 1-11.
13. Krämer, U. (2000) Cadmium for all meals – plants with an unusual appetite. *New Phytologist* **145**, 1-5.
14. Boyd, R. S. (2004) Ecology of metal hyperaccumulation. *New Phytologist* **162**, 563-567.
15. Krämer, U., Cotter-Howells, J. D., Charnock, J. M., Baker, A. J. M. and Smith, J. A. C. (1996) Free histidine as a metal chelator in plants that accumulate nickel. *Nature* **379**, 635-638.
16. Clemens, S., Palmgren, M. G. and Krämer, U. (2002). A long way ahead: understanding and engineering plant metal accumulation. *Trends in Plant Science* **7**:309-315.

17. Severne, B. C. (1974). Nickel accumulation by *Hypanthus floribundus*. *Nature* **248**: 807-808.

18. Mesjasz-Przybylowicz, J., Balkwill, K., Przybylowicz, W. J. and Annegarn, H. J. (1994). Proton microprobe and X-ray fluorescence investigations of nickel distribution in serpentine flora from South Africa. *Nuclear Instruments and Methods in Physics Research B* **89**: 208-212.

19. Psaras G. K., Constantinidis, Th., Cotsopoulos, B. and Manetas, Y. (2000). Relative abundance of nickel in the leaf epidermis of eight hyperaccumulators: evidence that the metal is excluded from both guard cells and trichomes. *Annals of Botany* **86**: 73-78.

20. Küpper, H., Lombi, E., Zhao, F. J., Wieshammer, G. and McGrath, S. P. (2001). Cellular compartmentation of nickel in the hyperaccumulators *Alyssum lesbiacum, Alyssum bertolonii* and *Thlaspi goesingense*. *Journal of Experimental Botany* **52**: 2291-2300.

21. Bidwell, S. D., Crawford, S. A., Woodrow, I. E., Sommer-Knudsen, J., Marshall, A. T. (2004). Sub-cellular localization of Ni in the hyperaccumulator, *Hypanthus floribundus* (Lindley) F. Muell. *Plant, Cell and Environment* **27**: 705-716.

22. Boyd, R. S. and Martens, S. N. (1998). The significance of metal hyperaccumulation for biotic interactions. *Chemoecology* **8**: 1-7.

23. Boyd, R. S. (1998). Hyperaccumulation as a plant defense strategy. In: *Plants that hyperaccumulate heavy metals* (ed. Brooks, R. R.), p. 181-201. New York, USA: CAB International.

24. Brooks, R. R. (1998). Geobotany and hyperaccumulators. In: Plants that hyperaccumulate heavy metals (ed. Brooks, R. R.), p. 55-94. New York, USA: CAB International.

25. Brooks, R. R., Chambers, M. F., Nicks, L. J. and Robinson, B. H. (1998). Phytomining. *Trends in Plant Sciences* **3**: 359-362.

26. Reeves, R. D. and Adigüzel, N. (2004). Rare plants and nickel accumulators from Turkish serpentine soils, with special reference to *Centaurea* species. *Turkish Journal of Botany* **28**: 147-153.

9

POLLUTION IN THE RED-BLOOMED HIMALAYAN DAL LAKE OF KASHMIR

Shafiq-ur-Rehman

ABSTRACT

Pollution is as old as civilization itself. The problem of pollution of water was not as serious in the past as it is now. Due to human interference, pollution of natural water resources by sewage or plant nutrients causes problems throughout the world. In recent years, the Dal Lake of the Himalayan Kashmir Valley has suffered with formation of a rare phenomenon of red-bloom of a new species discovered as Euglena shafiqii. In a series of examination performed on the pollution scenario of the Dal Lake, we also studied the values of toxic heavy metals (Fe, Cd, Cr, Cu, Mn, Ni and Zn) along with other physical and chemical features like temperature, hydrogen ion concentration, dissolved oxygen, T-alkalinity, nitrate-N, ammonium-N, phosphate-P, chloride, sodium and potassium in the waters affected by the red-bloom in different regions of the Dal Lake receiving anthropogenic waste effluents. Most of the overloaded chemical features were found high in the lake. Moreover, the levels of these events were much higher in the basins of the bloom-affected waters. The nutrients have been found important factor for the periodicity and aggregation of Euglena shafiqii, since during the bloom period these nutrients were reduced, thus believed to be utilized by the organisms for growth. The contents of Fe were high in the lake waters. Heavy metals concentrations were higher in the bloom-affected regions which receive wastes from hotels and settlements. Amongst heavy metals Mn levels were highest followed by Zn, Cr, Ni, Cu and Cd in waters at the same regions.

1. INTRODUCTION

Human civilizations have been intimately associated with water. And even today the association of water with anthropogenic activities, such as socio-economic, cultural, agricultural, industrial or technological developments, is found to be of great significance. The natural water resources, like the river Ganges in the Indian subcontinent or the Lake Dal in the western Himalayan Valley of Kashmir are symbolized as the centre of cultural, religious and national importance. The Dal Lake, being a centre of Kashmir civilization, has been closely associated with the socio-economic reform of the Himalayan Valley. Besides, the Dal Lake structures as a charming face of the beautiful body of the paradise for tourists. The Emperor Jahangir was as much impressed with the charming beauty as he recited a couplet of the *Ghazel*,

If there is a paradise on earth,
It is here, it is here, it is here.

The increasing human population has also been causing direct or indirect pressure onto the Dal through its activities in and around the lake and its catchment. The area within and around the lake has been encroached for human settlements, agricultural activities and for tourism industry. At present, there exist more than 2000 house boats on the waters at the bank of the Dal Lake. The maximum damage to the lake at the western side has been caused by continuous installation of new floating gardens for vegetable production. Almost all the old floating gardens have been turned into permanent agriculture activity and /or subsequent settlements. A recent study indicated that these floating gardens cover an area of 7.5 km^2 within the Dal Lake [1]. According to a recent report more than 2000 new settlements were erected in the Dal Lake in recent years [2]. Besides, deforestation of surrounding hills and agricultural activity in catchments of the Dal Lake have allowed huge erosion of the soil and subsequent deposition into the lake basins. It has been, however, estimated that about 36×10^6 m^3 of silt is annually poured into the Dal Lake [3].

Pollution is as old as civilization. Though it does not follow any mathematical roots for its growth, it does depend on the increase in population. The problem of pollution of water was not as serious as in the past as it is now [4]. The enrichment of waters by sewage or plant nutrients (a process called eutrophication) causes major problems throughout the world, such as blocking of vital water ways, making water hard to treat for drinking supplies, decreasing oxygen levels making fish stocks harder to support, reducing diversity of fauna, and lowering the amenity value of waters. The phenomenon stimulated by these events can well be quoted with the recent outbreak of the red-bloom of *Euglena shafiqii* [5] in the Dal Lake. And these events have prompted to undertake this study. It was, therefore, considered important to determine some important influential physico-chemical features and characteristics of the composition of some metals in the waters of the bloom Dal Lake.

2. MATERIALS AND METHODS

2.1 Study Area

The Dal Lake is situated at 34° 04' - 34° 11' N, 74° 48' - 74° 53' E in the north-east township of Srinagar in the heart of Kashmir valley in western Himalaya (1583 m above the sea level). The lake has a total area of about 20 km^2 of which approximately 12 km^2 is the total open water spread area. The lake is shallow in nature with a maximum depth of 3 m. It has a little water-inflow through Telbal nullah in the north and water-outflow channel at the Dal gate in the south that discharges water into the Jhelum River. An additional outflow channel is situated at the western side of the lake that discharges water into the Nagin Lake. A total of 277 km^2 area around the Dal Lake falls under the catchment area surveyed in 1989 [6]. It comprises 148 km^2 of Dachigam sanctuary, 80 km^2 of Telebal region, 47 km^2 of Hillside area and 21 km^2 of Srinagar old township area. A remote sensing study using Landsat data and aerial photographs (colour and colour-infrared) of 80 km^2 catchment area close to the Dal Lake revealed approximately 54% area covered for township settlement and other infrastructure while the rest of 46% area utilized for agricultural practices [1]. Due to construction of foreshore wall, road, and a hotel-cum-international conventional complex, some pockets of stagnant waters were resulted in the Dal Lake along its eastern side. These stagnant water pockets frequently receive large quantities of wastes from surrounding human settlements, agricultural lands and hotels that caused severe damages to the lake ecosystem. Consequently, these stagnant water bodies of the Dal Lake developed a rare phenomenon of the red-bloom (Shafiq-ur-Rehman, 1991) for the first time in the last week of June 1991, which spread to a vast area of the Dal Lake. The red-bloom was found to be a new and recognised as *Euglena shafiqii* [5]. The red-bloom of *E. shafiqii* continuously appeared every year during summer.

The study stations for the bloom-affected-region (BAR) were marked as P-1, P-2 and P-3 at the region of stagnant water bodies. The areas of the lake which did not experience the red-bloom were marked with no-bloom-region (NBR) as N-1 and N-2 (Figure 1). A suitable volume of water samples (at least 10 samples for physicochemical and 6 for metals) was collected from each test site during the month of June (before bloom formation), and July/August (after bloom formation) in 1994 and 1995.

2.2 Methodology

The water samples were centrifuged at 4500 r.p.m. for 10 min. Water temperature was determined at each test site. The selected physico-chemical parameters, such as pH, conductivity, dissolved oxygen, total alkalinity and chloride, ammonia-N, nitrate-N and phosphate-P, sodium and potassium in the samples of water, were determined as per standard methodology [7,8].

All glass wares and plastic wares used in this study were acid washed in high grade HNO_3 rinsed three with double distilled deionised water, prior to use. The samples of water were acidified with HNO_3 for metal analysis. The concentrations of metals (Cd, Cr, Cu, Fe, Mn, Ni, and Zn) were measured at their respective wave length (228.8 nm, 357.9 nm, 324.7 nm, 248.3 nm, 279.59 nm, 232.8 nm, and 213.9 nm) using hollow cathode lamps on flame atomic absorption spectrophotometer (Video 11, Thermo Jarrel Ash Corp, USA). A mixture of compressed air as oxidant and acetylene as fuel was employed. The reading on each sample was recorded in triplicate. The chemicals and reagents used in this study were of highly purified or analytical grade.

Statistical analysis of data was performed by ANOVA and descriptive statistics (Minitab, 11.12 version, Minitab Inc., U.S.A. and S-PLUS 2000) for physico-chemical and metal parameters. Significant differences were evaluated using *P* having value less than 0.05.

3. RESULTS

3.1 Physicochemical Features

The physical and chemical characteristics of the Dal Lake waters both from no-bloom regions (NBR, N-1 and N-2) and bloom-affected regions (BAR, P-1, P-2 and P-3) are given in Table 1. In brief, all the physical and chemical parameters showed high values in the lake waters, except dissolved oxygen which were found to be in decreased levels. Before the blooming, the BAR waters were observed to have higher values ($P<0.05$) of

Table 1 *Values of different physico-chemical parameters of Dal Lake waters samples (0 m).*

Parameters	*Non-Bloom Region (NBR) at N-1, N-2*	*Bloom-Affected Region (BAR) at P-1, P-2, P-3 Before*	*After*	*ANOVA*
(N)	*Mean ± S.D., (Range)* *(a)*	*Mand ± S.D., (Ragne)* *(b)*	*Mean ± S.D., (Range)* *(c)*	*Significant differences (P < 0.05) between (*) groups (a, b, or c)*
Water Temp. (°C)	26.00 ± 5.70, (14.0-31.0)	27.00 ± 5.00, (17.0-32.0)	29.20 ± 3.53, (17.0-32.0)	— a * c —
pH (at 25 °C)	7.80 ± 0.42, (7.2-8.4)	8.14 ± 0.45, (7.4-8.7)	8.20 ± 0.47, (7.4-8.9)	a * b a * c —
Cond. (μs,25 °C)	160.00 ± 29.01, (100-190)	136.00 ± 27.13, (100-175)	133.00 ± 18.02, (110-160)	a * b a * c —
Diss. Oxy. (mg.l^{-1})	13.10 ± 3.30, (7-16)	9.05 ± 2.04, (6-12)	6.88 ± 1.70, (4.5-9)	a * b a * c b * c
T. Alk. (mg.l^{-1})	91.00 ± 16.43, (70-120)	110.25 ± 20.55, (80-140)	126.25 ± 20.58, (90-150)	a * b a * c b * c
Nitrate-N (μg.l^{-1})	248.00 ± 80.00, (160-400)	280.00 ± 80.00, (180-400)	210.00 ± 44.00, (160-300)	— — b * c
Amm.-N (μg.l^{-1})	37.95 ± 19.32, (10-62)	48.00 ± 18.74, (18-70)	24.25 ± 10.79, (10-40)	— a * c b * c
Phosph.-P (μg.l^{-1})	73.15 ± 22.90, (45-110)	92.25 ± 19.34, (60-116)	42.25 ± 17.36, (25-70)	a * b a * c b * c
Chloride (mg.l^{-1})	42.85 ± 13.63, (24-60)	141.60 ± 52.09, (52-202)	107.90 ± 47.82, (35-180)	a * b a * c b * c
Sodium (mg.l^{-1})	4.65 ± 3.60, (0-10)	11.65 ± 5.63, (2-20)	10.30 ± 5.02, (1-18)	a * b a * c —
Potassium (mg.l^{-1})	3.95 ± 2.70, (0-7)	9.35 ± 3.77, (2-14)	7.10 ± 3.89, (1-12)	a * b a * c b * c

Number of samples 10-25(N)

temperature, pH, T-alkalinity, phosphate-P, chloride, sodium and potassium as compared to the NBR waters. On the other hand, the conductivity and dissolved oxygen showed declined levels ($P<0.05$) in the bloom affected waters (before the blooming) as compared to the NBR waters. After the appearance of the bloom in the same regions (i.e., in BAR during bloom period), however, the levels of dissolved oxygen, nitrate-N, ammonium-N, phosphate-P and chloride contents were significantly reduced ($P<0.05$), where as, the contents of sodium and potassium remained to be unaffected.

3.2 Metals Concentrations

Table 2 summarises the ANOVA of the metals, Fe, Cd, Cr, Cu, Mn, Ni, and Zn, distribution in different regions of the Dal Lake. Summary of means of these metals and their levels of significant differences are provided in the Table 3. Figures 2 to 8 represent concentrations of different metals analysed in waters collected from the NBR and BAR of the Dal Lake. Briefly, the concentrations of iron were highest and equally distributed in all the water samples collected from different regions of the lake (Figure 2). Cadmium concentrations were also found equally distributed in all the regions, except in site P-1 which showed 50% fewer concentration than other sites or regions (Figure 3). The concentrations of copper were found to be higher with 100% at site P-2 and lower with 50% at site P-1 as compared to the sites of NBR (Figure 4). Nickel concentrations were, however, differentially distributed in Dal Lake waters; it were lower by 75% in site P-1 and 50% in site P-3 but were higher by 150% in site P-2 as compared to waters of the NBR (Figure 5). The concentrations of chromium were higher in the BAR (by 200% in P-2 and 50% in P-1 and P-3) than the NBR (Figure 6). Similarly, manganese concentrations in lake waters were higher in the BAR, showing 1200% in P-2, 400% in P-1 and 100% in P-3 above than the levels in the NBR (Figure 7). Zinc concentrations were found to be highest in lake waters at site P-2 followed by P-1 Fig. 8.

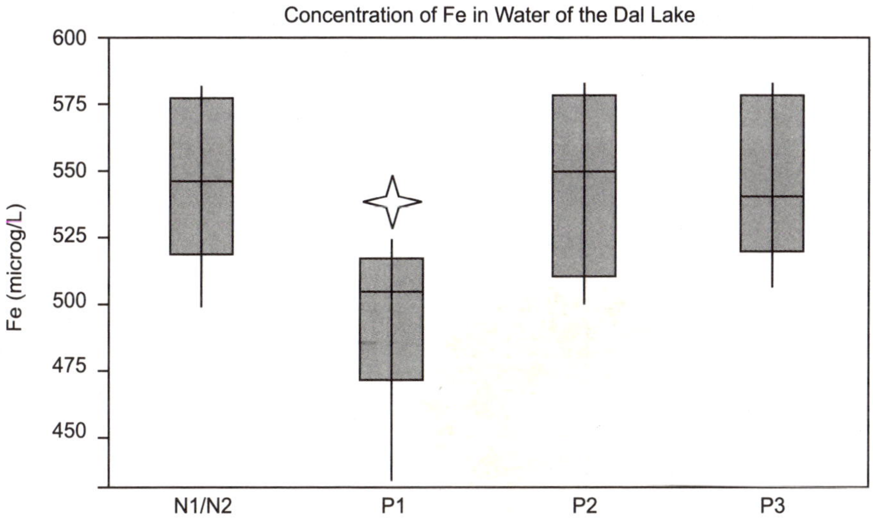

Fig. 2 *Summary in the box-whisker-plot represents statistical feature of concentration of Fe in different regions of the Dal Lake waters. Star shows significant difference (P at least<0.05) compared to N1/N2. For ANOVA and descriptive statistics refer to Tables 2 and 3.*

Table 2 *Summary of ANOVA for metals contents in the Dal lake.*

Source of Variation	DF	Fe	Cd	Cr	Cu	Ms Mn	for Ni	Metal Zn
Regions	3	3750.00*	37.50**	16000.00**	237.50**	17750.50**	2437.50**	9800.00**
Error	20	1035.00	4.80	69.90	17.80	95.10	19.30	234.00

* and** represent for 5% and 1% level of significance, respectively.

Table 3 *Summary of means of metal sand their differences.*

Regions	No. of Samples	Fe	Cd	Cr	Cu	Mn	Ni	Zn
1 = N1/N2 (NBR)	6	550^a	10^a	40^b	10^a	10^a	20^b	30^a
2 = P1 (BAR)	6	500^b	5^b	60^a	5^a	50^b	5^a	90^b
3 = P2 (BAR)	6	550^a	10^a	50^c	20^b	130^c	50^c	150^c
4 = P3 (BAR)	6	550^a	10^a	60^a	10^a	20^a	10^a	30^a
LSD (.05)		18.574	1.265	4.827	2.436	5.630	2.536	10.392
SE of maan differnece		38.745	2.639	10.069	5.081	11.745	5.291	21.678

Means at per have same superscript

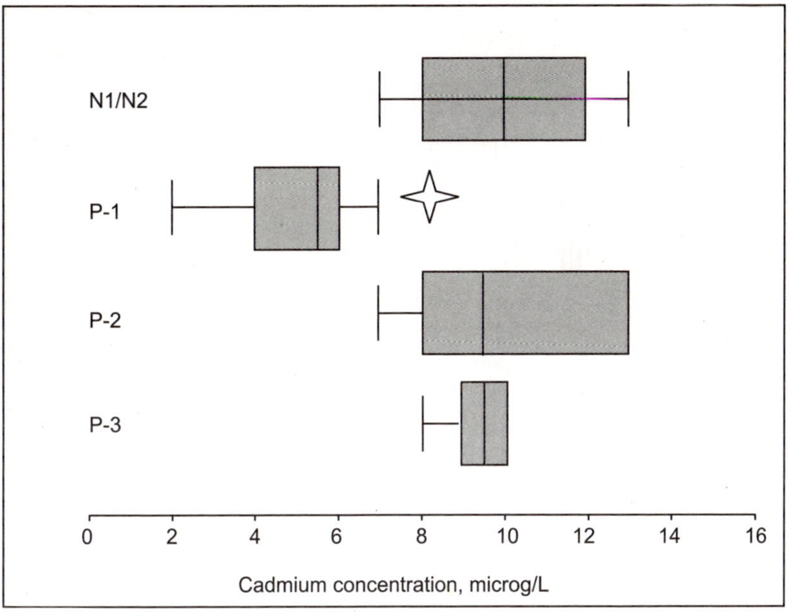

Fig. 3 *Summary in the box-whisker-plot represents statistical feature of cconcentrations of Cd in different regions of the Dal Lake waters. Star shows significant difference (P at least<0.05) as compared to N1/N2. For ANOVA and descriptive statistics refer to Tables 2 and 3.*

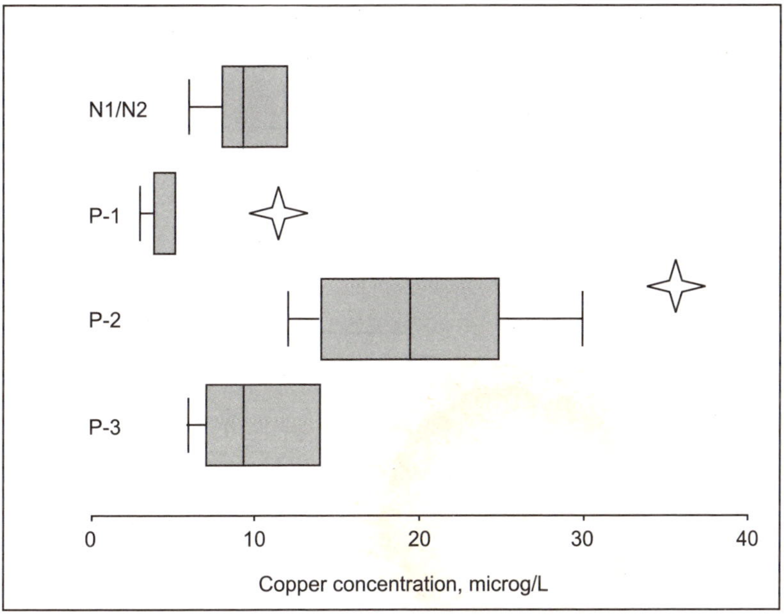

Fig. 4 *Summary in the box-whisker-plot represents statistical feature of cconcentrations of Cu in different regions of the Dal Lake waters. Star shows significant difference (P at least<0.05) as compared to N1/N2. For ANOVA and descriptive statistics refer to Tables 2 and 3.*

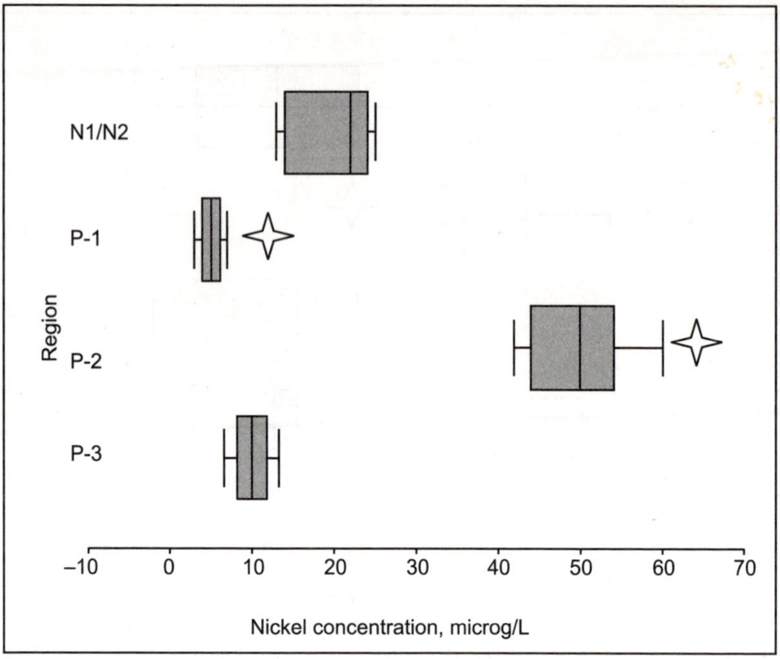

Fig. 5 *Summary in the box-whisker-plot represents statistical feature of concentrations of Ni in different regions of the Dal Lake waters. Star shows significant difference (P at least<0.05) as compared to N1/N2. For ANOVA and descriptive statistics refer to Tables 2 and 3.*

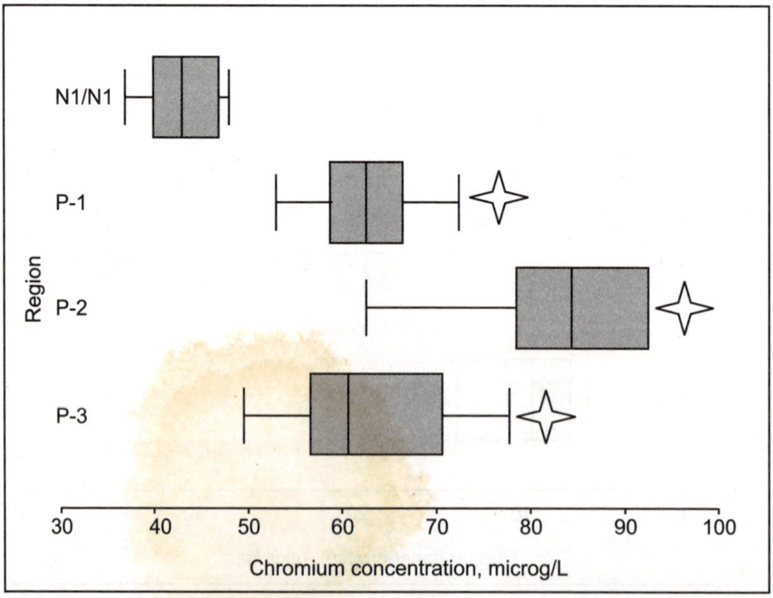

Fig. 6 *Summary in the box-whisker-plot represents statistical feature of concentrations of Cr in different regions of the Dal Lake waters. Star shows significant difference (P at least<0.05) as compared to N1/N2. For ANOVA and descriptive statistics refer to Tables 2 and 3.*

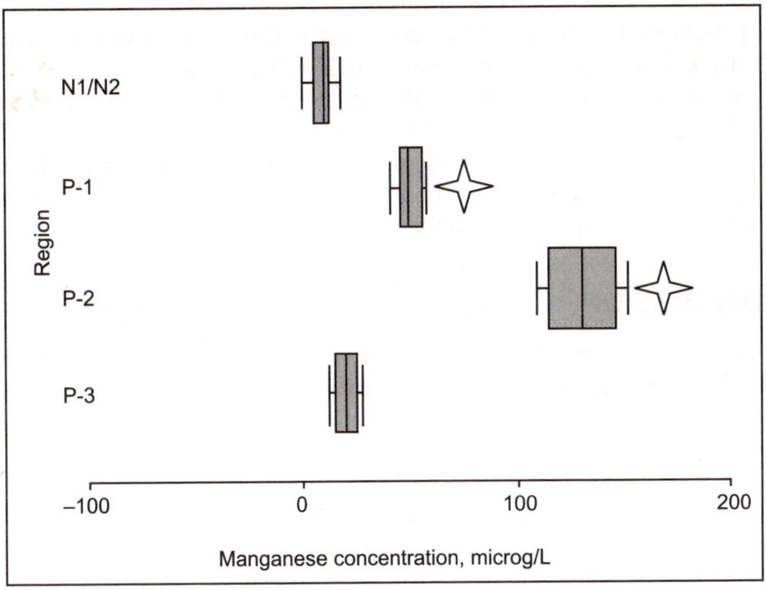

Fig. 7 *Summary in the box-whisker-plot represents statistical feature of concentrations of Mn in different regions of the Dal Lake waters. Star shows significant difference (P at least<0.05) as compared to N1/N2. For ANOVA and descriptive statistics refer to Tables 2 and 3.*

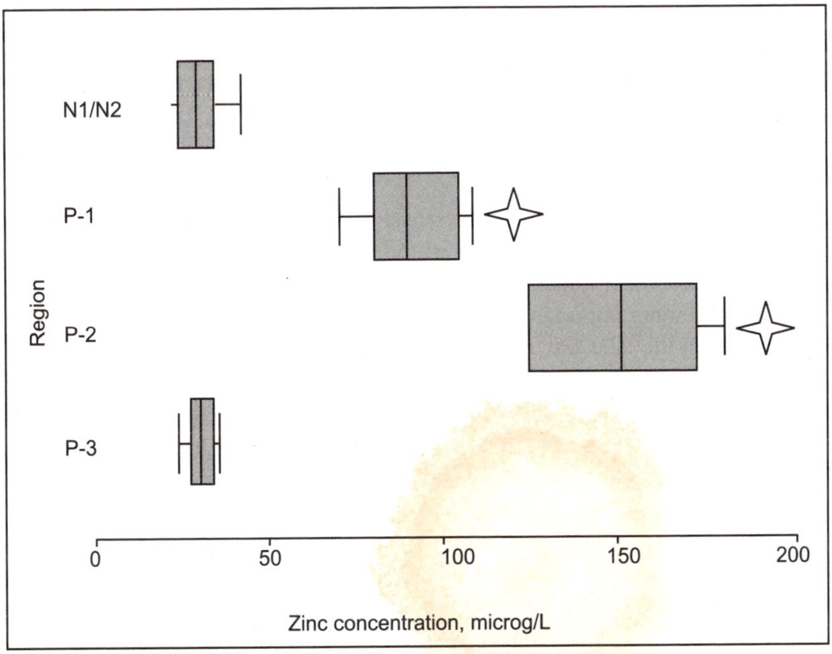

Fig. 8 *Summary in the box-whisker-plot represents statistical feature of concentrations of Zn in different regions of the Dal Lake waters. Star shows significant difference (P at least<0.05) as compared to N1/N2. For ANOVA and descriptive statistics refer to Tables 2 and 3.*

We have accessed these metal findings in the Box–Whisker Plots (Figure 2-8) for a clear presentation of the data to understand at a glance the statistical feature. It is evident that Cd concentration in the study sites P-3, P-2 and N-1/N-2 are at par, while P-1 shows its least value. Moreover, P-2 and N-1/N-2 show maximum variability, whereas P-3 exhibits least variability. Median of P-1 site is around 6 while the rest are having their median values around 10μg L^{-1}. It is, therefore, evident from the graphic summary that P-2 is higher in almost all the metals, whilst P-1 site is comparatively showing fewer values of Cd, Ni, and Cu.

4. DISCUSSION

Increasing evidences attribute that human interference with the nature is responsible capitally for causing serious damages to human environment. Tremendous growths in population, urbanization, economic exploitation through industrialization, modern agricultural practices and clearance of forests have painted a big question mark on the face of the green world and its very survival. Our natural water resources viz rivers and lakes, have already fallen victim to anthropogenic interference world wide. In Kashmir as well, over exploitation of the Lake Dal and its catchment for economic gain and settlements has ultimately threatened to the survival of this precious natural water resource of the Kashmir valley. These events in turn adversely affected the biodiversity, caused pollution and increasing cultural eutrophication of the lake. The living example emerged as the rare formation of the red-bloom of the *Euglena shafiqii* in the Dal Lake [5]. I found that trends in high temperature, few events of heavy precipitation before blooming, high pH values and increased levels of nutrients in summer gave rise to favourable conditions for the periodicity and growth of the red-bloom. Hammer [9] has also shown that the time of appearance and sequence of bloom formation in lake of Saskatchewan, Canada, were closely influenced by temperature. Our studies have also supported the view that the growth of many species of *Euglena* was favoured by high temperature [5,10,11].

Generally the water of the Dal Lake is basic in nature, however, the bloom-affected regions receiving anthropogenic wastes were found to be more basic. Moreover, those regions showed elevated trends of total alkalinity. Conversely, the dissolved oxygen, which is of paramount importance to all living organisms, showed decreasing order of its distribution in the lake waters before red-blooming. During the bloom period, the levels of dissolved oxygen have dropped further. Munawar [12] and Singh [13] observed lesser dissolved oxygen as favourable condition for *Euglena* growth. It is well documented that the pollution of lakes by sewage or fertilizers brings them to eutrophic state, which is of ecological, environmental and health concern. These not only cause damages to lake ecosystem but human health as well since the lake water is used for human consumption. The available nutrients, particularly phosphorus and nitrogen, have been widely acknowledged as having large considerable significance in the process of eutrophication. Nitrate itself is not toxic except at large doses. Under certain circumstances nitrate can be reduced to nitrite *in vivo*; it is this process that leads to concern regarding ingestion of nitrate, which causes blue baby syndrome and cancer. In the Dal Lake, however, the levels of phosphate-P, nitrate-N and ammonium-N were observed to be high. Moreover, before the development of the bloom, high levels of these nutrients were determined in the water. Interestingly, the concentrations of these nutrients were reduced in water during the bloom period, which seems to have been utilized by the organisms for growth and aggregation. It is well understood that the nitrogen and also phosphorus are the most important constituents of the organismal matter and hence are required by the biota. Sodium is present in all natural waters that impart them with softness in contrast to hardness. This element has a minor role in aquatic systems. Nonetheless, sodium contributes an important role in the ion-transport-energy-process. Sodium concentrations were high in the bloom-affected lake waters. Potassium is an important element that plays a vital role in the mechanism of fresh water ecology. Potassium, even at fewer concentration (than sodium), has an important role as enzyme activator in the process of ion transport in order to stabilize the biochemical configuration and integrity of the membranes and its channels in the cell. Like sodium, the concentrations of potassium remained higher in the bloom-affected lake waters before the bloom formation, and were not affected by the blooming. In natural waters, chlorides are usually present in low concentration. However, it plays metabolically active role in photosynthesis in water and photophosphorylation in autotrops. Free chlorine, which is commonly used as disinfectant for drinking and wastes water, soon gets either converted into chlorides or after complexing with organic matters forms toxic compounds. Chloride levels in the lake waters were found high. Moreover, the bloom-affected waters of the lake exhibited much higher values of chloride, indicating a potential hazard of chloride pollution in the Dal. Moreover, after the formation and subsequent presence of the red-bloom caused the chloride level to decrease as much as 24 per cent. In summer, the nutrients levels in the Dal Lake were high, particularly much higher in the bloom-affected lake basins, which receive untreated waste waters. It is, however, observed that the formation and presence of the red-bloom in these basins reduces the levels of the nutrients indicating their utilization by the organism *Euglena shafiqii* for growth and aggregation.

Based on regional geology, the spring and stream waters in Kashmir valley generally contain high levels of iron. Moreover, other culminating factors like urbanization, deforestation and agricultural activities in the catchment area with extensive land use and subsequent loss of soil, are the triggering contributors of the iron load in the lake ecosystem. The contents of iron in the Dal Lake were found to be equally distributed among all the regions and basins. However, the levels of iron in water have been related to the distribution of *euglenoids* [14]. Whereas, according to the present study, the contents of iron can not be related as a sole factor for the appearance, distribution, growth or aggregation of the red-bloom of *Euglena shafiqii*. As such, in the presence of high concentrations of iron, some other events such as high nutrient load, increased heavy metal contents, reduced levels of dissolved oxygen and stagnancy of water having minimum physical stress, elevated pH values and high temperature may play a vital role as co-factors for the growth and distribution of the red-bloom. These factors are probably the gift of the anthropogenic pollution from our so-called unchecked developments which are visible in and around the Dal Lake.

The trend of heavy metal concentrations in the lake waters had been fluctuating from region to region. The concentrations of metals were, however, higher in the regions of stagnant waters affected with the red-bloom. Moreover, amongst the bloom-affected waters, the concentrations of the heavy metals were found to be maximum at the site P-2 followed by P-1. Manganese, zinc and chromium appeared to be highest in concentrations in the both basins. The lake basins of Gagribal and Boddal, (sites P-1, and P-2/ P-3) which were separated from each other as two stagnant water pockets due to erection of the Centaur Hotel Complex, have become the receivers of untreated waste discharges of the hotels, human settlements and agricultural fields. Consequently, these regions have been suffering from huge carpeting of the red-bloom of *Euglena shafiqii* [5] owing to the pollution of both the metals and non-metals. The pollution of these elements could be hazardous as well to the human population due to consumption of contaminated food items, like fish and aquatic vegetables (like, *nadru* or lotus root, *sighara* or water chestnut, etc.) and drinking water supplied from the Dal Lake.

Conclusion

The present study concludes that the environmental condition of the Himalayan Dal Lake, which symbolizes as a socio-economic reform and national importance, is deteriorating. The water chemistry of the lake is modified. The evidences show that the lake water is basic in characteristic which bears decreased dissolved oxygen and raised alkalinity. The chemical nutrients, that contribute eutrophication of water bodies such as phosphorous, nitrate-N, ammonium-N and potassium, were high in contents. Since the Kashmir geology impresses its water with high iron levels, the similar expression of iron was also apparent in the lake waters. The heavy metals, Cd, Cr, Cu, Mn, Ni, and Zn were accumulated in abundance in the stagnant water areas. However, their regional distribution differed across the polluted zones. The human activities on the lake catchments generate waste effluents/municipal sewage from human settlements, hotels, and agricultural fields and orchards, which are directly dumped into the lake. These human events seemed to be directly responsible the chemical pollution of the lake and formation of the novel red-bloom of *Euglena shafiqii* carpeted over these polluted areas of the Dal Lake.

REFERENCES

1. Palria, S., Singh, T. S., Muley, M. V., Chakraborty, M., Tamilarasan V. and Kawosa, M. A. (1987) Water quality mapping in the Dal lake and the Wular lake of Jammu and Kashmir using lansat images and aerial photographs. Technical Report, Scientific note of Space Applications Centre (ISRO), Ahmedabad and Dept. of Ecology, Environment Science and Technology, Srinagar, India. 1-42.
2. Bukhari, F. (1994) 2000 new settlements in three years in Dal lake, 40 kanals of water body being filled at Brari Nambal. *Greater Kashmir*, August 11. 1.
3. Enex (1978) Study of the Pollution of Dal Lake, Srinagar, Kashmir, India. A report prepared for the Common wealth Fund for Technical Cooperation by Enex of New Zealand Inc.
4. Mall, I. D., Upadhyay S. N. and Sharma, Y. C. (1996) A review on economic treatment of wastewaters and effluent by adsorption. *Intl. J. Env. Studies* **51**, 77.
5. Shafiq-ur-Rehman. (1998) A red bloom of *Euglena shafiqii*, a new species in Dal lake, Srinagar, Kashmir. Water, air and soil Pollution. **108**, 69-82.
6. Kaul, B. L. (1992) Ecodegradation of Dal lake. In: Himalayan environment, man and the economic activities. (ed. Raina, J. L), Pointer Publishers, Jaipur, India. 226-236.
7. Adoni, A. D. (1985) Workbook on limnology. Indian MAB Committee, Department of Environment, Govt. of India, New Delhi, India. 1-216.

8. APHA. (1985) Standard Method of the Examination of Water and Waste Waters. APHA, AWWA and WPCF, Washington, D.C., USA.

9. Hammer, U. T. (1964) The succession of 'bloom' species of blue-green algae and some casual factors. Verhandlungen der internationale vereinigung für theoretische und angene wandte limnologie. **15**, 820-836.

10. Gonzalves E. A. and Joshi, D. B. (1946) Fresh water algae near Bombay. Bombay Natural History Society. **46**, 154-176.

11. Seenayya G. and Subbaraju, N. (1972) First international symposium on taxonomy and biology of blue-green algae. University of Madras, Madras, India. 52-57.

12. Munawar, M. (1970) Limnological studies of freshwater ponds of Hyderabad, India II, The biocenose distribution of unicellular and colonial phytoplankton in polluted and unpolluted environment. Hydrobiologia. **36**, 105-128.

13. Singh, V. P. (1960) Phytoplankton ecology the inland water of U.P. Proceedings of Symposium on Algology. Indian Council of Agricultural Research, New Delhi. 243-271.

14. Rao, C. B. (1953) On the distribution of algae in a group of six small ponds. *J. Ecol.* **41**, 62-71.

10

PRODUCTION OF RADIOISOTOPES FOR POSITRON EMISSION TOMOGRAPHY

Alok Srivastava

ABSTRACT

Radioisotopes find application in fields as diversified as medicine, agriculture, food preservation, water resource management and environmental studies but it is the medical application which constitutes the most significant part. Radioisotopes are finding increasing applications in the form of diagnostic as well as therapeutic radiopharmaceuticals. The production of radioisotopes for application in positron emission radiography (PET) is briefly reviewed from the scientific and technological point of view.

1. INTRODUCTION

Positron Emission Tomography (PET) is a medical imaging technique used to produce a three dimensional image of the functional processes in the body with the aid of positron-emitting radionuclides or by the employment of positron-emitting radiopharmaceuticals i.e. tagged substance wherein the positron-emitting radionuclide has been chemically incorporated into a metabolically active molecule. This moves normally through the patient to accumulate in the targeted organ. As it decays it emits a positron which after travelling up to a few mm sometimes independently and sometimes as positronium (short lived bound state) encounters a free electron and promptly annihilates resulting in the simultaneous emission of two identifiable annihilation photons (gamma rays) in opposite directions. The detection of the two 511 kev photons at opposite detectors within about 10 ns coincidence regime gives rise to spatial data which is translated into computer image.

Radioisotopes are unstable nuclides which spontaneously disintegrate by emission of radiation. At present nearly 2000 isotopes are known to man out of which 275 are stable and the rest radioactive. Radioisotopes are produced artificially by changing the neutron to proton ratio of the stable isotopes. This transformation is achieved through nuclear reactions by exposing targets made of stable isotopes with projectiles of different charge, mass and energy. A nuclear reaction is a process in which two nuclei or nuclear particles collide, to produce products different to the initial reactants. In principle a reaction can involve more than two particles colliding, but such an event is exceptionally rare. If the particles collide and separate without changing (except possibly in energy levels), the process is called a collision rather than a reaction.

Neutron induced nuclear reactions lead to formation of neutron rich radioisotopes. These decay by beta emission whereas charged particle induced nuclear reactions produce neutron deficient radioisotopes which decay by electron capture or positron emission.

The radioisotopes used for Positron Emission Tomography (PET) belong to the latter type and require particle accelerators to produce them. There are various types of particle acclerators available in the market. The electrostatic accelerators like Vande Graaff and Cockcroft Walton generators were the first to be introduced and were later followed by Tandem Accelerators like the Linear Accelerator and the Pelletron. However, it was the Cyclotron which was introduced in the intervening period that became the most sought after accelerator for medical applications.

The cyclotron, originally developed for nuclear physics research was later simplified for the benefit of increasing medical applications, being the ideal source for many short-lived, neutron-deficient radioisotopes, and has become today a versatile tool in the hands of the radiopharmaceutical scientists. From the first hospital based cyclotron installed at the Hammersmith Hospital in London in the early forties the field of medical cyclotron has seen a very significant growth. It is interesting to note that at present nearly 250 medical cyclotrons exist worldwide and new ones are being added every other month. Belgium has the unique distinction of having the highest number of medical cyclotrons per capita. The Indian Government has also taken initiative in this direction and is in the process of installing five medical cyclotrons for PET application in different parts of the country.

The earlier cyclotrons were variable energy multiparticle machines but in the late seventies the Japanese made a breakthrough in terms of designing compact cyclotrons which were easy to operate and were capable of providing protons with sufficient energy to produce the vital PET radioisotopes namely ^{11}C, ^{13}N, ^{15}O and ^{18}F. In the last two decades different types of cyclotrons have been developed to meet the specific demands of the medical community. Table 1 shows details of such cyclotrons in terms of the types of charge particles they can accelerate and the maximum energy to which they can accelerate such particles.

Table 1 *Types of accelerators available today for radioisotope production*

Type	*Characteristics*	*Energy (MeV)*	*Major radioisotopes produced*
Class I	single particle (d)	<4	^{15}O
Class II	single particle (d)	≤11	^{11}C, ^{13}N, ^{15}O, ^{18}F
Class III	single or two particle (p, d)	≤20	^{11}C, ^{13}N, ^{15}O, ^{18}F,^{86}Y, ^{124}I, ^{123}I, ^{67}Ga, ^{111}In
Class IV	single or multi particle (p, d, ^3He, ^4He)	≤40	^{38}K, ^{73}Se, $^{75\text{-}77}Br$, ^{123}I, ^{81}Rb, ^{81}Kr, ^{67}Ga, ^{111}In, ^{201}Tl, ^{22}Na, ^{57}Co, ^{48}Ti, ^{68}Ge, ^{72}As, ^{140}Nd
Class V	single or multi particle (p, d, ^3He, ^4He)	≤100	^{28}Mg, ^{72}Se, ^{82}Sr, ^{117m}Sn, ^{123}I
Class VI	single particle (p)	≤100	^{26}Al, ^{67}Cu, ^{68}Ge, ^{82}Sr, etc.

It can be observed from Table 1 that the energy of the projectiles plays a very important role. More the energy more are the reaction channels which are likely to open up. Therefore, knowledge of nuclear reaction cross sections at different energies (excitation function) becomes very important due to its direct relevance on the radioisotope production routes optimization i.e on the maximization of the yield of the desired product and minimization of the yields of the radioactive impurities.

The expected yield of a product for a certain energy range can be evaluated using the following expression:

$$Y = N_L J / M * I(1-e) \int_{E1}^{E2} [(dE/d(px)]^{-1} \sigma(E) dE \qquad ...(1)$$

where Y is the activity (in Bq) of the product, N_L is the Avagadro number, H is the enrichment (isotopic abundance) of the target nuclide, M is the mass number of the target element, I is the projectile current, $[(dE/d(\rho x)]$ is the stopping power, $\sigma(E)$ is the cross section at energy E, λ is the decay constant of the product and t is the time of irradiation.

The rate of production is dependent on the number of bombarding particles, the number of target nuclei and the cross section of the reaction, the length of bombardment (the production of short lived nuclei reaches saturation rather quickly) time. Therefore in order to optimize the production of any PET radioisotope the following production considerations have be taken into account:

(i) The threshold energy for the desired reaction
(ii) The energy where the maximum yield occurs
(iii) The chemical form of the target nucleus
(iv) The physical form of the target nucleus
(v) The chemical form of the desired product
(vi) The physical form of the desired product
(vii) The ease of separation of product from the target

A very good example to illustrate the above mentioned considerations is to look at the case where ^{18}F is needed as fluoride. The two most widely used nuclear reactions for producing ^{18}F are the $^{20}Ne\,(d, \alpha)\,^{18}F$ and the $^{18}O\,(p, n)\,^{18}F$. Both reactions have low threshold energy (0-2 MeV) and the maximum yield occurs below 10 MeV. However, the (p,n) reaction has a much higher cross section so that the quantity of ^{18}F that can be produced is more than double for a thick target reaction using protons of about 12 MeV in comparison to deuterons of comparable energy. The yield is expressed in terms of millicuries per milliampere of current per hour (mCi/mAh).

Thus having selected the mode of production and the energy, the next issue is to examine the chemical and physical forms of the target and product. The most convenient target material for producing ^{18}F as fluoride is a water target enriched in ^{18}O. The resulting ^{18}F is produced directly as the fluoride ion which can be extracted from the target water by simple ion exchange column.

Considerable effort has been put in the development of target materials. The targets are required to withstand high beam currents in accelerators whether exposed to internal beam, extracted beam or parasitic beam. The power density effective on a target in some case reaches up to few kilowatts per square centimeter. Therefore efficient heat transfer consideration becomes inevitable in the design of targets. The other considerations which have to be kept in mind during the design of target are chemical reactivity, the ease of chemical separation of the radioactive product from it and the recovery of target material if necessary besides high yield of the product.

Presently the most important PET radioisotope ^{18}F which in the form of ^{18}FDG (Fluoro Deoxy Glucose) is used for brain imaging and some of the common Single Photon Emission Computed Tomography (SPECT) radioisotopes like ^{67}Ga, ^{111}In and ^{123}I are produced using highly enriched target isotopes. Table 2 gives a list of the frequently used target materials along with the form in which they are being used for the production of PET/SPECT radioisotopes.

Table 2 *Frequently used target materials used for production of PET/SPECT radioisotopes.*

Target material	*Reaction*	*Product*	$T_{1/2}$
13-*C* (enriched)	*p*, *n*	13-*N*	10 m
14-*N* (natural)	*p*, alpha	11-*C*	20.3 m
14-*N* (natural)	*d*, *n*	15-*Q*	2 m
18-*O* (enriched)	*p*, *n*	18-*F*	110 m
112-*Cd* (enriched)	*p*, 2*n*	11-*In*	2.8 h
82-*Kr* (enriched)	*p*, 2*n*	81-*Rb*/81*m*-*Kr*	4.6 h/13 s
58-*Ni* (enriched)	*p*, 2*p*	57-*Co*	271.7 d
68-*Zn* (natural)	*p*, 2*n*	67-*Ga*	3.3 d

It should also be remembered that the chemical processing of the irradiated target material has to be undertaken in a very careful and sophisticated way as radiation induced chemical reactions and nuclear recoil effects could lead to the production of the desired radioisotope in the form of different species some of which may not be desirable.

In the case of ^{11}C produced from nitrogen gas target it was observed that the product radioisotope occurred as $^{11}CO_2$ or ^{11}CO depending on the target filling, beam current used and the time of irradiation. On the other hand it was observed that in the case of production of ^{13}N using water target the product radioisotope ended up either as $^{13}NO_2^-$ or $^{13}NO_3^-$ depending on the radiation dose effective in the target.

The radioisotope production technology using cyclotrons has made great strides in the past decade and today one can directly purchase special purpose cyclotrons, high current irradiation targets and automated/remotely controlled chemical processing units.

However, there are still some PET radioisotopes which are not produced in accelerators but are produced using generators and have found application at centers which do not have a cyclotron. Two such PET radioisotopes are the positron emitters ^{68}Ga and ^{82}Rb which have half lives of 68 minutes and 1.3 minutes respectively. The former is used for labeling and PET attenuation correction whereas the latter is used for study of myocardial blood flow. The parents of these PET radioisotopes namely ^{68}Ge and ^{82}Sr are produced using special nuclear reactions in accelerators. Besides the above two mentioned ($^{68}Ge/^{68}Ga$ and $^{82}Sr/^{82}Rb$) generator systems two more generators namely $^{62}Zn/^{62}Cu$ and $^{122}Xe/^{122}I$ are also available commercially in the market. The former is clinically established but its application is limited due to the short half life of the parent ^{62}Zn ($T_{1/2}$ = 9.2 h) whereas the difficulty in the production of the later ^{122}Xe ($T_{1/2}$ = 20.1 h) has made its application even more limited.

The radioisotope production technology is now well established and many radioisotopes for diagnostic and therapeutic applications can be purchased directly from commercial manufacturers. The number of radioisotopes potentially interesting for medical applications is quite large and it is envisaged that many more will be made available in the near future.

Acknowledgement

The author is thankful to Prof. Dr. S.M.Qaim of Forschungzentrum Juelich, Germany for making available his review article on Cyclotron Production of Medical Radionuclides and for fruitful discussions on the subject matter. The fruitful comments from Dr.R.K.Sharma, Director of Defense Laboratory, Tezpur is gratefully acknowledged. He would also like to place in record the valuable information on PET he received from time to time from Prof.Dr.Ulrich Scherer of the Aachen University of Applied Sciences, Juelich, Germany, Prof. F. Roesch of Instituet fuer Kernchemie, Universitaet Mainz, Mainz, Germany and Dr. F. H.A.Schneeweiss of Forschungzentrum Juelich, Germany while he was working as a DAAD Professor in the Aachen University of Applied Sciences, Juelich, Germany.

REFERENCES

1. Pillai, M.R.A. and Ramamoorthy, N. (1999) *Cyclotrons and their Applications*. In: Ionizing Radiation Sources. Ed. Tomar, B. S., Indian association of nuclear chemists and allied scientists, Mumbai.
2. Qaim, S. M. (2003) *Cyclotron production of medical radionuclides*. In: Handbook of nuclear chemistry. Ed. Vertes, A., Nagy, S. and Klencsar, Z., Kluwer Academic Publishers, Dordrecht, The Netherlands.
3. Qaim, S. M. (2001) Nuclear data for medical applications. *Radiochim. Acta* 89
4. Directory of Cyclotrons (2002) International Atomic Energy Report IAEA-DCRP/CD, IAEA, Vienna.
5. Ruth, T. J. (2003) Accelerators available for isotope production. In: Handbook of radiopharmaceuticals. Ed. Wech, M. J. and Redvanly, C. S., Wiley, Chichester.
6. Scharf, W. (1986) Particle accelerators and their uses. Harwood Academics Publishers GmbH, Germany.
7. Friedlander, G., Kennedy, J. W., Miller, J. M. and Macias, E. S. (1981) Nuclear and Radiochemistry, John Wiley and Sons, New York.

11

STUDY OF RADIOTRACER APPLICATIONS IN BIO-NANO-GEO SCIENCES AT SAHA INSTITUTE OF NUCLEAR PHYSICS

Susanta Lahiri and Kamalika Roy

ABSTRACT

The radiochemistry group of Saha Institute of Nuclear Physics, Kolkata, is actively engaged in the study of different biological and geological problems with the main emphasis being on metal bio-accumulation by different bio-reagents like yeast, algae, mushroom, biopolymers etc. The studies show that certain bioreagents can be efficiently used for preconcentration of some metals. In the field of geo sciences efforts have been made to develop separation techniques for naturally occurring stable isotopes which are isobaric to some cosmogenic radionuclides. This state of art separations is a step forward in the direction of solving long standing astronomical puzzles. We have also reported for the first time synthesis of radioactive gold nano particles with green (environmentally safe) chemical reagents.

1. INTRODUCTION

Radiotracers have made significant contributions in several areas like medicine, industry, agriculture and fundamental research. The applications of radiotracer techniques have considerable environmental impact, and potential for expansion is enormous. Apart from exploring the mechanism of various chemical reactions, radiotracers are widely used in studying the biological systems of various organisms starting from the single celled bacteria to higher plants and animals. Certain radioisotopes also find use in medical applications. They are either used as therapeutic radiopharmaceuticals (e.g., ^{90}Y, ^{186}Re, etc.) or for diagnostic purposes (e.g., ^{99m}Tc, ^{89}Zr, ^{68}Ga, ^{201}Tl, etc). Radiotracers in the form of natural and artificial radionuclides are employed to identify and model transport processes in diverse geological systems like water bodies, wetlands and coastal marine environments.

Radioisotope studies coupled with Accelerator based Mass Spectrometry (AMS) can help in estimating several geological events. For example, measurement of ^{36}Cl/^{35}Cl ratio may be used for ground water dating. The possible use of ^{36}Cl for dating old ground waters is defined by the competition between decay of the initial cosmic ^{36}Cl input (N_0) and the build-up of ^{36}Cl (N_t) by underground neutrons [1]:

$$N_t = N_o e^{-\lambda t} + N_{eq}(1 - e^{-\lambda t})$$

Where $N_0 e^{-\lambda t}$ denotes decay of cosmic input and N_{eq} $(1- e^{-\lambda t})$ the intra formational build-up (λ is the decay constant of ^{36}Cl and t is time for build up of ^{36}Cl).

The area of nano science research has made a big impact in the present scenario of scientific research. The research on radiotracer related nano materials is still in its infancy. Radioisotope (^{99m}Tc) in conjunction with carbon nano tubes has been recognized as new lung imaging medical bio-tracer [2]. The most accepted mechanism for the synthesis of metal nano particles by radiation involves a two-step process, i.e. nucleation and successive growth of the particles.

During last few years, a considerable time of radiochemistry research at Saha Institute of Nuclear Physics, Kolkata has been focused on the application of radiotracers in the area of bio- and geo- sciences, details of which are discussed in the following paragraphs.

2. RADIOTRACERS IN BIOLOGICAL SCIENCES

In view of increasing interest in cost effective and environmentally sustainable removal procedures of toxic metals and radionuclides, the use of hazardous chemicals is being restricted and replaced by greener alternatives. Various bio-reagents like Baker's yeast, certain phytoplankton, proteins and other biomolecules derived from seeds, etc., are drawing attention of the analytical chemists. Amongst various eukaryotic organisms *yeasts* play an important role as a model of eukaryotic cells for biochemical and physiochemical experiments. The possibility of using Baker's yeast *Saccharomyces cerevisiae* as sequestering agent for some long-lived radionuclides, ^{152}Eu, ^{137}Cs and ^{60}Co has been explored. The auxotrophy dependent sensitivity of particular strains of yeast to particular metal ion has been studied [3]. Four different strains of yeast, AP22 (MATα leu2 his3 trp1 ura3), CCFY (MATα leu2 his3 ade2 ura3 trp1 can1), 699 (MATα leu2 his3 ura3 trp1 ade2) and F399 (MATα gal7 lys2 tyr1 leu2 his4 thr4 MAL2 ade6 arg4 ura4) are taken for bioaccumulation studies with the above mentioned radioisotopes. It has been found that Baker's yeast, independent of its auxotrophy, is good bioaccumulator for europium from trace to macro level while accumulation of cobalt strongly depends on the histidine auxotrophy. In all of the above cases yeast shows no interest to accumulate long-lived ^{137}Cs radionuclides. It is therefore evident that the mechanism of uptake of the rare earth element, Eu follows a different biosynthetic pathway than that of the transition elements, or the alkali metal elements. In another set of experiment, two different strains of *Saccharomyces cerevisiae* (CCFY and AP22) were used to study the preconcentration of precious metal gold under different experimental conditions [4]. A quantitative relation between the total amount of gold uptake with the number of yeast cells has been established. The results show that the yeast cells can be used as environmentally benign extractant for gold in place of the conventional chemical extracting agents.

Certain bio-reagents e.g., phytoplankton species, have been shown to be effective scavengers of toxic metals. Various algal genera, viz.: *Lyngbya* and *Phormidium* from cyanophyceae; *Rhizoclonium* and *Chaetomorpha* from chlorophyceae and *Catenella* from rhodophyceae have been found to accumulate the radionuclides ^{152}Eu, ^{137}Cs and

^{60}Co to a considerable extent [5]. *Chaetomorpha* has the capacity of accumulating ^{152}Eu selectively from a mixture of three different radionuclides and can therefore be introduced as a bio-reagent for separation of long-lived radionuclides from the waste disposal sites. The bioaccumulation of ^{198}Au radionuclide, by *Rhizoclonium* a member of Chlorophyceae has also been studied [6]. Accumulation of gold on *Rhizoclonium* is almost pH independent in the lower pH range and is slightly higher in basic pH. To have an idea about the binding sites of gold on *Rhizoclonium* the gold accumulated biomass was analyzed. The entire ^{198}Au has been found in the cellulose part and a small portion in the protein part. This implies that algae might have reduced initially present Au(III) to Au(I), followed by Au(I) to Au(0) which is sequestered in cellulose fibers.

The biosorption of heavy and toxic radionuclides by three algal genera, *Spirulina*, *Oedogonium*, *Catenella* from different taxonomic groups was studied using "Tracer Packet" technique [7]. The tracer packet of heavy and toxic metals containing ^{197}Hg, $^{198\text{-}201}Tl$, $^{199\text{-}201}Pb$, ^{204}Bi and $^{204,205}Po$ in their no-carrier-added state was produced by irradiating a gold metal foil with medium energy ^{7}Li and ^{12}C beams successively in a 12MV Pelletron. *Spirulina* was found to be most suitable for metal removal at basic pH [8]. The results provide valuable information on the environmental risk assessment of many kinds of radionuclides using algae as bio-monitors.

Pinnate diatoms, (Elongated single-celled algae, mostly photosynthetic, that form silica cell walls, can be solitary or chain-forming, range in size from 2 microns to several millimeters and found in marine and aquatic systems in the top layer of water column or on/in the bottom sediments.) *Nitzscia obtusa* and *Navicula minima* also show bioaccumulation of ^{198}Au radionuclide in acidic to neutral pH. Maximum accumulation was observed with 1mgkg^{-1} to sub mgkg^{-1} gold concentration. The accumulation by the living cells was hindered when gold concentration was higher than 1mgkg^{-1} [9].

The common edible mushroom species, *Pleurotus citrinopileatus* was found to accumulate long-lived radioisotopes of cesium, ^{134}Cs ($T_{1/2}$ = 2.06 y) and ^{137}Cs ($T_{1/2}$ = 30.02 y) in controlled laboratory conditions. A large fraction of the total cesium was accumulated in the cap portion of the mushroom. The protein and fat fractions were extracted after accumulation of radiocesium. It was found that most of the radiocesium accumulation occurred in the protein fraction. The mushroom species normally white in color turned completely black after accumulation of radiocesium. These black mushrooms were produced upto fourth generation by tissue culture method and the successors remained black without addition of any further radioactive cesium. This indicates a permanent mutation of the experimental species. This observation may be helpful in the understanding of the process of biomagnification of ^{137}Cs radionuclide through food chain to human being. The black mushroom may also find application as a radio indicator for fallout and contaminated zones [10].

Study of metal protein interaction with different seed proteins have been carried out in multielemental environment. Binding ability of mercury, thallium, lead and bismuth with *Erythrina variegata* seed protein have been investigated using tracer packet technique. Due to lack of standard methods, inter-comparisons have been made applying three different approaches, like trichloroacetic acid precipitation, isoelectric precipitation and dialysis of protein after incubation with metals. All three approaches were in good agreement for binding of Hg, Tl and Bi radionuclides with the seed protein.

Hg and Bi showed high binding affinity with protein while Tl showed minimum interaction [9]. Proteins extracted from *Swietenia mahagoni* (L.) jacq. (Family Meliaceae) seed have been studied for the binding ability of various micronutrient metals, using the radioisotopes ^{61}Cu, ^{62}Zn, ^{66}Ga, ^{71}As, ^{69}Ge and ^{73}Se by tracer packet technique. It has been found that no-carrier-added copper and zinc radionuclides bind with the protein, while accumulation of other radionuclides by the protein is negligible [12].

Alginate, a natural biopolymer extracted mainly from brown seaweed and various bacteria can be used as adsorbing material. It may be converted into hydrogels via cross linking with divalent calcium ions bound to two carboxyl groups as adjacent alginate molecules. Adsorption behavior of no-carrier-added $^{197-200}$Pb, $^{197-200}$Tl and ^{197}Hg radionuclides to the calcium alginate beads has been studied. It has been found that over a wide range of pH 2-7, calcium alginate (CA) can remove quantitatively lead and mercury from the solution. The adsorbed radionuclides can be recovered from the CA beads using suitable reagents, e.g., 100% Tl and 80% Pb was desorbed by 0.1M HCl [13].

A list of bioreagents accumulating different metals at the optimized experimental conditions are given in Table 1.

Table 1 *Metal bioaccumulation by various bioreagents*

Bioreagent	*Experimental condition*	*Metal accumulation*	
		Trace	*Macro concentration*
Yeast: *(Saccharomyces cerevisiae)*	*pH~1-5*	*Eu*	*Trace to 0.1 mgkg^{-1}*
	pH~3	*Au*	*Trace to 0.1 mgkg^{-1} 100%)*
Algae: *Lyngbya and Phormidium*	*pH~7*	*^{152}Eu ^{60}Co*	*trace*
Rhizoclonium, Chaetomorpha	*pH~7*	*^{152}Eu ^{60}Co*	
Catenella	*pH~7*	*^{152}Eu^{60}Co*	
Rhizoclonium	*Independent of pH*	*^{198}Au*	
Spirulina	*pH 5*		
pH 8	*^{204}Po*		
	^{197}Hg		
	$^{199-201}$Pb		
	^{204}Bi		
Oedogonium	*pH 6*	*$^{199-201}$Pb*	
	pH 7	*^{197}Hg*	
Catenella from *Rhodophyceae*	*pH 7*	*^{204}Po*	
	^{197}Hg		
	$^{199-201}$Pb		
	^{204}Bi		
	$^{198-201}$Tl		
Diatom: *Nitzscia obtusa* and			
Navicula minima	*pH3*	*Au*	*Trace to 1.0 mgkg^{-1}*
Mushroom: *Pleurotus citrinopileatus*		*^{134}Cs*	
Seed protein: *Erythrina variegata*	*pH 7*	*^{197}Hg*	
	^{204}Bi		
Swietenia mahagoni	*pH 7*	*^{61}Cu*	
	^{62}Zn		
Biopolymer: Calcium alginate	*pH~2-7*	*$^{197-200}$Pb*	
		^{197}Hg	

Apart from the biological work mentioned in last few years the main thrust has been in developing methods of production and separation of no-carrier-added radionuclides useful for nuclear medicinal application. In nuclear medicine no-carrier-added radionuclides (i.e., radionuclides with high specific activity) with moderate radiological half life are always preferred. The radionuclide should be chemically and radioisotopically pure. For the first time heavy ion induced production of proton rich radionuclides have been proposed. The proton rich radionuclides are some time also positron emitter and find their suitability in PET studies. Some of the alternative radionuclides have been proposed for diagnostic and therapeutic applications which have better properties than those currently used in nuclear medicine. Table 2 shows some examples of the proposed radioisotopes useful for nuclear medicine [14, 15].

Table 2 *Alternative radionuclides proposed by SINP radiochemistry group useful for clinical applications*

	Radionuclides	*Half life*	*Decay mode*	*Principal g-energies*
Conventional	144 Ce	28.4 d	β–	133.5
	141 Ce	32.5 d	β–	145.4
Alternative	135 Ce	17.7 h	EC	265.2 300.0
Conventional	152 Eu	13.5 y	EC, β–	121.7, 344.2
	154 Eu	8.6 y	EC, β–	123.1
Alternative	145 Eu	5.9 d	EC	653.4, 893.7
	146 Eu	4.6 d	EC	633.1
	174 Eu	24.1 d	EC	121.2
	149 Eu	93.1 d	EC	277.1
Conventioanl	153 Gd	240.4 d	EC	97.4, 103.2
Alternative	147 Gd	38.1h	EC	229.3, 396.0
	149 Gd	9.3 d	EC	149, 298.6
Conventional	166 Ho	1.1d	β–	80.5
Alternative	160 Ho	5.0 h	IT, EC	197, 645.2, 728.2
Converntaional	167 Tm	9.2 d	EC	207.8
Alternative	163 Tm	1.8 h	EC	104.3, 241.5
	165 Tm	30.1	EC	242.8, 296.0
Converntaional	169 Yb	32	EC	109.8, 130.5, 177
Alternative	164 Yb	1.3 h	EC	390.6, 445.4

3. RADIOTRACERS IN EXPLORING GEOSCIENCES

Nature has kept her own signature using appropriate radiotracers during the last million and billion years at the time of every major event in our Universe. That may be creation of our solar system billions years BP (before present), explosion of nearby supernova million years BP or at the time of creation of the Earth, in this universe. However, these long-lived radionuclides, specially the supernova (SN) radionuclides, or radionuclides produced in a particular process have to be searched with great care as they are highly scanty in nature. Accelerator Mass Spectrometry (AMS), a highly sophisticated technique and capable of detecting down to 10^{-15} atom ratio, can only detect such radionuclides. The measurements in this ultra trace amounts by AMS unravels many intricacies in physical and biological sciences. AMS deals with the explanation of vast cosmic and galactic phenomena on one side and on the other side it also deals with DNA binding in the leaving systems. Above all, the AMS technique heralds in a new era of exciting research in varied number of experimental fields.

However, prior to detection chemical separations are required for ion source preparation of samples, or for separation of isobaric interferences especially from the naturally occurring stable isobars. For example, the radioisotopes ^{146}Sm (100My) and ^{182}Hf (9My) are believed to be formed by pure r-process during a supernova explosion and therefore the search of minute traces of ^{146}Sm and ^{182}Hf in the earth crust is of great interest. Only Accelerator Mass Spectrometry (AMS) is well suited for detecting such low levels of ^{182}Hf or ^{146}Sm. Any attempt to measure ^{182}Hf and ^{146}Sm by AMS must ensure that the sample is free from its naturally occurring stable isobar ^{182}W and ^{146}Nd respectively. A simple method for separation of tungsten and hafnium has been developed using radiometric simulation followed by checking the decontamination of tungsten from Hf in a synthetic sample by AMS [16]. Cosmologists believe that our solar system was formed from the remains of one or more supernovae. Because samarium-146 has a half-life of 100 million years, it is one of the few radioisotopes still left from the formation of the solar system. By measuring the amount of samarium-146 in the earth's crust, it is believed one can find out more about supernovae and about the conditions in which the solar system was formed. The main obstacle for measuring ^{146}Sm is its stable isobar ^{146}Nd, which is at least a million times more abundant than the infinitesimal amount of samarium-146. To overcome this problem, a liquid-liquid extraction method is developed which can separate out the samarium from the neodymium [17].

The other candidate radionuclide for unraveling the history formation of earth is ^{53}Mn ($T_{1/2}$=3.7My), which is cosmogenic in nature. The basic knowledge of ^{53}Mn production rates at various altitudes and locations is important for a study of earth surface processes that occurred during the last few half-lives of ^{53}Mn, ~10 My, the period not readily accessible with the other cosmogenic nuclides in common use. The abundance of ^{53}Mn in the earth crust is in the ratios of ^{53}Mn/^{55}Mn ~ $1/1\text{x}10^{14}$. Recent technical advances have made it possible to determine these ultra-trace ratios of ^{53}Mn by means of AMS. However the main interference in the detection of ^{53}Mn by AMS is its stable isobar ^{53}Cr, which is much more abundant in nature. Liquid-liquid extraction (LLX) method for separation of manganese from chromium has been developed. The developed method has been successfully verified for natural samples like crust, sediment and lava collected from the Pacific [18]. This way radiotracers help to study the basic processes in geological sciences.

4. RADIOTRACERS IN EXPLORING NANO-SCIENCES

The application of radiotracers in the field of nano-bio technology is still in its infancy. Biological systems have shown remarkable property of reducing cationic species into neutral ones through reduction by amino, carboxyl and hydroxyl moieties at the cell membrane sites. In such cases the neutral particles have been found to be in the nano range.

For the first time radioactive ^{198}Au nanoparticles have been reported. Moreover, the radioactive gold nanoparticles have been synthesized using a green chemical solvent polyethylene glycol (PEG 4000) [19]. The PEG rich phase containing radioactive gold nanoparticles has been characterized by UV-visible spectroscopy. The actual size in terms of diameter was determined using transmission electron microscopy (TEM) and has been seen to be in the range of 15-20nm. The developed procedure using an aqueous biphasic system is an environmentally safe technique for producing radioactive gold nanoparticles. ^{198}Au is short lived ($T_{1/2}$= 2.69d) and has high intensity gamma energy at 411keV; thus the synthesized radioactive gold nanoparticles may be useful for *in-vivo* and *in vitro* applications where both the properties of nano material as well as high energy gamma ray will be useful. The experiments with bulk amount of gold spiked with ^{198}Au suggest that even upto 50mM concentration gold nanoparticles are formed by the described method. Therefore the method can be extrapolated for much higher radioactivity to perform imaging based *in-vivo* tracking, organ specific delivery using antibodies on the gold nanoparticle surface, which may be useful for simultaneous detection of the organs and destroying of the abnormal cells.

Acknowledgement

One of the authors (K. Roy) expresses sincere thanks to the Council of Scientific and Industrial Research (CSIR) for providing research fellowship.

REFERENCES

1. Fontes, J. C., Brissaud, I. and Michelot, J. L. (1984) Hydrological implications of deep production of chlorine-36. *Nucl. Instr. Meth.* **B 5,** 303-307.
2. Chan, H.B.S., Ellis, B.L., Sharma, H.L., Frost, W., Caps, V., Shields, R.A., Tsang, S.C. (2004) Carbon-encapsulated radioactive Tc-99m nanoparticles. *Adv. Mater.* **16,** 144-147.
3. Roy, K., Sinha, P., and Lahiri, S. (2005) Selective bioaccumulation of ^{152}Eu by the yeast, *Saccharomyces cerevisiae.* Nuclear and Radiochemistry Symp., NUCAR 2005, Guru Nanak Dev University, Amritsar, 15-18 March, pp 577-578.
4. Roy, K., Sinha, P., and Lahiri, S. (2006) Accumulation studies of gold using Baker's yeast, *Saccharomyces cerevisiae,* In: Application of radiotracers in chemical, environmental and biological sciences. Eds. Lahiri, S., Nayak, D. and Mukherjee, A. Saha Institute of Nuclear Physics, pp 225-226.
5. Saha, A. K., Chakraborty, N., Nayak, D., Lahiri, S., Pal, R. (2006) Bioaccumulation of ^{152}Eu, ^{137}Cs, ^{60}Co using algae as bio-reagent, In: Application of radiotracers in chemical, environmental and biological sciences. Eds. Lahiri, S., Nayak, D. and Mukherjee, A. Saha Institute of Nuclear Physics, pp 221-222.
6. Nayak, D., Nag, M., Banerjee, S., Pal, R., Laskar, S., and Lahiri, S. (2006) Preconcentration of ^{198}Au in Rhizoclonium – A green alga. *J. Radioanal. Nucl. Chem.* **268,** 337-340.
7. Nayak D. and Lahiri, S. (2002) Production of tracer packet of heavy and toxic elements. *J. Radioanal. Nucl. Chem.* **254**, 619-621.
8. Nayak D., Lahiri, S., Mukhopadhyay, A. and Pal, R. (2003) Application of tracer packet technique to the study of the bio-sorption of heavy and toxic metal radionuclides by algae. *J. Radioanal. Nucl. Chem.* **256**, 535-539.

9. Chakraborty, N., Pal, R., Ramaswami, A., Nayak, D. and Lahiri, S. (2006) Diatom – a potential bio-accumulator of gold. *J. Radioanal. Nucl. Chem.* **270**, 645-649.

10. Mukhopadhyay, B., Nag, M., Laskar, S. and Lahiri, S. (2007) Accumulation of radiocaesium by *Pleurotus citrinopileatus* species of edible mushroom variety under controlled laboratory condition. *J. Radioanal. Nucl. Chem.* **273** (in press).

11. Nayak, D., Datta Samanta, T., Laskar S. and Lahiri, S. (2007) Application of tracer packet technique for studying metal-protein interactions with *Erythrina variegata* Linn. seed proteins, *J. Radioanal. Nucl. Chem.* **271**, 387-390.

12. Nag, M., Laskar, S., Ramaswami, A. and Lahiri, S. (2004) Application of tracer packet technique for studying metal protein interaction with *Swietenia mahagoni* (L.) Jacq. seed proteins. Discussion meeting on Application of radiotracers in chemical, environmental and biological Sciences (ARCEBS 04), April 15-16, 2004, Saha Institute of Nuclear Physics, Kolkata.

13. Nayak, D. and Lahiri, S. (2006) Biosorption of toxic, heavy, no-carrier-added radionuclides by calcium alginate beads. *J. Radioanal. Nucl. Chem.* **267**, 59-65.

14. Lahiri S. and Nayak, D. (2001) Production of medically important alternative radiolanthanides by heavy ion activation. *Physics Teacher* **43,** 44-49.

15. Lahiri, S. (2004) Production and separation of proton rich radionuclides by heavy ion activation. 6th International Conference on Nuclear and Radiochemistry (NRC-6), August 29- September 03, Aachen, Germany

16. Maji, S., Lahiri, S., Wierczinski, B., and Korschinek, G. (2006) Separation of trace level hafnium from tungsten: A step forward to solve astronomical puzzle. *Anal. Chem.* **78,** 2302-2305.

17. Maji, S., Lahiri, S., Wierczinski, B., and Korschinek, G. (2006) Separation of samarium and neodymium: a prerequisite for getting signals from nuclear synthesis. *Analyst* **131**, 1332-1334.

18. Lahiri, S., Nayak, D., and Korschinek, G. (2006) Separation of no-carrier-added ^{52}Mn from bulk chromium: A simulation study for AMS measurement of ^{53}Mn. *Anal. Chem.* **78**, 7517-7521.

19. Roy, K. and Lahiri, S. (2006) A green method for synthesis of radioactive gold nanoparticles. *Green Chem.* **8**, 1063–1066.

12

ARSENICOSIS AND THE ARSENIC-SELENIUM CONNECTION

Julian E. Spallholz and L. Mallory Boylan

ABSTRACT

Arsenic contamination of the domestic drinking water from shallow aquifers in Bangladesh available to villagers often exceeds the newest WHO standards of >10 ng As/L and the Bangladesh standard of >50 ng As/L. An estimated 9.2 million shallow tube wells in Bangladesh deliver water to 97% of the rural population with an estimated population of 57 million people at risk of arsenicosis. The contamination of drinking water by arsenic extends to W. Bengal, India and Nepal. The same shallow aquifers used for domestic water are also used to irrigate food crops, particularly rice. Irrigation adds arsenic to soils and increases the exposure of the population to additional arsenic via the foods consumed. Selenium, an essential trace mineral found in soils, is absorbed by plants entering the human food chain. We suggested that a low dietary intake of selenium may be contributing to the arsenicosis problem in Bangladesh. Selenium is a natural antidote to arsenic by accelerating its excretion and sequestration by complexation. It is also an antioxidant as a component of the enzyme glutathione peroxidase that may counteract the prooxidant effects of arsenic that contributes to arsenicosis and cancer. Analysis of agricultural soil samples from Bangladesh by flurometry, ICP-AES and NAA shows them to be very high in arsenic and biologically deficient in selenium. A low dietary intake of selenium from soils via foods maybe contributing to arsenicosis in the Ganges-Brahamaputa delta.

1. THE ARSENIC PROBLEM

Arsenic (As) is a non-metal found in the environment combined with sulfur as sulfides, orpiment and realgar (As_2S_3), and often with iron as arsenopyrites. Arsenic is also found in the environment as oxides; as the trioxide, (As_2O_3), the arsenates, (AsO_4), and arsenites (AsO_2). The arsenites and arsenates are often contaminants of domestic drinking waters. Levels of arsenic in drinking waters > 10 ppb (ng/L) exceed the established WHO limit for As in drinking water. Levels of As of 50 ppb in water have been established by some countries as the upper limit permitted in drinking water. Ingestion of As > 50 ppb from drinking water may lead to arsenicosis; arsenic toxicity characterized by melanosis, keratosis and cancers of the skin. The US Environmental Protection Agency classifies As as a known human carcinogen, a teratogen and suspected cardiovascular toxin. In addition to cancers of the skin, cancers of the liver, lung and bladder are recognized as being caused by arsenic. In animals, UV light and As ingestion are synergistic in their generation of skin cancer.

Arsenicosis and associated cancers occur in many parts of the world when arsenic concentrations are found in domestic drinking water that exceed upper regulatory limits. Major occurrences of arsenicosis are found in Taiwan, Chile, Mexico, China, Mongolia, portions of Europe and the United States. The worst human exposure to arsenic is geographically found in Bangladesh, W. Bengal, India and Nepal where many of the indigenous population are exposed to As levels in drinking water that exceed 50 ppb. The arsenic originates in mostly shallow underground aquifers brought to the surfaces in villages by hand pumped tube wells, the As having originated in the Himalayan mountains. Much of the Ganges-Brahmaputra delta is made from these As containing alluvial sediments constituting the source of much of the arsenic in well water.

1.1 The Suggested Arsenic Solutions

Prior to 1970's the primary source of drinking water in Bangladesh was from surface water and dug ponds. The high incidence of gastro-intestinal infections prompted the World Bank and UNICEF to embark on a massive public health program of drilling wells to provide clean water. Unbeknownst and with the best of intentions to improve public health the tube wells drilled by the World Bank and UNICEF often tapped into aquifers containing arsenic laden waters or into aquifers that overtime would release increasing amounts of arsenic. The first reports of arsenicosis and the discovery of arsenic in water in Bangladesh began in 1993. Within a few years the scope of the arsenic problem in the groundwater within one of the most densely populated places in the world became fully apparent. Millions of people, solely by misfortune, within the Ganges-Brahmaputra delta were at risk for arsenicosis.

The primary solution to obtain remediation of the regional arsenic crisis is by providing alternative sources of uncontaminated water and economical/environmental methods to treat contaminated water reducing exposure to arsenic. Various approaches to provide people with potable drinking water have been attempted. The initial approach was the identification of arsenic contaminated wells. More than 9 million wells have been reported to have been tested for arsenic and contaminated well heads have been painted red while non-contaminated well heads are painted green. Non-contaminated wells painted green over time often become contaminated when retested for As in water complicating the use of the initial non-contaminated wells. Filtration devices both relatively simple using sand and iron filings or nails to bind the arsenic have been used along with more sophisticated filtration devices. Both have generally proved to be beneficial but the disposal of the filtered arsenic and maintenance of the more sophisticated filtering devices has been problematic. Capture of rainwater has been an alternative attempt to replace arsenic laden well water. Costs for the collectors and the periods of the lack of rainfall hinder the collection of significant amounts of water for large populations. Finally, there has been some movement back to dug wells and the implementation of biological filtration and purification of that water to replace As contaminated tube well water.

Another way to possibly reduce arsenic toxicity from poisoning people from marginally contaminated well water source are to improve the diets of people. Several studies have reported that arsenicosis impacts the people with the poorest of poor diets. Diets higher in protein quality and antioxidants have been reported to improve people's resistance to arsenicosis. We had suggested several years to UNICEF that arsenicosis and associated cancers might be related to inadequate dietary intake of one specific essential dietary nutrient, selenium (Se).

1.2 The Working Hypothesis of the Selenium Solution

The rationale at the time (1998) upon first learning of the arsenic crisis in suggesting that dietary selenium inadequacy might be related in part to expression of arsenicosis was as follows:

1. Arsenic in experimental animals was known at the time and since the 1930's to counteract selenium toxicity.
2. Soils in many parts of the world are known to be low in selenium, and generally selenium is low in soils with high rainfall. The greater Ganges-Brahmaputra delta experiences annual monsoons and associated flooding leaching minerals, including selenium.
3. There was an extensive 286 page compendium published in 1996 on "The Geography of the Soils of Bangladesh". Some micronutrients are mentioned, but there is no reference to selenium within the text. The selenium content of Bangladeshi soils was not and is not well known.
4. Selenium is an essential dietary trace mineral not required by plants but it is associated predominantly with dietary protein, animal protein. Most dietary selenium ingested by people is selenomethionine. The diets of the very poor often are marginal and inadequate in total calories and protein.
5. One of selenium's predominate functions is that of a component of an antioxidant enzyme, glutathione peroxidase. Several arsenic species are prooxidants and therefore excessive ingestion from highly contaminated well water might further reduce the body burden of the postulated dietary inadequate selenium.
6. Other antioxidants might help reduce some arsenicosis symptoms but they could not replace the requirement for selenium being dietarily assaulted and having accelerated excretion by the elevated levels of arsenic ingestion.

Why selenium for arsenicosis prevention? It has been experimentally understood for sometime that selenium will counter the toxicity of several minerals; cadmium, copper, mercury, silver, lead and arsenic. This is accomplished by the formation of insoluble selenides of these minerals in vivo and as described below for arsenic, both complex with glutathione. As selenium, especially as selenides, are toxic themselves, nature in effect causes two toxins to annihilate each others toxicity.

1.3 What do we Now Know in 2006 About Selenium and Arsenicosis?

During the period following our attendance at the 3rd International Symposium on Arsenic at the Dhaka Community Hospital in May 2000, we were able to obtain a few agricultural soil samples from Jessore, Bangladesh. These samples were measured for selenium by the fluorimetric method which is sensitive to approximately 20 ng total Se. The soils averaged 0.051 ug Se/g which according to Oldfield [1] bordered exactly on soils with a Se deficient status (<0.05 ug Se/g). Our colleague Dr. Linda Smith, a geologist from the University of Texas, Dallas, having discovered the source of arsenic as being the Himalayan Mountains, at our request checked her outcropping rock samples from Nepal for selenium and found no detectable (<1.0 ug/g) selenium. Thus the high arsenic and low selenium origins of the Ganges-Bahamaputra soil and aquifers seemed to have been identified. At Texas Tech we continued to be interested in refining our understanding of selenium in soils, in foods, and in consumption from diets by people affected by arsenic.

Rice and common plant food samples along with fish were later obtained by us from Bangladesh and analyzed once again by the fluorimetric method of selenium analysis. The various Bangladesh rice samples averaged 0.132 mg Se/g (dry weight), with a range of 0.021-0.170 mg Se/g. In comparison, US produced rice is reported to average about 0.460 mg Se/g. The other plant foods were most often very low on a dried weight basis while fish usually possessed 1.0 mg Se/g or slightly more by dry weight. Using approximate rural Bangladesh dietary food intake records over a week, an estimate was made that some Bangladeshi adults were likely ingesting about 26 mg Se/day. This is below the WHO recommended adult minimum selenium intake of 40 mg Se/day. This value is comparable to the measured dietary intake of adult women in Nepal of 23 mg Se/day but greater than the people within the Keshan disease area of China (10 mg Se/day). Our adult estimate is just barely higher than the recent reported selenium dietary intake for Bangladeshi year-old infants consuming complimentary foods (11 mg Se/.day) and breast milk (14 mg Se/day) for a total of 25 mg Se/day made by Kimmons et al. [2]. Such values are questionable for the Bangladeshi selenium values were estimated based upon USDA food composition data and we suspect the foods and breast milk to measure less than the values expressed in US food tables. The authors acknowledge such differences might exist from the actual Bangladesh food values for Se and the Se value calculated. The present situation is that the dietary intake of selenium among people of the arsenic affected areas of Bangladesh has not been measured accurately and it should be done.

1.4 Selenium Counteracts Arsenic Toxicity

As previously noted, arsenic had been used experimentally in the 1930's to counteract and prevent selenium toxicity. In the 1960's and 70's it was noted experimentally in animals that arsenic accelerated the biliary excretion of selenium. Additional research in the 1980's and 1990's would show several experimental examples that the reverse toxic annihilation was also true, selenium added to cells in culture and to experimental animals reduced arsenic toxicity and increased its biliary excretion. In 2000, Gailer et al. demonstrated that at last one direct metabolic link between As-Se toxic annilation and increased biliary excretion was an As-Se complex with glutathione [3]. Likely formed in liver and passed to into bile as had been show years before by Levander [4], the excreted complex is the seleno-bis(S-glutathione) arsenium ion. This Se/As complex would be later identified in the bile of rabbits following injection of arsenite and selenate. Here, within this complex and the formation of insoluble selenides, is the likely explanation why arsenic toxicity counteracted selenium toxicity in the 1930's and why selenium latter would be shown to counteract systemic arsenic toxicity including the reduction of arsenic enhanced UV inducted skin cancer by selenium. All of these mechanistic interactions of selenium and arsenic assessing detoxification have been recently reviewed by Zeng et al. [5].

1.5 The Epidemiology of Selenium within the Ganges-Brahamaputra Delta

While extensive evaluation of water contamination of arsenic from tube wells have been made, and epidemiologic studies of the arsenicosis affected population has been reported in much detail, very little is known about the possible natural antidote to arsenic toxicity, selenium, in the area. Very little assessment has been made of the selenium in soils, in its distribution in foods and the human selenium status with and without arsenicosis. Some reports from W. Bengal India have found little or no associated changes in body selenium in people with and without arsenicosis. Other reports have found associated relationships between ingested arsenic and excretion of selenium in urine. In the Murshidabad district of West Bengal where arsenic-contaminated groundwater (mean: 0.11 mg/l, n=34) is the main source for drinking water, selenium in the diets of children were 37% less than the recommended amount but were considered adequate for males and slightly better than average for females. In view of the high arsenic ingestion from water we suggest that this is an example where the reported concentrations of Se in diets are likely inadequate for the population. Our own, admittingly limited data, suggests that low soluble selenium in soils, generally low to very low selenium in rice and other plant foods, but higher amounts of selenium in fish provides limited, marginally adequate human selenium nutriture. The data also suggests to us a human dietary selenium intake likely to be found marginal in the face of the reality of arsenic ingestion from contaminated water and its associated toxicity. Recent data on the concentration of both arsenic and selenium in hair is suggestive of high exposure of people to arsenic and lesser amounts of selenium in comparison to North Americans. The Chinese experience in Mongolia with selenium supplementation of Se-yeast over a 6-month period has shown a reduction in the visual symptoms of arsenicosis. While we hope to use inorganic selenium rather than an organic selenium supplement to affect arsenic excretion in a human supplemental trial measuring arsenic accumulation in hair and visual symptoms of arsenicosis, [http://www.bangladesh-selenium.org/] over the next couple of years, the ultimate experimentation will occur in children and young adults with no symptoms of arsenicosis taking selenium supplements as a preventative measure of protection, not treatment, of arsenic toxicity. While there is no substitute for arsenic free water improvement of the diets of people and possible selenium supplements might help to reverse the course of a tremendous environmental human catastrophe.

Acknowledgements

The authors gratefully acknowledge the cooperation of their many collogues that have provided samples for analysis, Dr. M.M. Rhaman in Bangladesh, Dr. David Robertson and his associates at the University of Missouri and Dr. Mohammed Aluddin at Cook

College in New York for their analytical support, the graduate students of Texas Tech that have helped with selenium and other mineral analyses, Richard Rigdon and the TraceElement Lab of Lubbock and Paul la Porte of the University of Chicago for his tireless efforts at helping to secure research funding and the assistance with his many colleagues working in Bangladesh to help find answers to the problem of arsenicosis. All these people are found at www.bangladesh-selenium.org/ , many with their pictures and we thank you all. We have come a long way in understanding the selenium situation, how it interacts with arsenic and yet we have a long way yet to go. May your enthusiasm and commitment carry us along the road of discovery. Some of the work here was supported by grants from ASPEN, the US Department of Energy and Exxon laboratories of Dhaka, Bangladesh. Travel to and from New Delhi has been supported in part by Texas Tech University and PharmaSe, Inc., www.pharmase.com. Travel within India was so graciously arranged by Dr. Alok Srivastava, Panjab University and supported by the Humboldt Foundation.

REFERENCES

1. Oldfield, J. E. (1992) *Selenium world atlas*. Selenium-Tellurium Development Assn. Grimbergen, Belgium. Revised on CD, 2002.
2. Kimmons, J. E., Dewey, K. G., Haque, E., Chakraborty, J., Osendarp, S. J. M., Brown, K. H. (2005) Low nutrient intakes among infants in rural Bangladesh are attributable to low intake and micronutrient density of complementary foods. *J. Nutr.* **135**, 444-451.
3. Gailer, J., George, G. N., Pickering, I. J., Prince, R. C., Ringwald, S. C., Pemberton, J. E., Glass, R. S., Younis, H. S., DeYoung, D. and Aposhian, V. (2005) A metabolic link between arsenite and selenite: The seleno-bis(S-glutathionyl) arsinium ion. *J. Am. Chem Soc.* **122**, 4637-4639.
4. Levander, O. A. and Baumann, C. A. (1966) Selenium metabolism VI. Effect of arsenic on the excretion of selenium in bile. *Tox. Appl. Pharm.* **9**, 106-115.
5. Zeng, H., Uthus, E. O., Combs, Jr. G. F. (2005) Mechanistic aspects of the interaction between selenium and arsenic. *J. Inorg. Biochem.* **99**, 1269-74.

13

BIOGEOCHEMISTRY OF LACUSTRINE SEDIMENT: AN ARCHIVE OF PALAEOCLIMATE VARIABLITY

Brijraj K. Das

ABSTRACT

The Indian monsoon is one major weather systems on the earth and, as evident by the unusually high rainfall in recent years, variations in the intensity have widespread socio-economic impact. Though the dedicated efforts by the scientists have been made over the centuries, a comprehensive picture of monsoon variability, especially on societally relevant time scales, has not emerged. This is because monsoon system is very complicated and to understand it and to predict, high resolution, short term data are required. Scientists have identified such archives while working on continents and obtained encouraging results from the Himalaya on seasonal to decadal scale resolution. It is realized that multiple 'proxy data' can be used to infer past monsoon variability and this is necessary in the assessment and prediction of future climate change.

Since organic matter is precipitated from water as particles of various sizes, its distribution is greatly affected by hydrodynamic conditions. One of the most general distribution features of organic matter in sediments is that its concentration rises as the particle size of sediments decreases. The latter is largely dependent upon source rock, distance of transport, intensity of rainfall and erosion etc. Therefore, the biogeochemical investigation of lake sediments particularly on well dated core sample will unravel the history of climate variability in the past which may correspond to global climate phenomena on a large scale. The biogeochemical study carried out on lake sediment core samples of Himalya have shown good results on Holocene palaeoclimate changes. A few cases are cited in this paper.

1. INTRODUCTION

Predicting the behavior of the monsoon is rather crucial particularly in the absence of long term high resolution (seasonal to decadal) records, and complexity and multiplicity of factors influencing the Indian monsoon and the complex processes of coupling between different climate phenomena (e.g El Nino southern oscillation (ENSO), North Atlantic Oscillation (NAO), tropical mid-latitude interactions.

In addition to these, the green house gas (CO_2) forcing is becoming increasingly important in global warming and sea levels rise. Some scientists have reported that sea levels are rising twice as fast as they were 150 years ago and man made green house emissions are the prime cause which is accelerating the pace of climate change. The level of carbon dioxide in the atmosphere today is 31 percent higher than what it was at the start of industrial revolution about 250 years ago. During the next century, CO_2 levels are expected to rise 90 to 250 percent over pre-industrial revolution level. Burning of fossil fuels is the most important source of green house gases. The most common use of fossil fuels in the form of coal and natural oil (petroleum) are for power generation, industries and transportation, accounting for about 75 percent of all emissions. With the growth of global economy, there has been unprecedented increase in the consumption of fossil fuels.

The major pools of carbon are lithosphere (earth's crust), ocean, soil, atmosphere and biosphere. Depending upon the use, any or all of these can act as sink or sources of carbon. Deforestation is a critical environmental problem in the developing countries. It is estimated that deforestation contributes about two giga tones of carbon into the atmosphere in addition to depletion of biodiversity, loss of economic opportunities, and increase in social problems [1].

In the existing scenario to understand fully and predict monsoon variability 'proxy'data on seasonal and decadal scale may be necessary. The longer continuous records of monsoonal variability are available from the marine sediments which provide information on centennial scale resolution. However, the data from continental setting is lacking, though the continents have the potential of providing well dated high resolution archives on seasonal and decadal scale such as tree rings, varved lake sediments, sediment facies, speleothems, micro-fossils, grain size, clay mineralogy, inorganic and organic geochemistry (biogeochemistry), etc.

Lake Sediments

Lakes are ideal place of deposition of weathered material derived from the surroundings and transported from distance sources through rivers, streams and nalas. It is a local site of sedimentation strongly influenced by its surroundings and bears records of events, which takes place in the surrounding drainage basin. Lakes and their sediments change rapidly in response to rapid change in the drainage basin, even seasonal changes are recorded. The study of lake sediments helps in reconstructing depositional history, climatic changes with time, besides pollution due to natural and anthropogenic sources. Many lake studies have shown that regional climatic oscillations can be related to global climatic processes.

Since lakes are smaller water bodies than sea, sedimentation in lakes is less disturbed and is of finer resolution. Therefore, lake sediments preserve records of past environmental changes. As the mineralogy of lake sediments is little altered by post-depositional reactions i.e. diagenesis and that physical and biological mixing in sediments are small relative to total amount of sediments deposited in the lake, the mineralogical and chemical variations of the sediments with depth can provide information on the physical and chemical changes that have occurred in the life of the lake. Such informations are found useful in evaluating long-term climatic changes of the region and in helping to assess the effects of human activities on the lake [2].

The organic matter content of lake sediments participate in a variety of biochemical and geochemical processes. Benthic animals and microbes depend on organic matter for their nutrition. Because many components of organic matter are relatively easily oxidized forms of reduced carbon, the dissolved oxygen concentration of sediment pore water is controlled largely by the availability of oxidizable organic matter below the sediment/water interface. Besides, the redox reactions, in which oxidized inorganic species become reduced by interaction with organic matter, are common in sediments organic matter, therefore is a dynamic biogeochemical component of sediments [3].

Source of Organic Matter in Lake Sediments

The particulate detritus of plants that have lived in the waters and on the land surrounding a lake comprises the primary source of organic matter to the bottom deposits. Plants have been divided into two groups on the basis of their biochemical compositions; (i) nonvascular plants that lack woody and cellulosic tissues, such as simple algae, and (ii) vascular plants that have these tissues, such as grasses, shrubs, and trees. The relative contribution from these two plant groups is influenced strongly by lake morphology, watershed topography, and relative abundances of lake and watershed plants [3]. The latter types of plants (vascular) exist on land and in shallow parts of lakes as bottom rooted emergent vegetation. Organic matter in lake sediments covers the spectrum of being predominantly algal in some lakes to being land-derived in others. The latter is probably more common in the lakes.

Biogeochemistry of Lake Sediments

The lacustrine environment is a very complicated hydrological and biological system which is reflected in the isotopic composition of its sources of organic carbon fraction. The main sources of organic carbon in a lake are submerged plants such as microphytes, floating pond weeds or planktonic material, and terrestrial vegetation. Of these, the submerged plants tend to be isotopically heavier because their source of CO_2 is dissolved bicarbonate rather than atmospheric CO_2. The latter being heavier, results in a fixed carbon also being enriched.

Similarly, the $\delta^{13}C$ value of planktonic material reflects the planktonic source of carbon i.e the use of dissolved bicarbonate results in $\delta^{13}C$ enrichments and use of atmospheric CO_2 leads to an overlap of planktonic $\delta^{13}C$ with the values for terrestrial plants. The latter fall into the well known C3 and C4 categories, with characteristic average $\delta^{13}C$ values of –27 and –14 %o, respectively.

Mansar Lake: A Case Study

The Mansar Lake of lesser Himalaya in Jammu (Figs.1a and b) is a fresh water lake at an altitude of 660 m above msl in lower Siwalik Formation sitting over an anticlinal axis of a large anticlinorium known as Surin-Mastgarh anticlinorium (Fig.2) [4]. It is the deepest lake in the region having a maximum depth of 38.25 m in the center. It is a tectonic lake formed during Holocene, 10,000 year ago. The rain fall in the region is about 1500 mm per year. The maximum evaporation is 1 to 1.5 m a year in the lake [5].

Fig. 1a *Mansar Lake, situated in Lower Siwalik Formation,* (For color illustration refer p xvii)

Fig. 1b *A panoramic view of Mansar lake, showing vegetation along the banks,* (For Color illustration refer p xviii)

The investigation has been carried out on 30 m long core retrieved from this lake, which shows that $\delta^{13}C$ value at shallow depth of 6 m is relatively heavier and gradually depletes becoming lighter till 16 m depth (Fig.3) [6, 7]. The heavier $\delta^{13}C$ at shallow depth may be due to algae or submerged vegetation as they derive CO_2 from dissolved HCO_3^- and not from atmospheric CO_2. The C/N ratio at shallow depth ranges from 6 to 13 with a general variation between 8 to 11.5 (Fig. 3). This value is not typically similar to non-vascular aquatic plants as it exceeds more than 10. This ratio for such plants are expected to be between 4 to 10. The $\delta^{13}C$ value ranges between -25%o to – 23%o which shows relative depletion in 13C indicating greater contribution from terrestrial vegetation. This indicates autochthonous and allochthonous carbon sources to the sediment. From 16m to 22m depth is sand deposit and beyond 22 m depth the ä13C of organic matter is heavier reaching -19%o and a general range is between -

25‰ to -19‰ and C/N ratio ranges from 9 to 20. The $\delta^{13}C$ value deviate from the extreme value of -17‰ expected of aquatic plants using dissolved bicarbonate. The decrease in $\delta^{13}C$ values may be due to decrease in the supply of molecular CO_2 caused by more flushing and a relative increase in floating planktonic material. The total organic carbon in shallow sediment is very low (Fig.3) and increases many fold in deeper sediments indicating greater productivity during warm and wet climatic regime. The increased productivity resulted in increase carbon content of total sediments and more dissolved bicarbonate contribute to the lake carbon pool to keep pace with the requirements of productive lake. During cold phase opposite effect operated making the organic carbon relatively depleted in $\delta^{13}C$ and also causing a decrease in the absolute carbon content in the sediment.

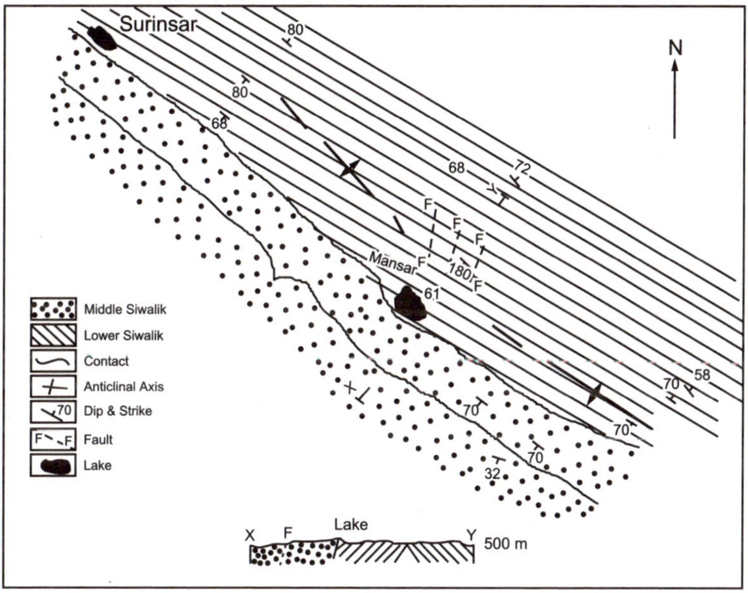

Fig. 2 *Geological and tectonic map of Mansar Lake showing lithology, anticlinal axis and faults (after Singh and Sharma, 1999)*

This suggests wetter and warm climatic regime during early Holocene followed by enhanced precipitation during which sands were deposited probably in middle Holocene and climate ameliorated to cold and dry conditions during late Holocene when terrestrial source of carbon increased. Similar climatic scenario, during Holocene, has emerged from the bio-geochemical studies of lake sediments of some Himachal Pradesh lakes carried out by the author.

Palaeoclimate studies carried out in Bhimtal-Naukuchiatal basin of Kumaun Himalaya by Kotlia et al. [8] has indicated arid conditions from ca.4,200 yr BP onwards and cool dry climate in Garhwal Himalaya on the basis of pollen studies from Alpine peat by Phadtare [9]. In Pakistan, the period from 3,900 Yr BP onwards has been interpreted as indicator of low precipitation and decreased run-off on the basis of varved sediment studies by von Rad et al. [10].

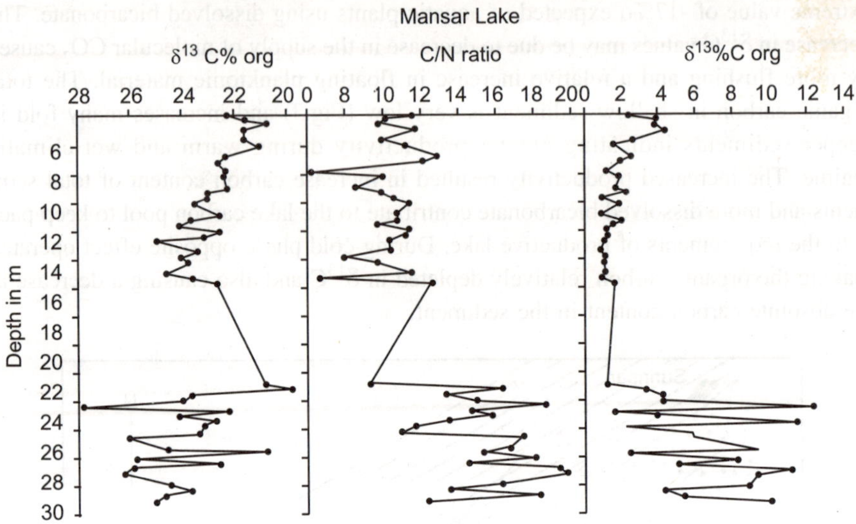

Fig. 3 *Vertical distribition of δ^{13}C, C/N ratio and organic carbon in sediments of Mansar Lake*

The drier climatic conditions from ca.4,500 yr BP are also recorded from the Nepal Himalaya on the basis of vegetation history in the lake Rara National Park by Yasuda and Tabata [11]. The maximum aridity for the whole Holocene in the western Tibet has been recorded at ca. 4,300 yr BP by Fontes et al. [12] and between ca. 4,000 to 3,000 yr BP by Gasse et al. [13]. Therefore, the climatic variability recorded in Mansar Lake sediment during Holocene is consistent with the records available from other areas including lesser Himalaya.

REFERENCES

1. Kushwaha, S.P.S. and Dadhwal, V.K. (2007) Carbon budget assessment in terrestrial ecosystems of India. International conference on *Asian monsoon variability during the past global changes*, pp. 58-60.
2. Jones, F. and Bowser, C.J. (1978) The mineralogy and related chemistry of lake sediments. In: *Lakes, Chemistry, Geology, and Physics*, (ed. Lerman, A.), Springer, New York, pp. 179-227.
3. Meyers, P.A. and Ishiwatari, R. (1993) Lacustrine organic geochemistry- an overview of indicators of organic matter sources and diagenesis in lake sediments. *Org. Geochem.* **20**, 7, 867-900.
4. Singh, R. and Sharma, V.K. (1999) Geoenvironmental appraisal of Mansar and Surinsar Lakes, Udhampur and Jammu districts. *Geol. Surv.Ind. Records* **131**, 19-24.
5. Rai, S.P., Kumar, V., Singh, O., Kumar, B. and Jain, S.K.. (2006) Bathymetry, sedimentation rate and physico-chemical characteristics of Mansar Lake in Himalayan Foot hills, Jammu & Kashmir. *Indian J. Geol. Soc. Ind.* **67**, 211-220.
6. Das, B.K., Schafer, P. and Jennerjahn, T.C. (1999) Biogeochemical studies in lesser Himalayan lakes of Jammu district, India. *SCOPE Sonderband*, **82**, 245-253.

7. Das, B.K. (2007) Climatic variability during Holocene- an example from lesser Himalayan lakes. International conference on *Asian monsoon variability during the past global changes*, pp. 10-12.

8. Kotlia, B.S., Bhalla, M.S., Mathur, P.D., Bhandari, S. and Chacko, S.T. (1997) Palaeoclimatic conditions in the upper Pleistocene and Holocene, Bhimtal-Naukuchiatal Lake basin in south-central Kumaun, North India. *Palaeogeogr. Palaeoclimatol. Palaeoecol.* **130**, 307-322.

9. Phadtare, N.R. (2000) Sharp decrease in summer monsoon strength 4000-3500 cal Yr BP in the Central Higher Himalaya of India based on pollen evidence from Alpine peat. *Quarter. Res.* **53**, 122-129.

10. Von Rad, U., Schaff, M., Michels, K.H., Schultz, H., Berger, W.H. and Sirocko, F. (1999) A 5000-year record of climatic change in varved sediments from the oxygen minimum zone of Pakistan. *Quarter. Res.***51**, 39-53.

11. Yasuda, Y. and Tabata, H. (1985) Vegetation and climatic changes in Nepal Himalayas II. A preliminary study of the Holocene vegetational history in the lake Rara National Park area, west Nepal. *Curr. Trends Geol* **6**, 13-17.

12. Fontes, J.C., Melieres, F., Gilbert, E., Qing L. and Gasse, F. (1993) Stable isotope and radiocarbon balances of two Tibetan lakes (Sumxi Co, Longmu Co) from 13,000 BP. *Quarter. Sci. Rev.* **12**, 875-889.

13. Gasse, F., Fontes, J.C., Van campo, E. and Wei, K. (1996) Holocene environmental changes in Bangong co basin (western Tibet). Part 4: Discussion and conclusions. *Palaeogeogr. Palaeoclimatol. Palaeoecol.* **120**, 79-92.

14

ENGINEERING INTERVENTIONS FOR MANAGEMENT OF MARGINALLY SALINE GROUNDWATERS

S. K. Kamra

ABSTRACT

Indiscriminate pumping is causing groundwater salinization and deterioration of pumped water quality due to upconing of saline water from lower zones in several coastal and irrigated regions of India. Under such conditions, it is imperative to selectively skim thin layers of fresh water accumulated over the saline groundwater through specially designed skimming structures and by enhancing groundwater recharge. In this paper, field appraisal of a few new technologies for skimming and recharging of fresh water in saline groundwater regions of Haryana, Andhra Pradesh, Gujarat and Tamil Nadu is presented. A clear perception emerging out of these studies is the need and importance of groundwater recharge in saline groundwater regions, either alone or in combination with skimming systems. Further individual farmer based skimming and recharge structures seem to have social viability and better chances of success and sustenance than the community/government, operated and maintained systems.

1. INTRODUCTION

The groundwater systems are the planet's largest freshwater resources and have storage capacity far more than all man-made surface reservoirs. About 60 percent of irrigation and 80 percent of drinking water requirements of India are met from groundwater sources. Proliferation of 17 million private open wells/ shallow tubewells and 90000 deep public tubewells has been vital for about seven- fold increase in India's irrigation potential from 7 to 50 million hectares (mha) over the last five decades. Due to easy access, operational convenience and private ownership, the groundwater development in the country has been quick but unregulated and has resulted in groundwater decline in more than 10% of the country's area. In the susceptible states of Andhra Pradesh, Bihar, Gujarat, Haryana, Karnataka, Kerala, Madhya Pradesh, Maharashtra, Punjab, Rajasthan, Tamil Nadu and Uttar Pradesh, declining ground water levels are forcing gradual deepening of open wells/ tubewells causing escalation in the cost of pumping. At current rate of development, about two fifth of the country's area is likely to be over-exploited (groundwater abstraction > 85% of annual replenishment) by 2017.

Excess pumping in the arid/semi-arid irrigated areas and coastal regions, containing limited depths of fresh water zones overlying native saline groundwater, results also in unwarranted increase in solute concentrations and deterioration of ground water quality. This occurs due to the phenomenon of *upconing* that moves the interface of fresh and saline waters up to within the domain of well screen. Inferior pumped-water quality and degradation of the aquifers in inland irrigated regions slow down or abandon groundwater development and consequently result in waterlogging and secondary soil salinization as observed in a major part of the southwest Haryana, Rajasthan and Punjab states. Under these conditions, it is imperative not to disturb saline water but to selectively skim fresh water accumulated due to recharge from rainfall, irrigation and/or canal seepage over the native saline groundwater through specially designed skimming structures and by enhancing groundwater recharge.

In this paper, basic features of different skimming structures are discussed in the context of Indian conditions. Further, field experiences on certain newly introduced technologies for skimming and recharging of fresh water in saline groundwater regions of Haryana, Andhra Pradesh, Gujarat and Tamil Nadu under a National Agricultural Technology Project are reviewed. In the end, salient observations and emerging issues relating to skimming and recharging technologies for saline groundwater regions are presented.

2. GROUNDWATER SKIMMING STRUCTURES

Various skimming well configurations such as single, multi-strainer, radial collector and scavenger wells (Fig.1) are possible to selectively abstract thin layers of fresh water overlying saline groundwater. The basic concept of all skimming structures is to modify the flow lines in such a way to maximize horizontal contribution of aquifer zones of acceptable quality to pumped water [11]. A single well (Fig. 1a) is used in unconfined aquifers in most parts of India. While using these wells in saline groundwater regions, well penetration is kept deep into the fresh water layer with a large gap between the bottom of the well and the fresh-saline water interface. These types of tubewell drainage projects have been executed at Masitawali in IGNP [4], Ghaggar depressional areas in Rajasthan and in Fatehabad branch area of Haryana.

A multi-strainer well (Fig. 1b), with relatively shallower penetration than single well, can be used with reduced *upconing* in fresh water layers of restricted depth. The system consists of closely spaced interconnected wells, each of low capacity, pumped by a central suction pump. Such structures are being extensively used close to canals/distributaries in Punjab in India [10] and in the Indus plains of Pakistan [7,11]. There are sporadic reports on the use of these systems in marginally saline regions of Haryana, Rajasthan, Andhra Pradesh and Tamil Nadu in India. Air leakage and priming problems have been reported from most of these studies.

Radial collector wells consisting of an open well and input radial drains on one or more sides (Fig. 1c) involve shallower penetration than a single vertical well operating at the same discharge. Since the radial drains collect water from shallow depths, *upconing* of saline water from lower depths is restricted. Large diameter open skimming wells experimented at Hisar in Haryana [6] and at Luni- Ki-Dhani in Rajasthan [4] and a drain line-sump based *Doruvu* technology embarked on a large scale in coastal sandy soils of Andhra Pradesh [9] are the local variants of radial collector wells. Scavenger wells (Fig. 1d) involve simultaneous abstraction of fresh and saline waters through two wells having screens in different quality zones, for controlling the rise of interface. The scavenger wells have been tested in the lower Indus basin of Pakistan [11] and have shown their potential in skimming of fresh water. Geological, hydrological and geo-chemical characteristics of the aquifers need to be studied in an integrated way to study the hydraulics and evaluate the performance of these skimming structures.

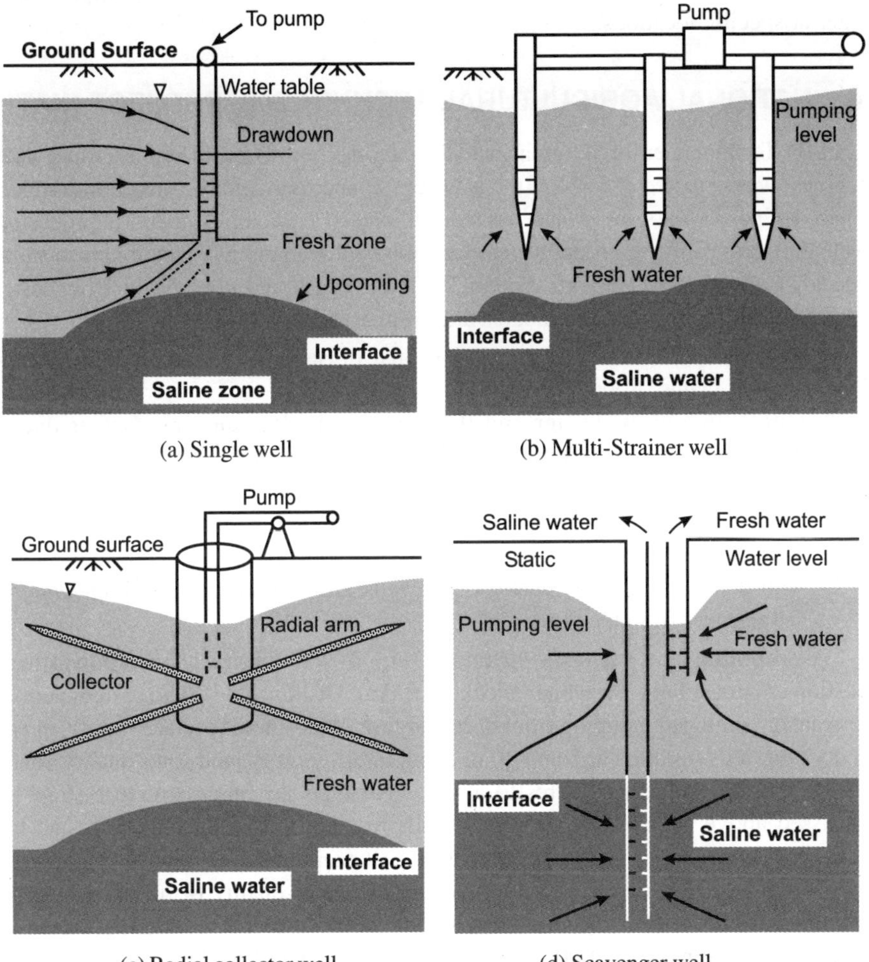

Fig. 1 *Different types of fresh water skimming wells.*, (For color illustration refer p xviii)

3. GROUNDWATER RECHARGE STRUCTURES

Central Ground Water Board [1, 2] and other agencies including research institutes, universities and NGOs have undertaken a large number of studies on induced recharge through percolation tanks, check dams, sub-surface dykes, underground *Bhundaras*, injection wells and roof water harvesting etc in different parts of the country. A broad categorization of these systems based on hydro-geological conditions and source of water to be recharged has been presented [3]. These projects have been/are being implemented in declining watertable locations having surplus monsoon runoff and sufficient sub-surface storage. Some projects have also been executed in polluted ground water areas to improve quality through dilution. Similar studies in shallow/saline groundwater regions, often considered unsuitable for groundwater recharge, are limited. Selective pumping (skimming) in these areas can create void space to be utilized for enhancing groundwater recharge using monsoon runoff or surplus canal water at specific locations.

4. NATIONAL AGRICULTURAL TECHNOLOGY PROJECT (NATP)

A NATP 'Technologies for Skimming and Recharging Fresh Water in Saline Groundwater Regions' was operated at CSSRI (the lead center) and four collaborating centers. Over a period of past 5 years, the project has made impressive accomplishments in proposing, evaluating and evolving groundwater skimming and recharging technologies in saline groundwater regions of Haryana, Andhra Pradesh, Gujarat and Tamil Nadu. Brief design features of all promising technologies are summarized below.

(a) **Haryana.** The north- western Indian State of Haryana is a part of Indo-Gangetic alluvial plains. About two third of the geographical area is currently underlain with saline groundwater and the situation is deteriorating further due to disproportionate pumping vis-a-vis groundwater recharge. In most of the marginally saline groundwater regions, low discharge shallow cavity wells are used for irrigation, which are inexpensive pumping structures not requiring strainer. Deep tubewells are not feasible due to increasing groundwater salinity with depth while many shallow tubewells are abandoned due to *upconing* of saline water from the deeper layers.

A skimming cum recharge structure (Fig. 2) was constructed at a downstream location at village Jagsi-Sarfabad (Block: Saffidon, Distrcit: Jind). The system permits separate or combined pumping from or recharging (with filtered surface runoff) of two cavity tubewells installed at 7 and 40 m depth in different groundwater quality zones. The system is similar in features to a scavenger type skimming structure (Fig. 1d) discussed above but consists of cavity wells instead of strainer tubewells and has additional facilities for groundwater recharge. It facilitates increased availability of good water in upper cavity and improves groundwater quality of deeper cavity for possible irrigation use. The total cost of the skimming and recharging system is Rs. 50000 including Rs. 20000 for the recharge unit. The technology is applicable on a community basis and is to be located at a downstream site where excess runoff is available for recharging.

At another site, **a pumping cum recharging structure and a recharge shaft** [8] have been constructed. These are individual farmer based systems suitable for falling/marginally saline groundwater regions. The pumping cum recharge structure facilitates groundwater recharge of a tubewell through gravel packed bore hole using rainfall or excess canal water after passing through a double chamber recharge filter. The cost of the pumping cum recharge structure is Rs. 20000 which includes Rs. 8000 for the recharge filter. The recharge shaft serves the sole purpose of groundwater recharge and is to be located close to any operational or abandoned tubewell. It consists of a gravel-packed filled borehole to carry filtered recharge water to favourable aquifer zones. It costs Rs. 9000 [8] that can be further reduced with improved designs and materials of the filter and gravel pack.

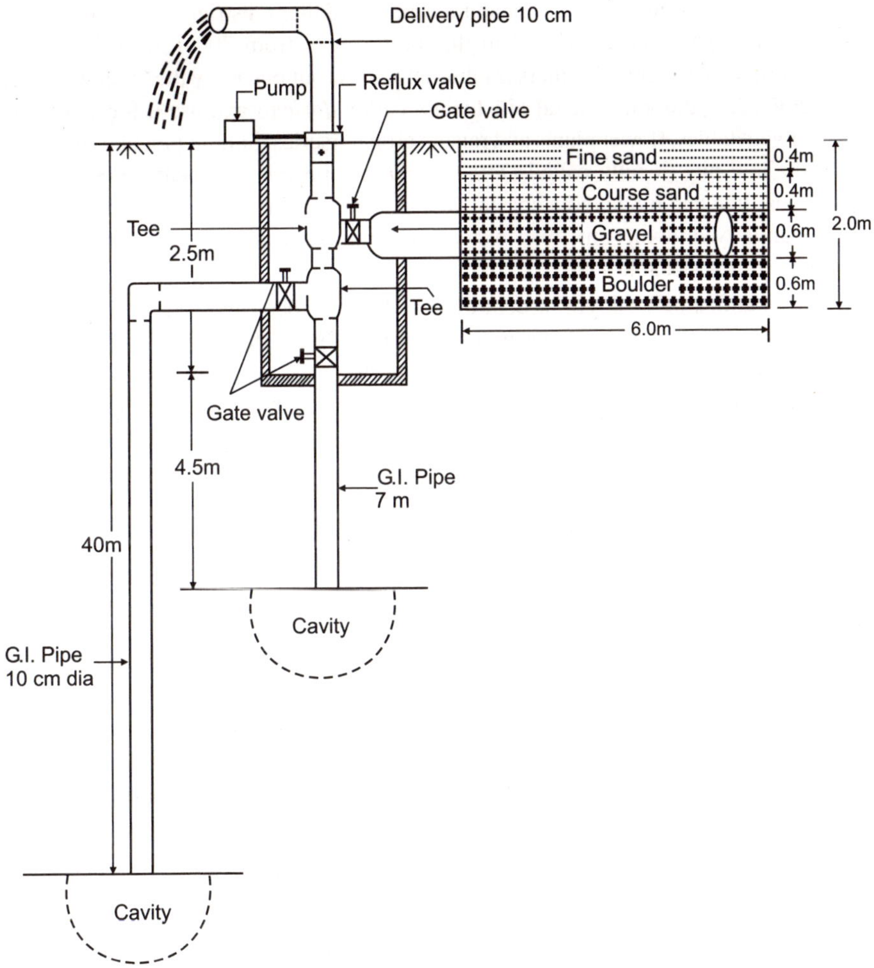

Fig. 2 *Groundwater Skimming cum Recharging Structure at Jagsi/ Sarfabad (Distt. Jind, Haryana)*

Hydraulic evaluation of these structures [5] indicated general improvement in the groundwater regime of area due to combined effect of the natural and imposed recharge interventions despite low recharge rates under shallow groundwater conditions. Pilot studies on pressurized injection, clogging and incorporation of inexpensive filters in the existing or abandoned cavity/strainer tubewells are recommended to enhance groundwater recharge in declining and marginally saline groundwater regions of

Haryana and adjoining alluvial states. Individual farmer-oriented recharging schemes are socially viable and have better chances of success than the community/government sponsored large recharge systems.

(b) **Coastal Sandy Soils of Andhra Pradesh and Tamil Nadu.** In Andhra Pradesh, about 1.74 lakh ha coastal sandy soils are characterized by good quality water floating over saline ground water at shallow depths which cannot be extracted with conventional tube wells. These soils occur in a 10 km wide and 972 km long strip extending from Ichapuram in Srikakulam district to Tada in Nellore district. The annual rainfall in this belt ranges from 700-1200 mm with an average of 855 mm. Farmers traditionally draw out manually the fresh water that collects in dug out conical pits locally called *doruvus* and use it for cultivation of vegetables, flower plants and raising crop nurseries. Similar practices involving shallow pits (locally called *Oothu Kuzhi*) are prevalent in nearly 600 km long coast line and 6.8 lakh ha coastal area in Tamil Nadu. The traditional method of manually drawing out fresh water from dug out conical pits is laborious and consumes sizeable land.

Radial collector well type structures (Fig. 1c) have been evolved for skimming of fresh water floating over saline groundwater for sandy coastal regions in these states. The system consists of an open well (called sump) of 1.2-1.5 m diameter with radial subsurface drains (at 2.4-4.0 m depth and of 30–40 m length on two sides for skimming of fresh groundwater. The installation begins with a sump well constructed with RCC rings of 1.2 – 1.5 m diameter and 0.3 m height to a depth of about 5 m from ground surface. A trapezoidal trench of 30-40 m length, 5-7 m top wideth and 3 m bottom width is dug on either side of the sump with the help of a proclainer machine. The radial subsurface drains consisting of perforated stone ware pipes (0.10-0.15 internal diameter; 0.6 m length, having 18-21 perforations of 5 mm dia) are laid at a slope of 0.2% towards sump at 2.4-4.0 m depth in the excavated trench. These pipes are interconnected using cleaned coconut coir and are also covered with 8-12 mm size gravel filter and also with the sump well. Since the radial drains are installed under highly unstable conditions due to sandy substratum and shallow groundwater, iron gutters are used to facilitate installation. Installation of radial drains in sandy soils is very tedious and contributes to more than 40% of the total cost of skimming wells. Horizontal drilling techniques in sandy soils need to be standardized to reduce this cost. The system yields a discharge of 5 to 15 lit/sec or more depending upon number of arms and nature of sand and can operate in combination with sprinkler/drip irrigation system. Pumping tests indicated 90 m as the safe spacing between two skimming wells. Such structures can meet *rabi* crop water demand of 2 ha area using sprinklers, have 1.7 benefit-cost ratio and 30% internal rate of return [8]. The technology has been widely adopted in Guntur and Prakasam districts of Andhra Pradesh, introduced in Nagapattinam district of Tamil Nadu and has vast scope for coastal sandy regions of other states. Geo-hydrological and geophysical studies conducted in these coastal areas indicate the existence of a sand layer of less than 8-10 m followed by a clay layer saturated with saline water.

Multiple filter point type skimming structures (Fig. 1b) tested at a number of sites in these districts, indicated limited scope for skimming due to insufficient water in the flow domain.

High cost of skimming wells at Rs. 30000 to Rs. 40000 and an equal amount for engine, sprinkler/drip system, is the major constraint for large-scale adoption. Governmental subsidiary for small and marginal farmers is mandatory. Horizontal drilling of drains in sandy soils needs standardization to reduce the cost of installation of these structures.

(c) **Groundwater Recharge in Saline Groundwater Regions of Gujarat and Tamil Nadu.** Groundwater recharge through farm and percolation ponds was evaluated for saline groundwater regions in rocky regions of Gujarat and Tamil Nadu [8]. The site Khapat in Porbander district of Gujarat (annual rainfall @ 500 mm) had 1.0–1.5 m thick clayey soil followed by sedimentary fractured aquifers. Two dug well at this site were recharged separately using runoff from rainfall harvested in two farm ponds after filtration through designed sand-gravel-boulder filter. The enhanced recharge during 2004 resulted in improved groundwater availability and significant reduction in salinity of water (from 9.50 to 0.90 dS m^{-1} and from 3.7 to 1.20 dS m^{-1}) in two wells. Due to groundwater recharge interventions, the cropping pattern in the area is changing from Jowar, Bajra to cotton and wheat in rabi and groundnut in summer resulting in notable improvements in farm incomes.

Similarly the recharge potential of percolation ponds (of about 3000 m^3 capacity) constructed in shallow clayey soils underlain by igneous fractured rocks using harvested runoff was evaluated at two sites (Aruppukottai, district: Virudhunagar; Ramanathapuram, district: Ramanathpuram) in Tamil Nadu. Recharge through these percolation ponds raised groundwater levels by 0.5 to 4 m and reduced its salinity by 0.5 to 3 ds m^{-1} in surrounding wells within 2-3 km distance. At Aruppukottai, studies on **recharge of an abandoned open well** using filtered excess runoff proved promising in improving the availability and quality of well water. The concept has considerable potential considering there are 5 lakh abandoned open/ tube wells in different parts of Tamil Nadu.

The research output of NATP project has shown the way for commissioning of larger groundwater skimming projects in coastal sandy regions of Andhra Pradesh and Tamil Nadu and artificial recharge projects in Haryana, Gujarat and Tamil Nadu. Further details on the features and performance evaluation of the above and a few additional groundwater skimming and recharge structures can be found elsewhere [8].

5. SALIENT OBSERVATIONS AND EMERGING ISSUES

1. Groundwater skimming and recharge technologies should be viewed in the context of the availability and accessibility of surplus water in time and space.
2. Radial collector type skimming wells have considerable scope in coastal sandy soils and need government subsidiary. Horizontal drilling techniques need to be explored to lay radial drains of skimming structures in coastal sands.
3. Individual farmer-oriented recharging schemes are socially viable and have better chance of success and sustenance than the community/government sponsored and maintained large recharge systems.

4. Incorporation of small and less costly recharge filters in the existing or abandoned dug wells/tubewells can contribute to significantly enhance groundwater recharge in the irrigation sector.
5. Pilot studies are needed on the design/adaptation of alternate filters for recharge structures in alluvial and hard rock regions and on pressurized well injection.
6. Possibilities of multiple uses of land and water in percolation ponds and tanks in western and south India should be explored since a large area of channels, bunds, and pond beds remain unutilized during non-recharging seasons.
7. Application of geophysical, geo-statistical and tracer techniques is vital for the selection of recharge sites and in understanding the mechanisms, demarcating the extent and quantifying the recharge rates and volumes over large areas. (Observation 7 is an important general conclusion but can not be supported by data within the framework of this article. Tracer studies were referred in results of coastal sandy soils of Andhra Pradesh and Tamil Nadu and we have results on these aspects for other areas as well [8]).

Note: Guidelines of groundwater structures for different areas are provided in Ref. [8]

REFERENCES

1. Central Ground Water Board. (2000) Guide on Artificial Recharge to Groundwater. p. 93
2. Central Ground Water Board. (2002) Master Plan for Artificial Recharge to Groundwater in India. 116 pp.
3. Chadha, D.K. (2002). State of art of artificial recharge applied on village level scheme in India. In: Management of Aquifer Recharge and Subsurface Storage: Making Better Use of Our Largest Reservoir (Ed. Tuinhof, A. and Heederik, J.P), Netherlands National Committee- International Association of Hydrogeologists (NNC-IAH), Wageningen, The Netherlands. Publication No. 4, 19-24.
4. Hooja, R., Shrinivas, V. and Sharma, G. (1995) Waterlogging and salinity problems n IGNP, Rajasthan. In: Proc. National Seminar on Reclamation and Management of Waterlogged Saline Soils (Eds. Rao, K. V. G. K., Agarwal, M. C., Singh, O. P., Oosterbaan, R. J.), CSSRI, Karnal and HAU, Hissar (Indo- Netherlands Collaborative Project), pp. 141-159.
5. Kamra, S. K., Anchal, V., Aswal, S. and Lal, K. (2006) Groundwater Recharge through Cavity Wells in Saline Groundwater Regions. In: Recharge Systems for Protecting and Enhancing Groundwater Resources. Proc. 5th International Symposium on Management of Aquifer Recharge (ISMAR5), Berlin (Germany), June 11-16, 2005, IHP-VI Series on Groundwater No. 13, 699- 704.
6. Kumar, R. and Singh, J. (1995) Drainage systems for groundwater management. In: Proc. National Seminar on Reclamation and Management of Waterlogged Saline Soils (Eds. Rao, K. V. G. K., Agarwal, M. C., Singh, O. P., Oosterbaan, R. J.), CSSRI, Karnal and HAU, Hissar (Indo- Netherlands Collaborative Project), pp. 50- 62.

7. Mazhar Saeed, M., Ashraf, M. and Asghar, M. N. (2003) Hydraulic and hydro-salinity behavour of skimming wells under different pumping regimes. *Agricultural Water Management*, **61**, 163- 177.
8. National Agricultural Technology Project (2006) *Technologies for Skimming and Recharging Fresh Water in Saline Groundwater Regions*, Final Report, Central Soil Salinity Research Institute, Karnal (Haryana), India, 39 pp (with 125 page appendix)
9. Raghu Babu, M., Rajendra Prasad, B. and Srikanth, I. (2004) *Subsurface skimming techniques for coastal sandy soils*. NATP Bulletin No.1/ 2004, Saline Water Scheme, Bapatla (Andhra Pradesh), India.
10. Shakya, S.K. (2002) *Agricultural drainage under actual farming conditions on watershed basis*. Department of Soil and Water Engineering, Punjab Agricultural University, Ludhiana (Punjab), India.
11. Sufi, A.B., Latif, M., Skogerboe, G.V. (1998) Simulating skimming well techniques for sustained exploitation of groundwater. Irrigation and Drainage Systems. **12**, 203- 226.

15

RIVER RESPONSE TO CLIMATE CHANGE: WHY, WHERE AND HOW?

Rajiv Sinha

ABSTRACT

Recent geological history of the earth is known for rapid and frequent climate changes. There is a serious concern about the nature and consequences of climate change as it has an enormous impact on agriculture, water resources and ecological system. There is also a widespread recognition that human activities are altering climatic and environmental conditions over very short time scales. Most analyses of these changes depend on the time series data restricted to a few decades or a century at the most and therefore the causative factors of these changes remain unresolved. A complete understanding of the processes and variability of climate changes would therefore require the use of "proxy data" for tracking the past climate changes for a much longer time scale. As a corollary, future climate change would also be a function of natural factors such as earth's orbital motion, insolation and anthropogenic factors such as greenhouse gas emissions. The assessment and prediction of future climate change would therefore be strongly linked to the understanding of past climates. The response of river systems to ongoing and future climate change has become a subject of intensive research. River landforms and deposits provide one of the most readily available continental records to study environmental change. The Indian sub-continent is host to several large river systems such as the Ganga, Brahmaputra, Narmada and Cauvery. The Gangetic plains of India represent the world's largest area of modern alluvial sedimentation and support a large population of 250 million people. Natural environmental changes therefore have a large impact on the social and economic development of the region. Natural environmental changes have been augmented by the human-caused changes manifested in terms of land cover changes, increased sediment yield in rivers, increased flood frequency and a general degradation of the eco-system.

1. INTRODUCTION

The planet Earth has witnessed century scale changes in temperature, and the global warming scenarios in the past century. This has been a matter of concern among the scientists as well as the society at large. As a consequence, the response of various components of the Earth system to ongoing and future climate changes has become a matter of intensive research. The reliability of predictive capability of the climate response models needs to be tested; and this, to some extent can be done by studying sedimentary records such as those produced by rivers which serve as important archives of the earth's climate change.

Climate change is caused by the combined effects of many forcing functions; and this, in turn, leads to complexities in the natural response times of the various components of the Earth's climate system. Climate scientists have often emphasized the large-scale responses such as (a) the presence of ice sheets at high latitudes, (b) changes in the strength of low-latitude monsoon circulations, (c) the reaction of ice-free land surfaces outside monsoon regions, (d) the flow of deep water, and (e) the movement of carbon among reservoirs in the climate system [1]. In the Indian context, these changes are largely related to the strength of low-latitude monsoon circulation. Monsoon variability has been investigated on various time-scales [2, 3]. The studies show that much of the variability over the past 1,50,000 years occurs at orbital periodicities "and all of the palaeo-climate time series show four monsoon maxima that occur during interglacial conditions and coincide with precession maxima and maxima of northern hemisphere summer radiation" [2].

Most parts of India belong to a monsoonal system that is influenced considerably by the Himalayan orogeny. Strong precipitation gradients occur from the arid regions of Thar towards the Himalaya in the northeast, and towards the western Indian Ocean in the southwest. The 250 million people living in this area depend on agriculture and are largely dependent on monsoonal precipitation. Future climatic changes are likely to have an enormous impact on food production and water availability in this region and deleterious changes could lead to the resettlement of millions of people. Evaluation of the realism of the predictions of future climate changes in this region has been hampered by the comparative lack of well-dated paleoclimate data from the Indian subcontinent.

Although the climatic records from oceans for the Quaternary period are plenty, there is a real dearth of records from continental settings for accurate and precise reconstruction of Quaternary paleoclimate. Particularly lacking is the response of large river systems to monsoonal fluctuations both in terms of hydrological response as well as sediment flux. It is well established that major climatic shifts driven by insolation changes have occurred in the last 20 ka and several studies have demonstrated that rivers respond to these climate changes. However, the sensitivity of the response to climate change is a function of the size of the system and the response of large river systems to climatic shifts is very complex and not fully understood. It is envisaged that there is a tight coupling of sedimentary responses from source area to catchment basin to coastal and marine depocenters. However, given the geomorphic diversity of the river system like the Ganga-Brahmaputra, it is not understood as to how the different components of the systems have responded. Whereas hydrology largely explains variability in Ganga-Brahmaputra system, it does not fully explain close source-to-sink linkages.

2. CLIMATIC FACTORS AND THEIR CONTROL ON RIVER SYSTEMS

The development of the river systems itself is strongly controlled by climatic effects which are generally manifested through their control upon erosion processes. Two most important climatic factors influencing the river systems include precipitation and temperature through their control on vegetation cover. An important control of precipitation in the Indian region is the Himalaya, which forms a barrier and forces the moisture-laden winds to ascend and thus cause precipitation. Climate variability in monsoonal settings demonstrates that much of the considerable variability over the past 150,000 years occurs at orbital periodicities. Four monsoon maxima during this period occur during interglacial conditions and they coincide with the precession maxima and minima of the northern hemisphere summer radiation. It has been suggested that climate-related signals propagate downstream and show tight coupling between source area, catchment basin, and coastal and marine depocenters of the river systems. However, the nature and amplitude of climate shift required to bring about a response in large river systems is not fully understood.

The present-day river systems of the Indian Subcontinent have fascinated scientists for centuries. More recently earth scientists have been developing their investigations of these special and difficult systems. Some of the difficulty is related to the size of the rivers and the sheer extent of the alluvial plains of northern India. These facts of scale must certainly have contributed to the sense of wonder and reverence felt for the rivers and the Himalayas on the part of most of the peoples living in this most heavily populated large natural systems on the surface of the Earth.

There are a number of reasons why work on this topic is developing so actively now. The general acceptance of the ideas of plate tectonics that became established in the 1960's has put the fluvial systems of the Sub-continent on a new basis. In the north they are seen as a response to the continent-continent collision that started to build the Himalayas in Early Tertiary times. The fluvial systems are a direct result of the uplift and erosion of mountains which are still active today, as convergence of known direction and rate is continuing. So these systems are still working today on a scale that has no rival anywhere else on the present Earth's surface. Indeed tracking of ocean chemistry through geological time shows the global importance of Himalayan uplift and its fluvial erosion.

The world of Quaternary climate research has been revolutionised by recent work on the chemical cycles that have been discovered in cores from the deep oceans and from the thick ice sheets of Greenland and Antarctica. Not only has this work impacted on understanding of the detailed chronology and mechanisms of climate change during the Ice Age and before, but it has resulted in a precision in dating oceanic events that could not have been foreseen twenty years ago. Unfortunately this precision is still sadly lacking in river deposits, and the challenge of relating the Sub-continent's monsoonal climate cycles as seen in the fluvial record, with known global oceanic events, is still a difficult one, though being addressed.

The Ganga River System: A Case Study

Recent studies in the Ganga system [4-7] have shown that changes in Southwest Monsoon precipitation have affected all parts of the system over the past ~100 ka, from the foothills to the delta. In Himalayan valleys north of the Eastern Ganga Plains, monsoon intensification has been linked to landslides and regolith removal, initially overwhelming the river's sediment transport capacity and causing aggradation, with subsequent erosion of material after about 7 ka [8]. Fluvial activity was greatly reduced during the Last Glacial Maximum [4, 9, 10] but subsequently intensified greatly. In consequence, sediment yield to the Ganga-Brahmaputra delta between 11 and 7 ka was more than double present rates [11], and large volumes of sediment were transported to the Bengal deep-sea fan after about 13 ka, during a period of rapid sea-level rise [12]. The subsidence rate also varied in the delta area, with especially rapid subsidence in the Sylhet Basin of Bangladesh [11].

An important role for climate is also indicated from studies of deep-sea fan systems in southern Asia [13]. Other factors that have affected discharge include meltwater fluctuations due to the growth and melting of Himalayan glaciers [14] and the activity of the Northeast Monsoon [15]. Additionally, there is local evidence for avulsion and river capture on the plains and in the headwaters [16-19], caused in some cases by tectonic activity at mountain exits [20] and by the intrinsic effect of sediment accumulation on megafans [21].

The relationship between discharge of fluid and sediment is complex, with feedback loops, non-linear responses, and response times that affect the system's sediment transport capacity in time and space [22]. Knox [23] noted that even modest climatic shifts may have huge effects on flood response and sediment yield. Within the Himalayan context, sediment pulses may reflect enhanced precipitation [8, 11], but may also be independent of precipitation. For example, the 1950 Assam earthquake released 47 billion m^3 of sediment, which caused unprecedented floods as a result of channel aggradation and generated a wave of aggradation and subsequent degradation along the Brahmaputra River over a period of decades [24]. Depending on the balance between fluid and sediment availability and the energy available for erosion, rivers of the Ganga Plains may aggrade their channels and floodplains, or may incise or become underfit, detaching from the adjacent floodplains [5].

Hydrological changes in large river systems could be global due to phenomena that are widespread such as past glaciation or current ENSO events. Otherwise they may be linked to spatial variability in precipitation, sediment availability, distribution of snowmelt and vegetational effects, as well as the crossing of response thresholds. However, all parts of southern Asia are affected by the huge driver of the monsoon, and may be expected to show some degree of parallel response. Goodbred [4] suggested that the Ganga system responds rapidly to fluctuations in fluid and sediment discharge in the headwaters, despite alluvial storage in the plains. He noted that sediment pulses recorded in the delta succession correspond in age with periods of sediment release in the headwaters, as well as with modeled precipitation maxima. Glacioeustatic changes in sea-level over the past 30 ka have clearly influenced the Ganga-Brahmaputra delta.

In the modern Ganga-Brahmaputra delta region, Goodbred and Kuehl [11] documented two buried valleys extending for at least 350 km inland from the modern shoreline valley systems filled with at least 90 m of coarse- to fine-grained sand, coarsening upstream and thinning laterally. A boulder-rich layer underlying the valley fills has a slope much higher than that of the modern delta, suggesting that rivers were graded down to more than 100 m below present sea-level [25]. A radiocarbon date of about 28 ka ^{14}C yr BP [25] from below the boulder layer and dates from alluvium that post-dates the Last Glacial Maximum suggest that the valleys filled during the lowstand of Marine Isotope Stage 2 and the succeeding transgression, as sea-level rise promoted accommodation and backfilling.

Several studies have indicated that relative sea level changes can impact large river systems and their deltas. As the sea level rises, freshwater/seawater interface is pushed landward, river gradient drops considerably and more accommodation space is created within the flooded reaches of the valley complex. Islam and Tooley (2004) suggested on the basis of lithostratigraphic and biostratigraphic data that the sedimentary environment within the Bengal basin from Early Mid-Holocene until Ca. 1750 BP was dominated by regional sea level movements [26]. They recognized five periods of marine to fluvial transitions during this period at a site (Panigati) ~100 km inland of the present-day tidal limit. Similarly, the Rhine extended some 800 km farther seaward from its present position during the LGM [27] and the terrace intersection of Pleniglacial terrace and Holocene deposits shifted landward by ~150 km as a result of Holocene sea level rise [28]. It has been demonstrated that the rate of sea level rise also affects the channel pattern in the near-coast fluvial area. In the Rhine-Meuse delta a faster rate gave rise to low-energy, narrow anastomosing river pattern (7500-4000 ^{14}C year BP) whereas a slower rate resulted in dominance of low sinuosity meandering rivers (post-4000 ^{14}C year BP) [29] although straight anastomosing channels continues to exist until present [30]. Low stream power was related to discharge and low gradients, and thus indirectly linked also to rapid sea level rise (Makaske, 2001). Studies in the East China Sea [31] have also shown that sea level changes during MIS 2 and 1, coupled with climate controlled water discharge, sediment supply and shelf physiography controlled the valley fills in the region. During the sea level rise in late MIS 2, the upper parts of the drowned lowstand fluvial deposits were reworked into a valley wide tidal bar and tidal sheet complex. This phase was followed by wetter climatic conditions during MIS 1 and a higher sediment delivery which resulted in burial and preservation of the low stand fluvial deposits to a depth below the tidal effects.

3. CONCLUDING REMARKS

The development and sustenance of large river systems is strongly influenced by the interplay of climate, tectonics and source area characteristics. Many of the modern large rivers are confined within the tropics in monsoonal settings, obviously due to the high precipitation in these regions. The coupled effect of tectonics and climate is quite critical to sustain the large rivers. Mountain uplift and orographic precipitation are closely interlinked. Focused denudation and erosion triggered by tectonics cause more uplift through isostatic effects, thereby sustaining the orographic precipitation (positive feedback system). Moreover, continued uplift and erosion provide continuity of slopes, resulting in high discharge and sediment transfer which, in turn, leads to valley enlargement.

Although several efforts have been made to understand the responses of river systems to climatic shifts in monsoonal regimes, high-resolution studies backed by high quality chronometric data are very limited. Site-specific studies on individual river systems must be combined with modeling studies. The realism of the models could be tested using the proxy-data of various climatic archives, in turn leading to better models with enhanced predictive capability for the future. Large river systems such as the Ganges, Brahmaputra, Narmada, Godavari, Krishna, Cauvery should receive attention in a mission mode to develop a comprehensive understanding of responses of river systems to climate changes in monsoon-dominated regions. This is indeed important in the context of our food (agricultural), water, and land use security. Big river systems are multi-component systems that have dynamic evolutionary histories over million year times scales, indeed tens of millions of years. Model-driven studies in different components of big river systems backed by high resolution stratigraphic data (including reliable multi-proxy data) would require to be spatially integrated. Such studies need strong inter-disciplinary teams of earth scientists who in turn would interface with climate and hydrological modelers.

REFERENCES

1. Ruddiman, W. F. (2001) Earth's Climate: past and Future. W.H. Freeman & Co. New York, p. 465.
2. Prell, W. L. and Kutzbach, J. E. (1987) Monsoon variability over the past 150,000 years. *J. Geophys. Res.* **92**, 8411-8425.
3. Overpeck, J., Anderson, D., Trumbore, S. and Prell, W. (1996) The Southwest Indian monsoon over the last 18000 years. *Climate Dynamics* **12,** 213-225.
4. Goodbred, S. L. J. (2003) Response of the Ganges dispersal system to climate change: a source-to-sink view since the last interstade. *Sedimentary Geology* **162,** 83-104.
5. Gibling, M. R., Tandon, S. K., Sinha, R. and Jain, M. (2005) Discontinuity -bounded alluvial sequences of the sounthern Gangetic plains, India; aggradation and degradation in response to monsoonal strength. *J. Sedimentary Res.* **75**, 369-385.
6. Sinha, R., Tandon, S. K., Gibling, M. R., Bhattacharjee, P. S. and Dasgupta, A. S. (2005) Late Quaternary geology and alluvial stratigraphy of the Ganga basin. *Himalayan Geology* **26**, 223-240.
7. Tandon, S. K., Gibling M. R., Sinha R., Singh V., Ghazanfari P., Dasgupta A., Jain M. and Jain V., (in press). *Alluvial valleys of the Gangetic Plains, India: causes and timing of incision*. In: Ed. Dalrymple, B., Incised Valleys in Time and Space, SEPM Spl vol. 85.
8. Pratt, B., Burbank, D. W., Heimsath, A. and Ojha, T. (2002) Impulsive alluviation during early Holocene strengthened monsoons, central Nepal Himalaya. *Geology* **30**, 911-914.

9. Cullen, J. L. (1981) Microfossil evidence for changing salinity patterns in the Bay of Bengal over the last 20 000 years. *Palaeogeography, Palaeoclimatology, Palaeoecology* **35**, 315-356.
10. Wiedicke, M., Kudrass, H.-R. and Hubscher, C. (1999) Oolitic beach barriers of the last Glacial sea-level lowstand at the outer Bengal shelf. *Marine Geology* **157**, 7-18.
11. Goodbred, S. L. and Kuehl, S. A. (2000) Enormous Ganges-Brahmaputra sediment discharge during strengthened early Holocene monsoon. *Geology* **28**, 1083-1086.
12. Weber, M. E., Wiedicke, M. H., Kudrass, H. R., Hubscher, C. and Erlenkeuser, H. (1997) Active growth of the Bengal Fan during sea-level rise and highstand. *Geology* **25**, 315-318.
13. Prins, M. A. and Postma, G. (2000) Effects of climate, sea level, and tectonics unraveled for last deglaciation turbidite records of the Arabian Sea. *Geology* **28,** 375-378.
14. Owen, L. A., Finkel, R. C. and Caffee, M. W. (2002) A note on the extent of glaciation throughout the Himalaya during the global Late Glacial Maximum. *Quater. Sci. Rev.* **21**, 147-157.
15. Enzel, Y., Ely, L. L., Mishra, S., Ramesh, R., Amit, R., Lazar, B., Rajaguru, S. N., Baker, V. R. and Sandler, A. (1999) High-resolution Holocene environmental changes in the Thar desert, Northwestern India. *Science* **284** 125-128.
16. Wilhelmy, H. (1969) The ancient river valley on the eastern border of the Indus Plain and the Sarasvati problem. *Zeitschrift für Geomorphologie* **8,** 76-93.
17. Parkash, B., Kumar, S., Rao, M. S., Giri, S. C., Kumar, C. S., Gupta, S. and Srivastava, P. (2000) Holocene tectonic movements and stress field in the western Gangetic plains. *Curr. Sci.* **79**, 438-449.
18. Roy, N. G. and Sinha, R. (2005) Alluvial geomorphology and confluence dynamics in the Gangetic plains, Farrukhabad-Kannauj area, Uttar Pradesh, India. *Curr. Sci.* **88**, 2000-2006.
19. Jain, V. and Sinha, R., 2005. Response of active tectonics on the alluvial Baghmati river, Himalayan foreland basin, eastern India. *Geomorphology* **70**, 339-356.
20. Gupta, S. (1997) Himalayan drainage patterns and the origin of fluvial megafans in the Ganga foreland basin. *Geology* **25**, 11-14.
21. Wells, N. A. and Dorr, J. N. (1987) Shifting of the Kosi river, northern India. *Geology* **15**, 204-207.
22. Bogaart, P. W., van Balen, R. T., Kasse, C. and Vandenberghe, J. (2003) Process-based modelling of fluvial system response to rapid climate change — I. model formulation and generic applications. *Quarter. Sci. Rev.* **22,** 2077-2095.
23. Knox, J. C. (1993) Large increases in flood magnitude in response to modest changes in climate. *Nature* **361**, 430-432.
24. Goswami, D.C. (1985) Brahmaputra River, Assam, India: Physiography, basin denudation, and channel aggradation. *Water Res. Res.* **21**, 959-978.
25. Umitsu, M. (1993) Late Quaternary sedimentary environments and landforms in the Ganges delta. *Sedimentary Geology* **83**, 177-186.
26. Islam, M. S. and Tooley, M. J. (2004) Coastal and sea level changes during the Holocene in Bangladesh. *Quarter. Intl.* **55**, 61-75.
27. Gibbard, P. L. (1988) The history of the great northwest European rivers during the past three million years. *Phil. Trans. R. Soc. Lond.* **318,** 559-602.

28. Van Dijk, G. J., Berendsen, H. J. A. and Roeleveld, W. (1991) Holocene water level development in The Netherlands' river area: implications for sea level reconstruction. *Geologie en Mijnbouw* **70,** 311-326.

29. Törnquist, T. E. (1993) Holocene alternation of meandering and anastomosing fluvial systems in the Rhine-Meuse delta (central Netherlands) controlled by sea level rise and subsoil erodibility. *J. Sed. Petrol.* **63**, 683-693.

30. Makaske, B. (2001) Anastomosing rivers: a review of their classification, origin and sedimentary products. *Earth-Science Rev*. **53**, 149-196.

31. Wellner, R. W. and Bartek, L. R. (2003) The effect of sea level, climate and shelf physiography on the development of incised valley complexes: a modern example from the East China Sea. *J. Sed. Res*. **73**, 926-940.

16

RECESSION RATE OF TRIBUTARY OF GANGOTRI GLACIER SYSTEM, GARHWAL HIMALAYA-INDIA

B. S. Kotlia, Rakesh K. Dumka, M. S. Miral, L. K. Joshi and K. Kumar

ABSTRACT

The recession rate of the tributary glaciers of the Gangotri glacier system, e.g., Thelu and Raktvarna glaciers in the Garhwal Himalaya has been calculated by GPS and toposheet (Survey of India toposheet no. 53 N/1, 1962, scale 1:50,000). In the last 42 years, The Thelu glacier has receded about 1288m at an approximate rate of 30.66m/year. Similarly, the Raktvarna has receded 1585 m with 37.73m of annual recession. At present, the Chaturangi glacier is found 293m away from the main trunk of the Gangotri glacier. Evidences of geomorphic landforms, both endogenic and exogenic in the valley also support the recession of glaciers in the recent past.

1. INTRODUCTION

Himalaya, the youngest and fragile mountain system of the earth has direct influence on climate, regional hydrology and environment of the subcontinent. The global climate change has a significant impact on the high mountain environment during the past few decades. The most visible signals of global warming are seen in ice melt and corresponding impact on high-mountain systems. The observation of the glacier retreat is one of the primary apparatus for climate system monitoring [1, 2]. The similar impact is also found in the Himalayan glaciers which are in the state of general retreat since 1850 [3] and because of this, the average air temperature of the Himalayan region has risen by 1°C since mid-1970's [4]. The trend in the recession of the Himalayan glaciers is expected to continue this century [5-8].

The Himalayan glaciers are retreating faster than the glaciers in other parts of the world [9]. Because of climatic and topographical factors, the rate of recession and amount of volume change are irregular for glaciers across the Himalayan arc. In the central sector of the Himalaya, only a few glaciers have been studied from recession point of view [10-19]. See Table 1 in the text.

Table 1 *Recession of Gangotri glacier*

Period	*Retreat(m)*	*Data Source*
1935-1956	10.16	[10]
1956-1971	27.33	[11]
1971-1977	30.84	[12]
1977-1990	8.082	[13]
1996-1999	25.33	[15]
2004-2005	12.10	[19]

Study Area

The Gangotri glacier (30°44' 35"- 30°56' 21" N and 79° 04' 41"–790 16' 34" E), situated in Uttarkashi district, Uttarakhand (Fig. 1), drains in NW direction. It is about 30.20 km long with width varying between 0.5 to 2.5 km. The total glacierized area of the catchment is 258.56 sq km, out of which the Gangotri system comprises 109.03 sq km [15]. The total spread of the ice volume is 39.18 km^3 [20]. The main trunk of the Gangotri glacier adjoins a number of small and large tributary glaciers (Fig. 1), important of which are Thelu (30° 57' 26.6"N- 79°.5' 43.9"E) and Raktavarn (30° 55' 45.0"N- 79°0.5' 50"E) tributary glaciers.

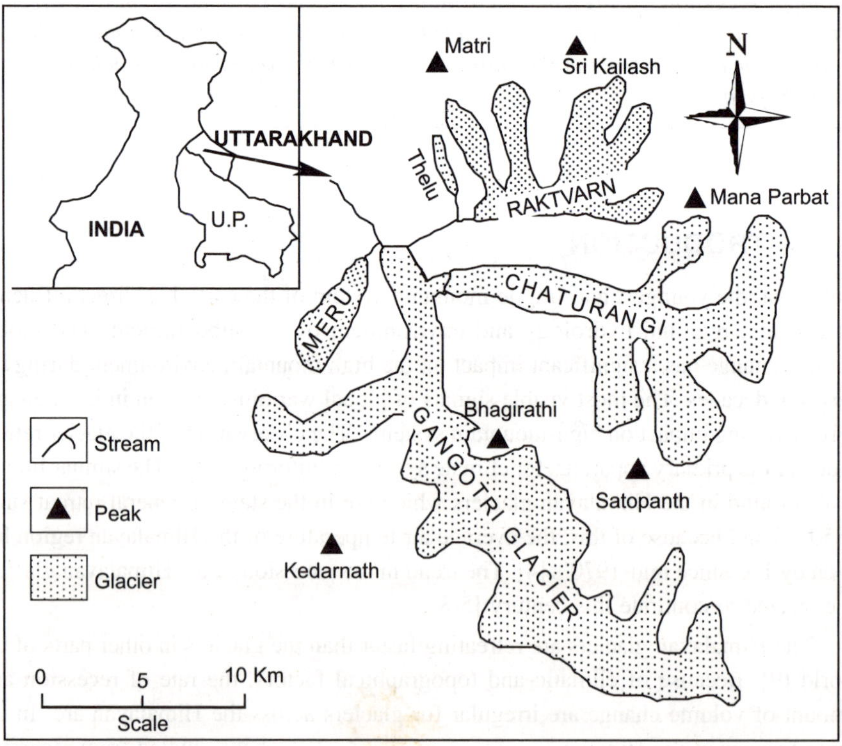

Fig. 1 *Gangotri glacier system with tributary glaciers*

A number of supra glacial lakes and thick pile of supraglacial moraines are observed on the surface of the Gangotri glacier system [21]. River Bhagirathi gushes out from the terminus of Gangotri (Gaumukh) glacier system. Beyond the terminus of Gangotri glacier, several other tributary glaciers feed the river Bhagirathi. The present study was carried out to estimate the recession rate of three important tributary glaciers, Thelu (5,318m) Raktvarn (4,350m) and Chaturangi respectively (4,157m).

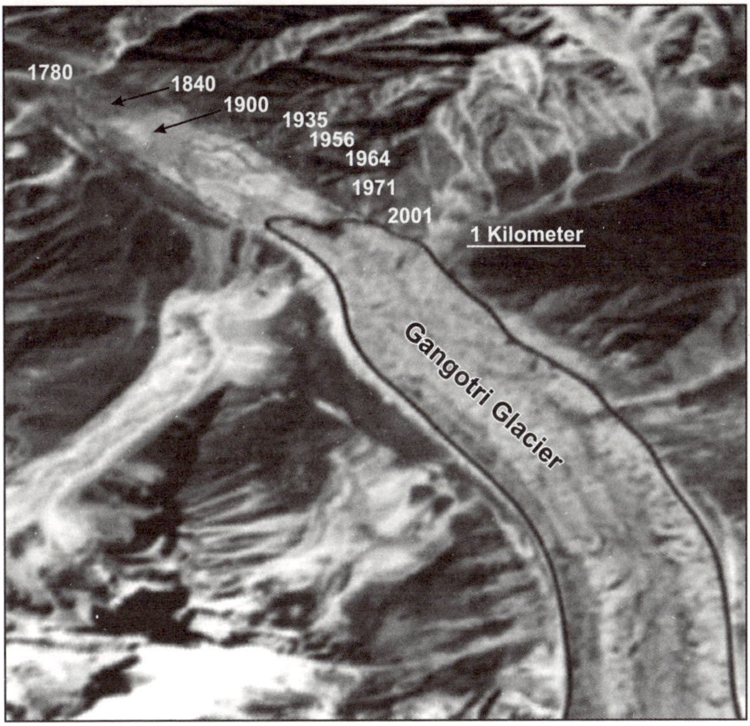

Fig. 2 *Retreat in Gangotri glacier in the last 220 years (source, Jeff Kargel, USGS),* (For color illustration refer p xix)

Methodology

Our measurements of receding rate were mainly focused on three tributary glaciers, Thelu, Raktavarn and Chaturangi. The melt water streams of these tributary glaciers adjoin the main Gangotri glacier. Our study was based on the survey of India toposheet (53/N/1, 1962), with GPS added investigation and direct field measurements in August 2004.

The average rate of recession is computed by comparing the position of 1962 toposheet and 2004 GPS positions (Figs. 3,4). The position of 1962 was Everest based spheroid, whereas, the position of 2004 is based on WGS 84. Since the direct comparison between these two may turn out to be problematic because of uncertainties, the coordinates of 1962 were converted into WGS 84 coordinates using software GEOTRANS, developed by National Imagery and Mapping Agency (NIMA), USA.

The converted coordinates of 1962 and coordinates of 2004 were transformed into the Cartesian coordinates and the distance between 1962 and 2004 was calculated using the software SKI-PRO (GPS data processing software). The position of 1962 was also marked in the field. This method was used for the Thelu and Raltavarna glaciers. However, the retreat rate of Chaturangi was calculated with help of measuring tape.

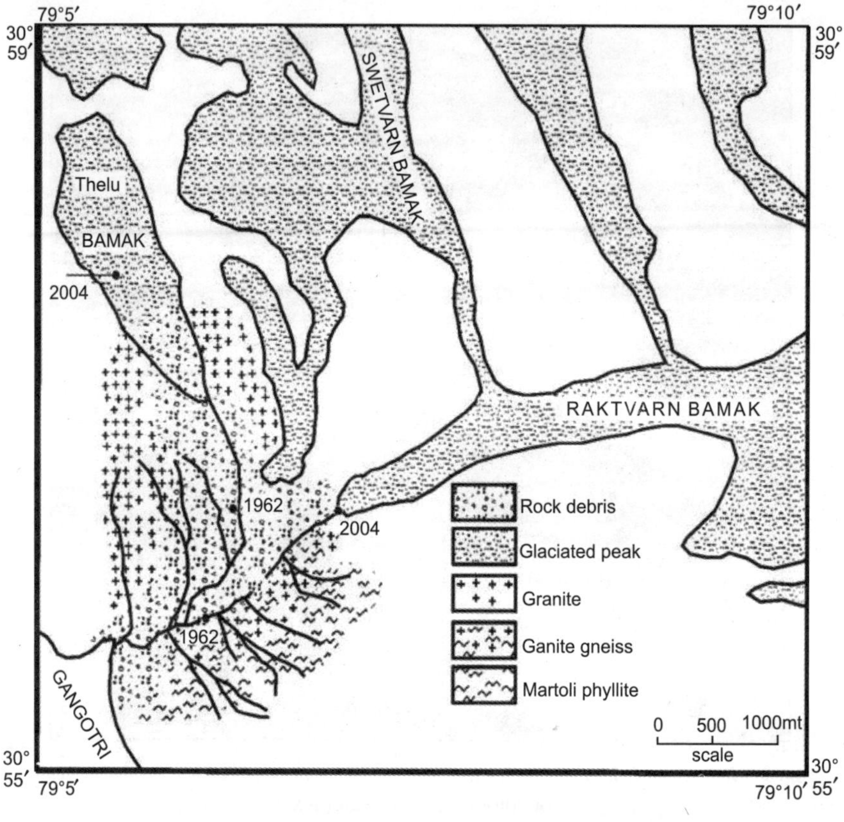

Fig. 3 *Geological map and topographic position (1962 and 2004) of Thelu and Raktvarna glaciers*

Results and Discussion

During the field investigation several longitudinal and transverse crevasses along with the supra glacial lakes were observed in upper part of the Thelu, Chaturangi and Raktvarna glaciers. The movement of a glacier generates moraines and rocky debris, pushed along by the glacier and deposited on its top and on its edges. The formation and the type of moraines, together with other landforms, provide the key morphological evidence to the previous positions of a glacier, particularly their frontal part of terminus which provides indications of the retreat. The field survey was carried out during the month of August 2004. The position of 1962 and 2004 were marked on the ground by above mentioned method. The difference between the Cartesian coordinates for both the year was calculated using SKI-PRO (Fig. 3). The calculated values reveal that the Thelu glacier is receded around 1288m (ca. 30.66m/y) between 1962 and 2004 with 356° geodetic azimuth. During this period, the Raktvarna glacier has receded around 1585m (37.73 m/y) with 78° geodetic azimuth. The terminus of Thelu glacier in 1962 was situated just 994m towards N 05° W of 1962's terminus of Raktvarna glacier, which in 2004 is around 2596m towards N 42° W).

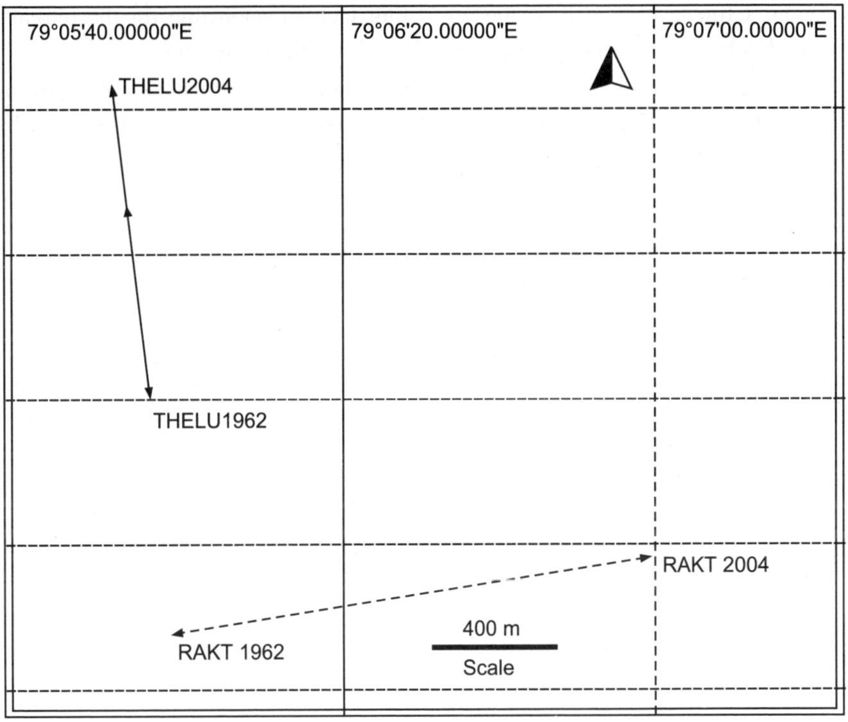

Fig. 4 *Snout position of Thelu and Raktvarna in 1962 and 2004*

Prior to 1971, the Chaturangi glacier had joined the Gangotri glacier [9], but has since retreated by over 100m [14]. Our field observations suggest that the retreat has amounted to about 393m. That means the terminus of Chaturangi retreats by 193m between 1996 and 2004.

Geomorphic evidences of recessional, terminal, lateral, medial, ground and surface moraine, taulus cones, boulders, bogs and chaotic assemblage of hummocky (mounded) moraines and dead ice with small ponds and lacustrine deposits seem to be characteristic features of the retreat. Slow ice movement and little precipitation explain the vast quantities of debris in the lower ablation zone, as there is a corresponding slow removal of great mass of rock waste occurring from very active weathering processes. Huge accumulation of glacial deposits are due to the continuous shrinkage of the glacier. Due to the glacio-fluvial action, the ground moraines, deposited by the old glaciers have been modified in the Thelu, Raktavarn and Chaturangi glacier valley. Huge boulder capped pillars of skeletal (residual) moraines are found also at Bhuggaddi 4.5 km behind the terminus of the Gangotri glacier. We have also observed that the rock falls all along the treks of the Thelu and Chaturangi glaciers are much more compared to other areas in the Gangotri system, probably indicating neotectonic activity along the Tethyan Fault. There are three prominent terraces below the Thelu that have cut by step faulting or cross faulting along the Tethyan fault.

The continuous retreat of Himalayan glaciers during the 20th century is still under processes. Climate obviously is the governing factor responsible for the retreat of glaciers. Such effect is also signified in the Gangotri glacier system.

The recession rate of the glaciers was estimated on the basis of coordinate differentiation of 1962 and 2004 (average of 42 years). There is a factor of error such as in the conversion of coordinates from Everest datum to WGS 84 datum as well as the error during GPS measurements due to the presence of high peaks on both sides of the tributary glaciers. However this is very less as compared to the recession and the method is useful for measuring the retreat rate of the glaciers.

Acknowledgements

The authors are grateful to the Department of Science and Technology for providing the fund for field work. The laboratory facilities provided by the G.B. Pant Institute of Himalayan Environment and Development, Kosi - Katarmal, Almora are thankfully acknowledged.

REFERENCES

1. Wood, F.B. (1990) Monitoring global climatic change: the case of greenhouse warming. *Bull. Amer. Meteorol. Soc.* **71**, 42 –52.
2. Haeberli, W., Cihlar, J. and Barry, R.G. (2000) Glacier monitoring within the global climate observing system. *Ann. Glaciol.* **31**, 241–246.
3. Mayweski, P. and Jaschke, P.A. (1979) Himalaya and Trans Himalayan glacier fluctuation since A. D. 1812. *Arctic Alp. Res.* **11**, 267-287.
4. Hasnain, S.I. (2000) Status of the Glacier Research in the HKH region. ICIMOD, Kathmandu, Nepal.
5. Ageta, Y. and Kadota, T. (1992) Predictions of changes of glacier mass balance in the Nepal Himalaya and Tibetan Plateau: a case study of air temperature increase for three glaciers. *Ann. Glaciol* **16**, 89-94.
6. Nakawo, M., Fujita, K., Ageta, Y., Shankar, K., Pokhrel, A.P. and Tandong, Y. (1997) Basic studies for assessing the impacts of the global warming on the Himalayan cryosphere, **1994-96**, *Bull. Glacier Res.* **15**, 53-58.
7. Hasnain, S.I. (1999) Himalayan Glaciers: Hydrology and Hydrochemistry. Allied Publ., Ltd., New Delhi.
8. Naito, N., Nakawo, M., Kadota, T. and Raymaond, C.F. (2000) Numerical simulation of recent shrinkage of Khumbu Glacier, Nepal Himalayas. *International Association of Hydrological Sciences Publication* 264 (Symposium at Seattle 2000 - Debris Covered Glaciers), pp. 245-264.

9. Vohra, C.P. (1978) Glacier resources of the Himalaya and their importance to environment studies. *Proceedings of the National Seminar on Resources Development and Environment in Himalaya Region,* DST, pp. 441-460.
10. Jangpangi, B.S. (1958) Study of some of the central Himalayan glaciers. *J. Sci. Ind. Res.* **17**, 91-93.
11. Vohra, C.P. (1971) Gangotri glaciers expedition. *Report geological survey of India.*
12. Puri, V. (1977) Geological Survey of India, *Ann. Gen. Rep,* 1 (5).
13. Puri, V. (1990) Geological Survey of India, *Ann. Gen. Rep,* 124 (8).
14. Sharma, M.C. and Owen, L.A. (1996) Quaternary glacial history of NW Garhwal, central Himalayas. *J.Quat. Sci. Rev.* **15**, 335–365.
15. Naithani A.K., Nainwal H.C., Sati K.K., Prashad C. (2001) Geomorphological evidences of retreat of the Gangotri glacier and its characteristics. *Curr. Sci.*, **80**, 87-94.
16. Kar, R. Ranhotra, P.S. Bhattacharyya A, and Sekar, B. (2002) Vegetation vis-a-vis climate and glacial fluctuations of the Gangotri glacier since 2000 years. *Curr.* Sci. **82**, 347-351.
17. Dhobhal, D.P., Gergan, J.T. and Thayyen, R.J. (2004) Recession and morphogeometrical changes of Dokriani glacier (1962-1995) Garhwal Himalaya, India. *Curr.Sci.* **86**, 692-696.
18. Srivastava, D. (2004) Recession of Gangotri glacier, Proc. Workshop on Gangotri glacier Lucknow, Special publication **80**, 21-32.
19. Kumar, K., Dumka, R. K., Miral, M. S., Satyal, G. S., 2007. Estimation of retreat rate of Gangotri glacier using rapid static and kinematic GPS survey. *Curr. Sci.* (submitted for publication)
20. Kaul, M.N. (1990) Glacial and fluvial geomorphology of Western Himalaya, *Concept Publishing Company*, 23-26.
21. Miral, M.S. (2001) Glacier research expedition, Gaumukh to Badrinath via Kalindikhal Pass (19,510 ft) *Envis bulletin*, Himalayan ecology and development **9**, 73-76.

17

MOLECULAR TECHNOLOGIES APPLICATION IN CANCER DIAGNOSIS FOR IMPROVED TREATMENT STRATEGY

A. Raghu Ram Rao and Amit Ahuja

ABSTRACT

Molecular technology is devoted to the study of synthesis, properties and applications of structures and materials with at least one critical dimension less than the scale of approximately 100 nm. Cancer is one of the most significant causes of morbidity and mortality in the world. Clinical outcome of cancer diagnosis is strongly related to the stage at which the malignancy is detected and therefore, early screening has become desirable. Here comes the role of various molecular technology tools for the early diagnosis of cancer. Various molecular technology tools used for cancer diagnosis are Nanocantilevers, Quantum Dots, Nanoshells, Dendrimers, Nanopores, Nanotubes, Genomic Profiling, Protein Profiling, and Protein Microarray. All these tools are based upon their interaction at the molecular level in the body and size plays an important role in their mode of action. Molecular technology holds a promising future in its hands for the cancer diagnosis and treatment and for many other diseases.

1. INTRODUCTION

Molecular technology is an emerging technology seeking to exploit distinct technological advances by controlling the structure of materials at a reduced dimension with the scale approaching individual molecules and their organized aggregates or supramolecular structures. Molecular technology is a broad area of science and technology devoted to studies of synthesis, properties and applications of structures and materials with at least one critical dimension less than the scale of approximately 100 nm. Application in life sciences research, particularly at the cell level sets the stage for an exciting role of molecular technology in healthcare. Novel nanomaterials are envisaged to have a major impact on a number of different relevant areas. As we know the quantum dots, nanowires and proteins, which are built from atoms, have similar average diameters. These basic structures can be organized to form larger structures such as viruses and nanoelectric circuits, which can be organized to form basic functional structures such as cells and computer chips.

Cancer is one of the most significant causes of morbidity and mortality in the world. Clinical outcome of cancer diagnosis is strongly related to the stage at which the malignancy is detected and therefore, early screening is desirable. Many cancers are detected at advanced stages, when invasion of surrounding tissues and metastasis to distant sites have already occurred, rendering standard treatment modes relatively ineffective. Standard treatment modes including chemotherapy and radiotherapy have consistent toxicities, but do not benefit all patients. An effort is under way to translate advances in molecular biology, bioinformatics, and computing into tools that may improve the diagnosis, staging, and treatment of patients with cancer. These tools range from tests that evaluate single genes for unique interindividual differences to chips that profile the expression of entire sets of genes or proteins. These emerging genomic and proteomic technologies may help to individualize and improve the clinical management of cancer.

Cancer is a leading cause of death worldwide. More than 10 million people are diagnosed with cancer every year. It is estimated that there will be 15 million new cases every year by 2020. Cancer causes 6 million deaths every year or 12% of deaths worldwide. In the United States, about 16 million new cancer cases have been diagnosed since 1990, where there were 553,768 cancer deaths and the overall costs for cancer treatment were estimated to be as high as $156.7 billion in 2001, and about 1.3 million new cases were diagnosed and more than half a million deaths were caused by cancer, i.e., one in every four deaths, in 2002.

The human and economic burden of cancer on our society continues to grow (Figs. 1, 2).

- Cancer now the #1 killer **[1]**
- 570,300 Americans will die of cancer this year
- 1,372,900 Americans this year will hear the words "you have cancer…"

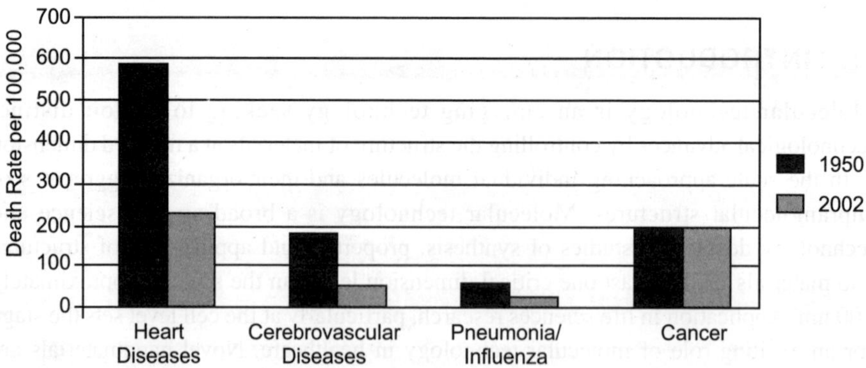

Fig. 1 *Mortality rates associated with some common diseases prevalent today [1],*

Complexity of Cancer Immunosupression In The Tumor Environment

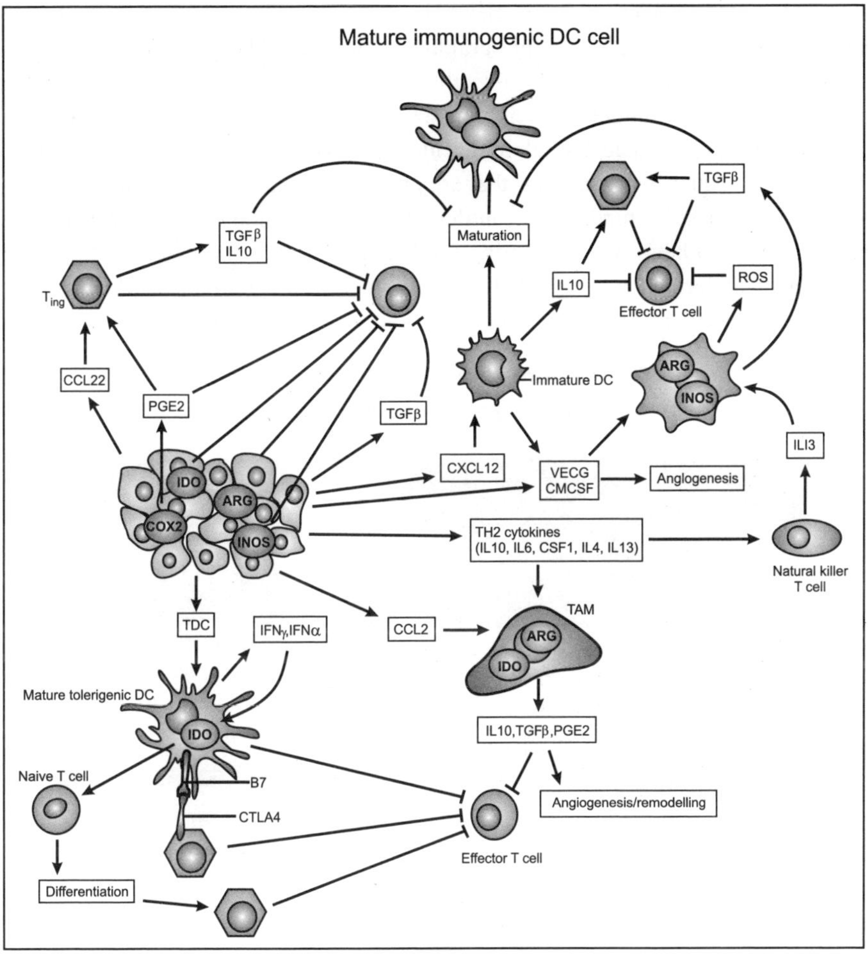

Fig. 2 *Interactions between the tumor and cells of the immune system that foster an immunosuppressive microenvironment [Figure taken from Ref. No. 2],*
(For color illustration refer p xix)

Cancer Detection and Diagnosis

Detection of cancer at early stages is a critical step in improving cancer treatment. The current diagnosis techniques are usually in tissue level, which have low efficiency and fail to detect invisible cancer cells. In order to successfully detect cancer at its earliest stages, we must be able to detect molecular changes even when they occur only in a small percentage of cells. The potential for nanostructures to enter and analyse single cells suggest they could meet this need

Various molecular technology tools used for cancer diagnosis

Nano-cantilever: [3, 4]

Nanoscale cantilevers, microscopic, flexible beams resembling a row of diving boards, are built using semiconductor lithographic techniques. These can be coated with molecules capable of binding specific substrates—DNA complementary to a specific gene sequence, for example. Such micron-sized devices, comprising many nanometer-sized cantilevers, can detect single molecules of DNA or protein. They may bind to altered DNA sequences or proteins that are present in certain types of cancer (Fig. 3). When these molecules bind to the cantilevers, surface tension changes, which causes the cantilevers to bend. By monitoring the bending of the cantilevers, scientists can tell whether molecules are present. This property will prove effective when cancer – associated molecules are present even at very low concentrations.

Cantilevers Can Make Cancer Tests Faster and More Efficient

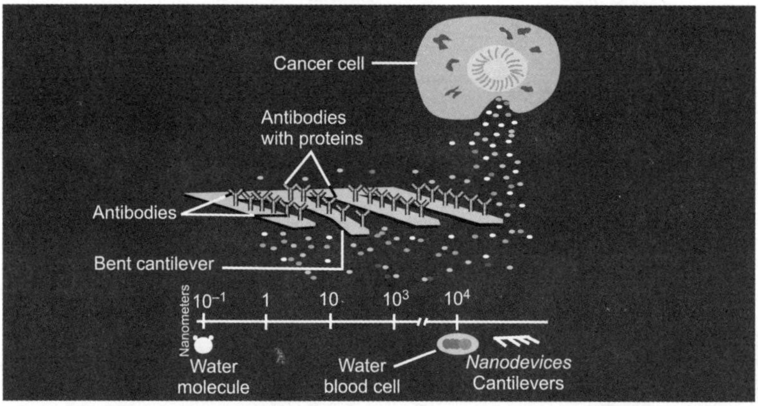

Fig. 3 *Role of Nanocantlevers in detecting cancer cells,*
(For color illustration refer p xx)

Quantum Dots: [5-9]

Quantum Dots (QDs) are fluorescent semiconductor nanocrystals that consists of one or more electrons. QDs are often referred to as artificial atoms. QDs are typically composed of atoms of elements from groups II and VI or groups III and V of the periodic table. They can be as small as 1 nm, typically between 2 and 10 nm. Their diameter corresponds between 10 and 50 atoms. The number of electrons inside a Quantum Dot ranges from one to several thousand. The determinant of the wavelength or colour of light depends upon the size, shape, material and coating of the crystal. Quantum Dots with the same material can emit different colors when their size is changed (Fig. 4).

Fig. 4 *Fluorescence induced by exposure to ultraviolet light in vials containing various sized Cadmium selenide (CdSe) quantum dots,* (For color illustration refer p xx)

Latex beads filled with these crystals can be designed to bind to specific DNA sequences. When the crystals are stimulated by UV light, each bead emits light that serves as a sort of spectral bar code, identifying a particular region of DNA. To detect cancer, scientists can design quantum dots that bind to sequences of DNA that are associated with the disease. When the quantum dots are stimulated with light, they emit their unique bar codes, or labels, making the critical, cancer-associated DNA sequences visible. The diversity of quantum dots will allow scientists to create many unique labels, which can identify numerous regions of DNA simultaneously (Fig. 5). This will be important in the detection of cancer, which results from the accumulation of many different changes within a cell.

Quantum Dots Can Find Cancer Signatures

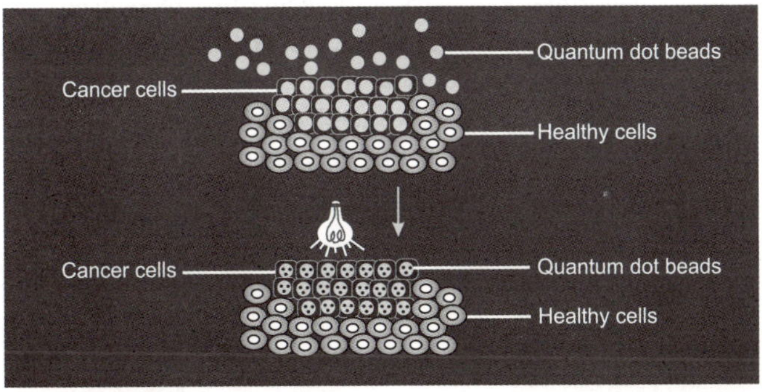

Fig. 5 *Quantum Dots can be used in the body, eliminating the need for biopsy* (For color illustration refer p xxi)

Nanoshells: [10-13]

Nanoshells are miniscule beads coated with gold. By manipulating the thickness of the layers making up the nanoshells, scientists can design these beads to absorb specific wavelengths of light. The most useful nanoshells are those that absorb near-infrared light, which can easily penetrate several centimeters of human tissue. The absorption of light by the nanoshells creates an intense heat that is lethal to cells.

Researchers can link nanoshells to antibodies that recognize cancer cells. Scientists envision letting these nanoshells seek out their cancerous targets, and then apply near-infrared light (Fig. 6). In laboratory cultures, the heat generated by the light-absorbing nanoshells has successfully killed tumor cells while leaving neighboring cells intact [13].

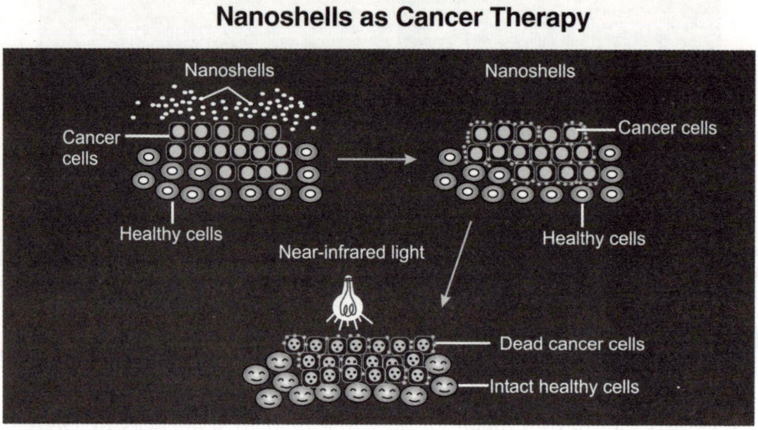

Fig. 6 *The use of nanoshells in cancer therapy,* (For color illustration refer p xxi)

Dendrimers: [14-16]

Dendrimers are man-made molecules about the size of an average protein, and have a branched shape (Fig. 7). This shape gives them a vast amount of surface area to which scientists can attach therapeutic agents or other biologically active molecules. Researchers aim eventually to create nanodevices that do much more than deliver treatment. A single dendrimer can carry a molecule that recognizes cancer cells, a therapeutic agent to kill those cells, and a molecule that recognizes the signals of cell death. Researchers hope to manipulate dendrimers to release their contents only in the presence of certain trigger molecules associated with cancer. Following drug release, the dendrimers may also report back whether they are successfully killing their targets.

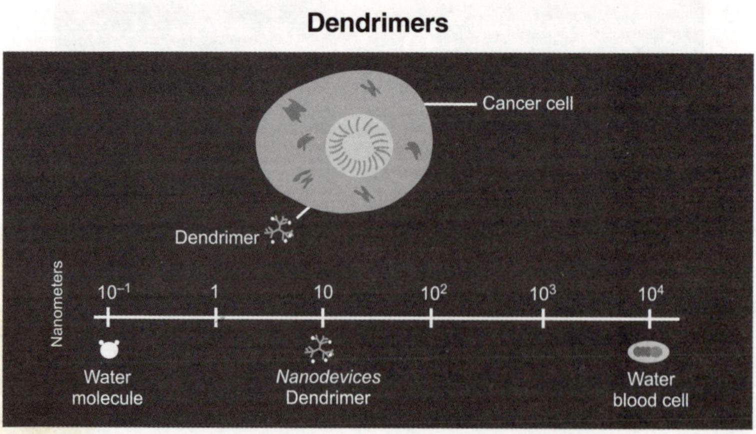

Fig. 7 *Comparison between the size of cancer cell and a dendrimer,* (For color illustration refer p xxii)

Nanopores: [17, 18]

Nanopores are tiny holes that allow DNA to pass through one strand at a time and make DNA sequencing more efficient. The shape and electrical properties of each base, or letter can be monitored as the DNA passes through the nanopore. Because these properties are unique for each of the four bases that make up the genetic code, scientists can use the passage of DNA through a nanopore to decipher the encoded information, including errors in the code known to be associated with cancer.

Nanotubes: [19-21]

Nanotubes are carbon rods about half the diameter of a molecule of DNA that cannot only detect the presence of altered genes, but may also help researchers pinpoint the exact location of those changes. To prepare DNA for nanotube analysis, a bulky molecule is attached to regions of the DNA that are associated with cancer (Fig. 8). Tags are designed to seek out specific mutations in the DNA and bind to them. Once the mutation has been tagged, a nanotube tip is used to trace the physical shape of DNA and pinpoint the mutated regions. The nanotube creates a map showing the shape of the DNA molecule, including the tags identifying important mutations. Since the location of mutations can influence the effects they have on a cell, these techniques will be important in predicting cancer.

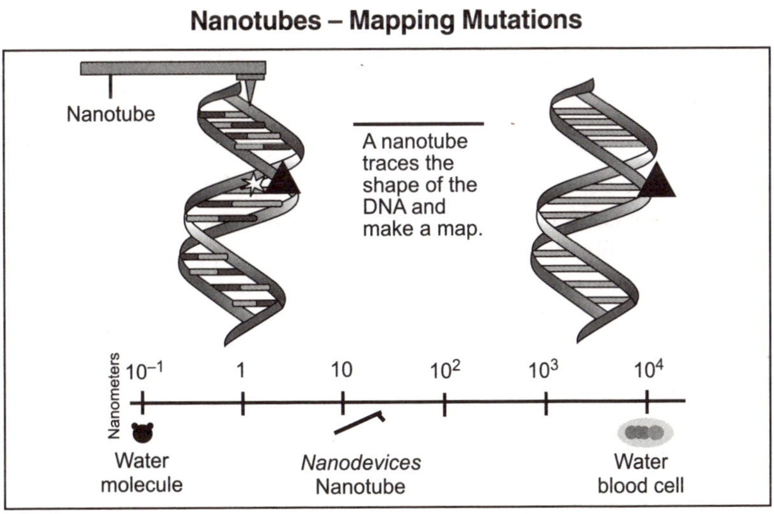

Fig. 8 ***Identification of mutations in the DNA by nanotube,***
(For color illustration refer p xxii)

Genomic Profiling

Profiling the expression of thousands of genes using microarray technology may also contribute to the clinical management of cancer. Microarrays have helped to improve cancer diagnosis by identifying both specific genes that contribute to cancer susceptibility and patterns of gene expression that distinguish different hematological malignancies. The use of genomic profiling has also led to an improved understanding of cancer pathogenesis, an important step in the development of novel targeted therapies. Identification of specific gene products in certain leukemias, as well as breast and ovarian cancers, has led to the development of specific drugs that target these tumors with relatively little toxicity to normal tissue. These drugs target proteins that are uniquely found or are relatively abundant in tumor cells, allowing for more specific targeting of tumor with less toxicity to normal tissues and, hence, fewer adverse effects. In addition to contributing to new drug development, genomic profiling has been used to identify patterns of gene expression that may predict the potential efficacy and toxicity of existing chemotherapeutic agents in individual patients.

DNA Arrays for Transcriptional Profiling: [22-32]

A DNA array consists of rows and rows of oligonucleotide strands, or complementary DNAs, lined up in dots on a miniature silicon chip or glass slide. Oligonucleotides can be synthesized in situ on the array, and comprise of short segments of DNA complementary to RNA transcripts of interest. Alternatively, cDNA arrays can be constructed containing immobilized sequences that may range greatly in length, but are usually much longer than the corresponding oligonucleotide. To use the arrays, RNA is extracted from tumor tissue, amplified and labeled with a fluorescent dye. The labeled RNA is hybridized to the array in the presence of a reference RNA, which is labeled with a different fluorescent dye. After hybridization, the fluorescence from each spot on the array provides a measure of the relative abundance of a given transcript in the two RNA populations, and so reflects the relative expression level of its corresponding gene. Expression levels can be compared across many samples, normal and pathological. With appropriate pattern recognition software, one can then identify groups of genes whose expression varies in a similar manner, and group cell lines or tumor tissue specimens with similar repertoires of expressed genes. Cluster algorithms are a powerful systematic tool because they can reveal hidden patterns of relationships that transcend the histological appearance of a tumor and lead to hypotheses about gene function and tumor pathology. *(Detailed protocol is given in Ref. No. 32)*

Protein Profiling

Clinical proteomics seeks to determine how changes in information flow at the cellular level, represented by alterations in complex patterns of protein activation and expression, relate to disease pathogenesis, and whether certain changes are predictive of disease development, progression, and response to specific treatment modalities.

Protein Microarray: [33-35]

Protein microarrays are a series of immobilized spots, each containing a bait molecule such as an antibody, a nucleic acid, a drug, or a recombinant protein or peptide, to which the patient's tissue or fluid is applied (Fig. 9). If proteins from the sample bind to the bait molecule, they maybe detected using antibody probes. Proteomic profiling with protein microarrays has been used in the research setting to assay levels and activation states of key signaling proteins with the goal of reclassifying human tumors to more accurately reflect the degree of malignancy and metastatic potential, and to determine molecular signatures indicative of early cancer [33].

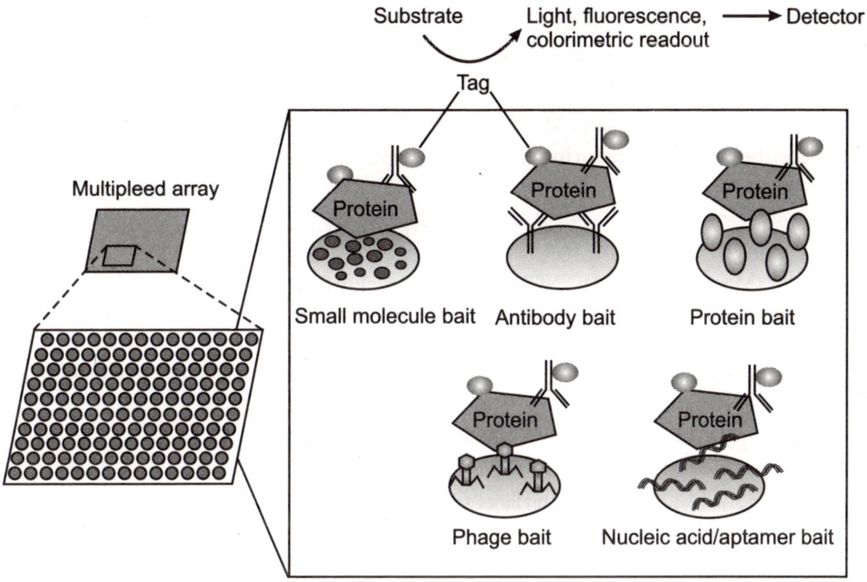

Fig. 9 ***Enlarged image of the Protein Microarray system,*** (For color illustration refer p xxiii)

High-throughput ELISA protein array systems comprise optically flat amino-terminal-modified glass plates which are arrayed with 14,000 addressable spots. Each spot detects a different protein partner through a fluorescently tagged secondary antibody. This format can detect specific proteins from within a cocktail of many analytes in a high-throughput and robust manner. The next step will be to extend antibody arrays to detect phosphorylated forms of proteins and to include binding molecules other than antibodies. In this way it will be possible in the future to conduct protein pathway profiling that reads out the global state of a signal transduction pathway in a tissue cell.

Future with Molecular Technology

A cancer patient visits his oncologist, gives a few drops of blood or a biopsy specimen, and finds that his genetic expression pattern shows that he has a certain subtype of disease. Meanwhile, another expression pattern predicts that his genetic profile should respond well to chemotherapy regimens A and B (with minimal side effects). During his treatment, protein expression patterns are used to make sure that his treatment is effectively disrupting the targeted cellular pathway in his tumor.

After treatment, more gene and protein expression patterns verify that the cancer is in remission.

Conclusion

Molecular Technology is thus an emerging field of medical science which could help us to fight more effectively with various killer diseases such as prostrate cancer, AIDS etc. It can also make medical diagnosis and treatment more effective and patient friendly. Molecular Technology offers possibilities for health breakthroughs, yet many of these developments seem so high-tech that it is hard to imagine their being used as health interventions among the poor. So we must work out such molecular devices which are sensitive and also cost effective.

The authors wish to thank Ms. Amandeep Kaur, Ms. Jasjit Kaur and Ms. Ramandish Arora for their help in preparing this manuscript.

REFERENCES

1. 1950 MortalityData -CDC/NCHS, NVSS, Mortality Revised. 2002 Mortality Data–NVSR-Death Final Data 2002–Volume 53, No. 5. Cost data from American Cancer Society Cancer Facts & Figures 2005.
2. Muller A.J. and Scherle P.A. (2006) Targeting the mechanisms of tumoral immune tolerance with small-molecule inhibitors. *Nature* **6**, 742.
3. Lechuga L.M. *et al.* (2006) A highly sensitive microsystem based on nanomechanical biosensors for genomics applications. *Sensors Actuators B: Chemical* **118**, 1-20.
4. Zhang J. *et al.* (2006) Rapid and label-free nanomechanical detection of biomarker transcripts in human RNA. *Nature Nanotechnol.* **1**, 214 – 220.
5. Nie S. (2004) Researchers Use Dots to Light Up Tumor Cells. *J. Am. Med. Assoc.* **292**, 1944-1945.
6. Alivisatos, A.P. (1996) Perspectives on the physical chemistry of semiconductor nanocrystals. *J. Phys. Chem.* **100**, 13226-13239.
7. Yin, Y. and Alivisatos A.P. (2005) Colloidal nanocrystal synthesis and the organic-inorganic interface. *Nature* **437**, 664-670.
8. Murphy C.J. and Coffer J.L. (2002) Quantum Dots: A Primer, *Appl. Spectrosc.* **56**, 16A-27A.
9. Sapra S. and Sarma D.D. (2004) Evolution of the electronic structure with size in II-VI semiconductor nanocrystals. *Phys. Rev. B* **69**, 125304-125312.
10. Khlebtsov B. *et al.* (2006) Optical amplification of photothermal therapy with gold nanoparticles and nanoclusters. *Nanotechnology* **17**, 5167-5179.
11. Yang Y. *et al.* (2005) Enumeration of DNA molecules bound to a nanomechanical oscillator. *Nanoletters* **5**, 925-929.
12. Huan X. *et al.* (2006) Cancer cell imaging and photothermal therapy in the near-infrared region by using gold nanorods. *J. Am. Chem. Soc.* **128**, 2115 -2120.
13. Lowery A.R. *et al.* (2006) Immunonanoshells for targeted photothermal ablation of tumor cells. *Intern. J. Nanomed.* **1**, 1-6.
14. Majoros I.J. *et al.* (2006) PAMAM dendrimer-based multifunctional conjugate for cancer therapy: synthesis, characterization, and functionality. *Biomacromolecules* **7**, 572-579.
15. Morgan M.T. *et al.* (2006) Dendrimer-encapsulated camptothecins: increased solubility, cellular uptake, and cellular retention affords enhanced anticancer activity in vitro. *Cancer Res.* **66**, 11913-21.
16. Myc A. *et al.* (2007) Dendrimer-based targeted delivery of an apoptotic sensor in cancer cells. *Biomacromolecules* **8**, 13-18.

17. Siwy Z. *et al.* (2005) Protein Biosensors Based on Biofunctionalized Conical Gold Nanotubes. *J. Am. Chem. Soc.* **127**, 5000-5001.
18. Lagerqvist J. *et al.* (2006) Fast DNA Sequencing via Transverse Electronic Transport. *Nanoletters* **6**, 779-782.
19. Liu Z., *et al.* (2007) Supramolecular Chemistry on Water- Soluble Carbon Nanotubes for Drug Loading and Delivery. *Nano* **1**, 50-56.
20. Shi, K., N.*W. et al.* (2005) Carbon nanotubes as multifunctional biological transporters and near-infrared agents for selective cancer cell destruction. *Proc. Natl. Acad. Sci. USA* **102**, 11600-11605.
21. Liu Z. *et al* (2007) *In vivo* biodistribution and highly efficient tumor targeting of carbon nanotubes in mice. *Nature Nanotechnology* **2**, 47-52.
22. Pease, A. C. *et al.* (1994) Light-generated oligonucleotide arrays for rapid DNA sequence analysis. *Proc. Natl Acad. Sci. USA* **91**, 5022–5026.
23. De Risi, J. *et al.* (1996) Use of a cDNA microarray to analyse gene expression patterns in human cancer. *Nature Genet.* **14**, 457–460.
24. Lipshutz, R. *et al.* (1999) High density synthetic oligonucleotide arrays. *Nature Genet.* **21**, S20–S24.
25. Gasson, S. *et al.* (1999) Mask less fabrication of light-directed oligonucleotide microarrays using a digital micro mirror array. *Nature Biotechnol.* **17**, 974–978.
26. Alizadeh, A. A. *et al.* (2000) Distinct types of diffuse large B-cell lymphoma identified by gene expression profiling. *Nature* **403**, 503–511.
27. Golub, T. R. *et al.* (1999) Molecular classification of cancer: class discovery and class prediction by gene expression monitoring. *Science* **286**, 531–537.
28. Ross, D. T. *et al.* (2000) Systematic variation in gene expression patterns in human cancer cell lines. *Nature Genet.* **24**, 227–235.
29. Perou, C. M. *et al.* (1999) Distinctive gene expression patterns in human mammary epithelial cells and breast cancers. *Proc. Natl Acad. Sci. USA* **96**, 9212–9217.
30. Perou, C. M. *et al.* (2000) Molecular portraits of human breast tumors. *Nature* **406**, 747–752.
31. Bittner, M. *et al.* (2000) Molecular classification of cutaneous malignant melanoma by gene expression profiling. *Nature* **406**, 536–540.
32. Sgroi, D. *et al* (1999) *In vivo* gene expression profile analysis. *Cancer Res.* **59**, 5656–5661.
33. Liotta L.A., Petricoin E.F. (2000) Molecular profiling of human cancer. *Nat Rev Genet.* **1**, 48-56.
34. Huang R. *et al.* (2004) Enhanced Protein Profiling Arrays with ELISA-Based Amplification for High-Throughput Molecular Changes of Tumor Patients' Plasma. *Clinical Cancer Research* **10**, 598-609.
35. Wulfkuhle J.D. *et al.* (2003) Signal pathway profiling of ovarian cancer from human tissue specimens using reverse-phase protein microarrays. *Proteomics* **3**, 2085-2090.

18

METAGENOMICS: A SPLENDID NOVEL APPROACH TO EXPLORE AND EXPLOIT MICROBIAL DIVERSITY

Pavan K. Pindi, N. Kishore, S. Ram Reddy, S. M Reddy, S. Chatterjee, I. Sherameti, R. Oelmueller and A. Varma

ABSTRACT

Microbial diversity is one of the greatest knowledge gaps in the biological sciences and remains largely unexplored and unexploited. Despite the improvements in cultivation methods, there has not been any significant change in our perception of microbial diversity. Molecular phylogenetic methods have recently provided a means for identifying the types of organisms that occur in microbial communities without the need for cultivation. However, they do not provide any idea about their ecophysiological functions. Metagenomics, the genomic analysis of a population of microorganisms that provides an access to the physiology and genetics of uncultured organisms, has emerged as a powerful technique in recent times. Metagenomic analysis involves isolating DNA from an environmental sample, cloning the DNA into a suitable vector, transforming the clones into a host bacterium and screening the resultant transformants. Plasmid, cosmid, fosmid and BACs as cloning vectors and E. coli as the surrogate host are employed to construct metagenomic libraries. Three types of screening methods, viz. sequence-based, function based and SIGEX are used to screen the libraries for the desired clones. Metagenomics has unlimited applications in basic biology and biotechnology. The long term goal of metagenomic analysis is to reconstruct the genomes of uncultured organisms, understand the functional dynamics in natural communities and also to exploit the culture-independent microorganisms for novel bioreactive compounds. Already, spectacular results have been achieved that exposed only the tip of iceberg. To date, metagenomics has uncovered a variety of novel genes that range from small genes coding for enzymes to complex gene clusters encoding antibiotics. Thus it has the potential to substantially impact white and red biotechnologies. Obviously, metagenomes have attracted the attention of academia and specialized pharmaceutical companies. Notwithstanding certain teething problems, it is sufficiently powerful to answer many intricate biological questions and yield products for solving real world problems.

1. INTRODUCTION

Soil microorganisms display a pivotal role in various biogeochemical cycles and are responsible for the cycling of organic compounds [1]. They also influence above ground ecosystems by contributing to plant nutrition, plant health, soil structure and soil fertility [2-6]. Prokaryotic microorganisms represent by far the largest reservoir of genetic diversity on earth. They outnumber all other organisms, have a larger biomass and make the planet livable by managing its biogeochemistry, cycling nutrients and breaking down wastes. Besides, microbes possess the highest potential for production of bioactive compounds, enzymes, polymers and most of the tools used in biotechnology [7]. Microbial world seems to offer a greater natural resource for molecular biology.

Microbial diversity (Fig. 1) in soils exceeds that of other environments. One gram of soil can contain up to 10 billion microorganisms of possibly thousands of different species [8]. Our present knowledge of microbial diversity in any environment springs from the studies made on cultivable microorganisms. Molecular analysis of environmental DNA has revealed substantial phylogenetic diversity with little or no

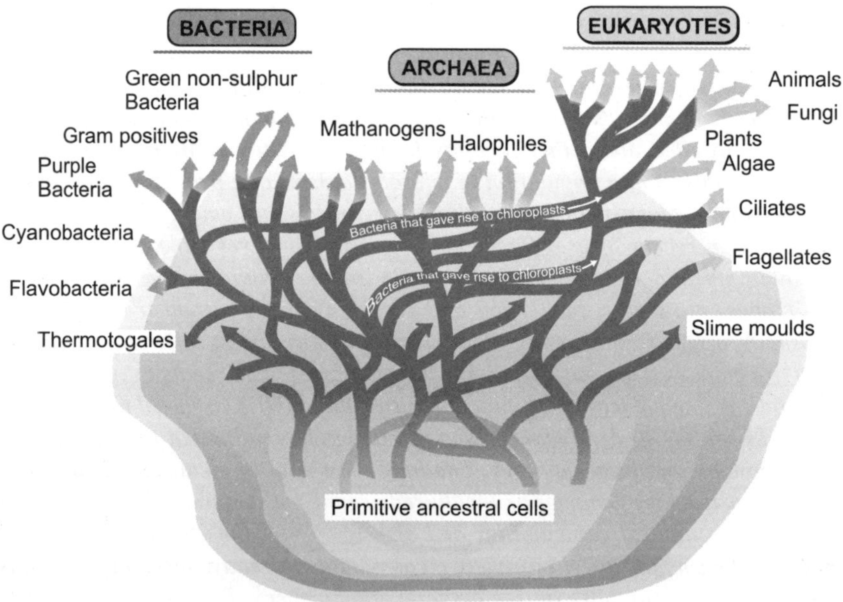

Fig. 1 *Microbial diversity,* (For color illustration refer p xxiii)

representation among organisms previously studied. Based on DNA-DNA reassociation studies, Torsvik *et al.* have estimated that there are 4000 different bacterial genomic units in 1 g of soil [9,10]. The genetic diversity of cultivable microorganisms from soil is about 200 fold less than this. Analysis of rRNA gene sequences has resulted in more than tripling of the number of identifiable bacterial divisions [11, 12]. At present, 52 bacterial phyla can be discerned; of which 26 are known to be as candidate phyla, i.e. they are known only from environmental rRNA clone sequences [13]. Although these approaches have provided some information on the identity and distribution of microbial species, molecular diversity alone does not reveal physiology, biochemistry or ecological functions. Present estimates suggest that >99 percent of the microorganisms in most environmental niches are not amenable to growth in pure culture, therefore, very little is known about their physiology and their role in their respective environments. In a study of Kellenberger, approximately 10^7 cells were counted in 1 g of soil, but only 0.1 percent of the cells were culturable [14]. Thus 99.9 percent of the genetic diversity present in the population was lost owing to difficulties in enriching and isolating microorganisms. These unrecovered microorganisms represent an unexplored reservoir of novel strains, which may produce novel natural products. In order to access this vast resource, one approach is to put an effort into the development of novel cultivation techniques [15-19]. In one study, a collection of 350 isolates of soil bacteria containing a high degree of phylogenetic novelty was generated by using simple solid media in petridishes. Even then, the number of strains obtained with these improved culture techniques falls far below the population expected from molecular predictions. Alternatively, polymerase chain reaction (PCR) amplification is used to access the genes from environmental samples. However, this method does not reveal the potential reservoir of undiscovered genes harbored by soil microbes, because PCR detection depends on primers that are based on known genes. Further, PCR could unveil new and different types of the known genes, but miss others entirely [20]. Yet another drawback is that, most often, there is no way to correlate the 16S rRNA sequence of a specific microbe with its ecophysiology.

As such, today, we have access only to 1 percent of the microbial population because it can be cultured. Clearly, the remaining 99 percent eludes our knowledge because it is culture independent. It is quite possible that this 99 percent of soil microbial population (uncultivated) is phenotypically and genetically different from the 1 percent of cultured population. It is important to study microbial diversity not only for basic scientific research, but also to understand the link between diversity and community structure and function. More importantly, the uncultured microbial strains are likely to harbor the genes that encode bioactive compounds. Then, a question worth million dollars remains to be answered: how to access the diversity of culture-independent microbes? Given the understanding that many organisms will not be coaxed readily into culture, molecular biologists have striven to devise methods to analyze the physiology and ecology of these diverse uncultured organisms which ultimately shaped in the emergence of a new powerful field, metagenomics. This review article is aimed at bringing out the salient features of metagenomics, its applications and its relevance to India (Table 1).

Table 1 *Abbreviations and glossary of words used in the text*

eDNA	–	Environmental DNA
BAC	–	Bacterial artificial chromosomes
SIGEX	–	Substrate induced gene expression
FACS	–	Fluorescence activated cell sorting

(Contd...)

GFP	–	Green fluorescent protein
ORF	–	Open reading frame
TGGE	–	Temperature gradient gel electrophoresis
DGGE	–	Denaturing gradient gel electrophoresis
PFGE	–	Pulsed field gel electrophoresis
RISA	–	Ribosomal intergenic spacer analysis
ARISA	–	Automated ribosomal intergenic analysis
Metagenomic DNA	:	DNA extracted directly from the environment, represents the composite DNA of microbiota of that environment.
Low-through put screening	:	A screening method in which each clone is tested separately for desired feature.
High-through put screening	:	A screening method in which a large number of clones are screened by automated method to separate desirable clones from the other clones.
Red biotechnology	:	Biotechnology in health care that uses substances produced in the human body to fight disease – medicines, vaccines, diagnostics and gene therapy.
White biotechnology	:	Industrial use of biotechnology, for example, to produce fine chemicals, biofuels, bioplastics, enzymes for use in detergents, food and feed.

2. METAGENOMICS

The word 'metagenomics' was first coined by Handelsman *et al.* [21]. The term is derived from the statistical concept of *meta*-analysis (the process of statistically combining separate analyses) and *genomics*- the comprehensive analysis of an organism's genetic material. The term "Metagenome" is used to describe the compound genome of whole microbiota, especially in the soil [22]. Metagenomics is the culture-independent genomic analysis of metagenome. It involves isolation of DNA from an environmental sample that contains an assemblage of organisms, cloning the DNA into a suitable vector, transforming the clones into a host bacterium and screening the resultant transformants. The clones can be screened for phylogenetic markers or "anchors" such as 16S rRNA and *rec* A or for other conserved genes by hybridization or multiplex PCR or for expression of certain traits or they can be sequenced randomly [23]. Notwithstanding the individual targeted approaches, the ultimate goal of metagenomics is to understand the uncultured world and to gain insight into groups of prokaryotes that are otherwise entirely unknown. In simplest terms it provides a window on the richness of the uncultured world (Table 2).

Table 2 *Chronology of metagenomics*

Year	*Major milestones in metagenomics*
1991	First publication of the major working steps
1995	Healy and coworkers publish the isolation of metagenome sequences from enrichment cultures
1998	Rondon and colleagues coin the term 'metagenome' and report on BAC clones

(Contd...)

1999	Major breakthrough in searches for genes with biotechnological value
2000	Report on a new type of phototrophy in the sea
2002	Report on overlapping archaeal cosmids
2003	Snap shot sequences of a phylogenetically diverse drinking water biofilm
2004	Near complete sequence of an acidophilic biofilm Release of Sargasso sea sequences
2005	Substrate induced gene expression (SIGEX) screening

Metagenomic analysis has several advantages over culture or PCR-based methods. For example, it (i) provides access to uncultured microorganisms, (ii) does not require prior knowledge of gene sequences, and (iii) recovers complete genes. Recent developments in metagenomics have drastically changed the perception of microbiologists for biodiversity. It redefined the concept of genome and accelerated the rate of gene discovery. The magnitude and significance of uncultured organisms were realized than ever before. Within a few years, the study of uncultured microorganisms has expanded beyond asking 'who is there' to include difficult questions like 'what are they doing?' 'what can they give us?'. The potential for application of metagenomics to biotechnology seems endless. The vast resource of thousands of genomes of microorganisms present in environments such as soil, waters and sediments can be exploited for biotechnology using this approach [24].

Applications of metagenomics largely depend on advancement of methods in library construction and analysis. Libraries constructed by cloning environmental DNA (eDNA) that include the genes of uncultured organisms are called "Metagenomic Libraries". The first metagenomic library was developed by Healy *et al.* [25]. Later De Long group constructed a gene library of prokaryotes from sea water [26]. Construction of libraries with DNA extracted from soil lagged due to difficulties associated with maintaining the integrity of DNA during its extraction and purification from soil matrix, but eventually produced analyses analogous to those from sea water [22, 27-32]. In the following sections methodologies followed for construction of metagenomic libraries and their analysis have been described (Fig 2).

3. METHODS FOR CONSTRUCTION OF METAGENOMIC LIBRARIES

3.1 Isolation and Purification of Metagenomic DNA

DNA extraction from the environment is a critical first step because it needs to reflect the existing genetic diversity. The choice of method depends on whether there is a requirement for small (<10 kb) or larger DNA segments. Large segments are required to capture operons encoding biosynthetic pathways. Furthermore, large inserts reduce the library size that in turn reduces the number of clones to be maintained and screened. However, cloning of high molecular weight DNA from soil is particularly difficult because of co-purification of humic acids with DNA. Commonly, polyvinyl pyrrolidone or hexadecyl methyl ammonium bromide is incorporated to reduce humic contamination. Two methods are generally adopted for isolation of DNA: the direct lysis method and the cell extraction method. However, direct analysis method has been reported to yield at least 10 fold more DNA than cell lysis method [33]. This

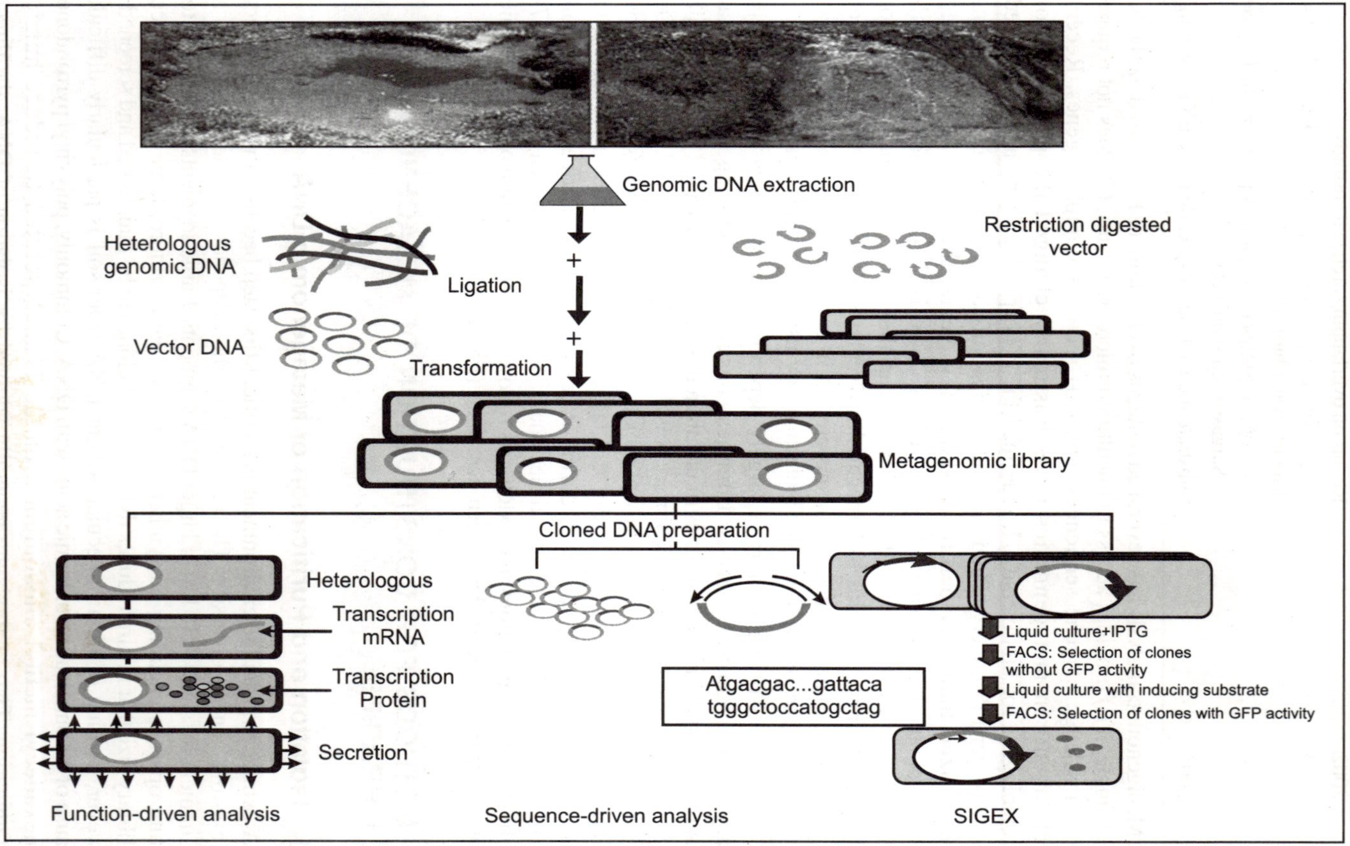

Fig. 2 *Metagenomic libraries,* (For color illustration refer p xxiv)

method is based on the initial extraction of extracellular DNA with alkaline buffer, followed by the direct lysis of the cells in the soil by chemical and mechanical means and then quantitative extraction of released DNA. For purification, multiple electrophoresis runs are commonly employed [22-34]. An optimized purification procedure has recently been reported [35].

3.2 Size Selection

Following purification, the metagenomic DNA is partially digested with restriction enzymes. Restriction digestion requires a starting DNA minimally three times longer than the desired insert. For example, if large inserts (≥ 100 kb) are desired, contiguous DNA of several hundred kilobases is required. However, the DNA of this size is quite vulnerable to shearing. To circumvent this problem, novel methods are continuously being explored that will cut down the number of steps in the lysis procedure and eliminate the need for restriction digestion. Quaiser *et al* prepared a soil metagenomic library in the fosmid vector pEpiFOS employing blunt end ligation [35]. Their library included 2.5×10^4 clones with insert sizes of 30 randomly chosen clones between 32.5 and 43.5 kb. Physical shearing and blunt end ligation also eliminate the bias introduced by restriction digests, and naturally decreases the amount of DNA to be handled.

3.3 Library Size

Metagenomic libraries can be classified into two groups with respect to average insert size: small insert libraries of less than 15 kb and large-insert libraries more than 40 kb. The advantages and disadvantages of these two types of libraries are précised in Table 3.

Table 3 *Advantages and disadvantage of small-insert and large-insert soil libraries*

Small-insert library (plasmids)	*Large-insert library (cosmids, fosmids, BACs)*
Advanges	
High copy number allows detection of weakly-expressed foreign genes	Large insert size
Expression of foreign genes from vector promoters is feasible	Small number of clones can be screened to obtain positive results
Cloning of sheared DNA or soil DNA contaminated with matrix substances is possible	Suitable for cloning of enzyme activities and pathways that are encoded by large gene clusters
Technically simple	Suitable for partial genomic characterization of uncultured soil microorganisms
Disadvantages	
Small insert size	Low copy-number might prevent detection of weakly-expressed foreign genes
Large number of clones must be screened to obtain positives	Limited expression of foreign genes by vector promoters
Not suitable for cloning of activities and pathways that are encoded by large gene clusters	Requires high-molecular soil DNA of high purity for library construction

3.4 Choice of Vector and Host

There is a limited choice of vectors for cloning metagenomic DNA. Most commonly used vectors are: plasmids, fosmids, cosmids and bacterial artificial chromosomes (BACs). Plasmids are ideal for small-insert libraries (less than 15 kb). Fosmids and cosmids are suitable for libraries with moderate size inserts [26, 34] (38-52 kb). These vectors have much greater cloning efficiencies than BACs. Nonetheless, BAC is the choice for cloning for various reasons. BACs based on the *E. coli* factor can carry large inserts (>300 kb). Even with large inserts they are stably maintained at one or two copies per cell. It has less potential for recombinations and for lethal expression of cloned genes. As it exists in supercoiled state, isolation and manipulations are easy [36]. A BAC vector that has been used to create soil metagenomic libraries is pBeloBAC 11 [37]. This vector has two selective markers for cloning purpose, the *lac* Z gene and chloramphenicol resistance gene. In addition, pBeloBac 11 has three unique cloning sites flanked by the T_7 SP6 promoters. *Not* I cuts on each side of the cloning site, excising the inserts and permitting the size determination by pulsed field gel electrophoresis (PFGE).

The most commonly used host for BAC cloning is DH1OB strain of *E. coli*. This strain contains mutations that block restriction of foreign DNA by endogenous restriction endonucleases, restriction of DNA containing methylated DNA and recombination. It takes up large DNA sufficient for >100 kb insert library construction [38]. The other advantages of cloning in *E. coli* include high transformation efficiency, the ability to express heterologous genes, the presence of simple primary metabolites for the production of heterologous secondary metabolites and ease of fermentation scale-up.

In order to expand the host machinery and metabolic background, some more hosts are also attempted. In this regard, two obvious host choices are *Streptomyces* and *Bacillus*. Several investigators are endeavoring to develop shuttle vectors. Sosio *et al.* have developed BACs that shuttle between *E. coli* and *Streptomyces* [39]. On the other hand, Handelsman *et al.* have developed the vector super BAC1 that shuttles between *E. coli* and *Bacillus* [40]. Quite surprisingly, so far, no eukaryotic host has been used to construct metagenomic soil libraries.

3.5 Analysis of Metagenomic Libraries

Analysis of the comprehensive metagenomic libraries calls for a thorough knowledge of the structural and functional aspects of a wide array of microbial genes and also the tools and methodologies required to be employed. Only then can an appropriate strategy be designed to realize the specific objectives. Strategies for the screening and analysis of metagenomic libraries have recently been summarized [24]. Of the many approaches, two, the function-driven analysis and sequence-driven analysis, have proved to be suitable for the extraction of biological information from metagenomic libraries. Very recently, a third high-throughput screening strategy, which is termed substrate induced gene expression (SIGEX) screening has been introduced for the identification and recovery of genes that encode catabolic pathways [41].

3.5.1 Function-driven Analysis

The function-driven analysis involves the identification of clones that express a desired trait, followed by characterization of the active clones by sequence and biochemical analysis. The principle underlying this analysis is to identify clones that express a function. Success demands faithful transcription and translation of the gene or genes of interest and secretion of the gene product, if the screen or assay requires it to be extracellular. The power of this approach is that it does not require that the genes of interest be recognizable by sequence analysis making it the only approach to metagenomics that has the potential to identify entirely new classes of genes for new or known functions. The frequency of metagenomic clones that express any given activity is low. For example, in search for lipolytic clones derived from German soil, only 1 in 7,30,000 clones showed activity [42]. Similarly, in a library of DNA from North American soil, 29 out of a total of 25,000 clones expressed hemolytic activity. The scarcity of active clones, therefore, necessitates development of efficient screening procedures.

Functional analysis can be achieved with low-throughput screens or high-throughput screens, or by selection. Low-throughput screening is based on characters such as visual detection of growth inhibition of indicator bacteria, and selection such as resistance to antibiotics or complementation based on necessary biochemical pathways (Table 4). Low-throughput screens can detect a wide range of activities, but their utility is limited by the low frequency of active clones. High-throughput screens do not require a selectable phenotype. One example of high-throughput screens is an intracellular biosensor [43].

Several selections based on functional analysis have been proved to be fruitful [28, 42-46]. One drawback of function based analysis is the dependence on expression of cloned genes by a foreign host. *E. coli* was successfully employed as the host during screening of soil-derived metagenomic DNA. However, a significant limitation, as many point out, is that many genes, perhaps most, will not be expressed in any particular host bacterium selected for cloning. To enhance the likelihood of detecting any particular gene or gene cluster other hosts can be used to express random soil genes. For instance, employment of *Streptomyces* as an alternative to *E. coli* yielded novel bioactive compounds [44, 47].

3.5.2 Sequence-driven Analysis

Sequence-driven analysis relies on the use of conserved DNA sequences to design hybridization probes or PCR primers to screen metagenomic libraries for clones that contain sequences of interest. It involves complete sequencing of clones containing phylogenetic anchors that indicate the taxonomic group that is the probable source of the DNA fragment.

Once a gene of interest is identified, phylogenetic anchors such as 16S rRNA gene and the archaeal DNA repair gene *rad* A can be searched in the flanking DNA to provide a link of phylogeny with the functional gene. Sequence analysis guided by the identification of phylogenetic markers is a powerful approach. It produced the first genomic sequence linked to 16S rRNA gene of an uncultured archaeon [26]. Sequencing of a clone isolated from sea water, through this approach, revealed the sensational information that rhodopsins are not confined to the Archaea, as was previously thought

Table 4 *Function driven analysis for industrially relevant enzymes and biocatalysts from metagenomic libraries*

Function	*Habitat*	*Library type*	*Average insert size (kb)*	*Number of clones screened*	*Library size (Mb)*	*Substrate*
Esterase/Lipase	Forest soil	Plasmid	8	67,000	536	Tributyrin
Esterase/Lipase	Forest soil	Fosmid	40	19,968	799	Tributyrin
Esterase/Lipase	Sandy ecosystem	Fosmid	30	29,884	903	Tributyrin
Esterase/Lipase	Sandy ecosystem	Fosmid	40	25,344	1,014	Tributyrin
Esterase/Lipase	Soil	Plasmid	6	286,000	1,716	Tributyrin
Esterase/Lipase	Soil	Plasmid	6	730,000	4,380	Triolein
Oxidation of polyols	Soil	Plasmid	3	900,000	2,700	1,2-Ethanediol, 1,2-propanediol, 2,3-butanediol
Alcohol oxidoreductase	Soil/ enrichment	Plasmid	4	400,000	1,600	Glycerol/1,2-propanediol
Amidase	Soil/ enrichment	Plasmid	5	193,000	965	D-phenylglycine-L-leucine
Amylase	Soil	Plasmid	5	80,000	400	Starch
Amylase	Soil	BAC	27	3,648	100	Starch
Biotin production	Soil/ excrement enrichment	Cosmid	35	50,000	1,750	Biotin-deficient medium

(*Contd...*)

Function	*Habitat*	*Library type*	*Average insert size (kb)*	*Number of clones screened*	*Library size (Mb)*	*Substrate*
Protease	Soil	Plasmid	10	100,000	1,000	Skimmed milk
Cellulase	Sediment enrichment	λ phage	6	310,000	1,860	Carboxymethyl-cellulose
Chitinase	Seawater	λ phage	5	825,000	4,125	Methyl-umbelliferyl-diacetylchito-bioside
Dehydratase	Soil/ sediment enrichment	Plasmid	4	560,000	2,240	Glycerol
4-Hydroxy butyrate conversion	Soil	Plasmid	6	930,000	5,580	4-Hydroxy butyrate
b-Lactamase	Soil	Plasmid	5	80,000	400	Ampicillin
Turbomycin	Soil	BAC	44.5	24,546	1,092	
	Soil	BAC	63	12,000	756	
	Soil	BAC	27	3,648	100	
	Soil	Fosmid	30	13,440	403	
	Soil	Cosmid	Nd	5,000	Nd	

c.f. 35, 40, 51, 52, 53

[48]. A promising application of phylogenetic anchor guided sequencing is to collect and sequence many genomic fragments from one taxon. The alternative to phylogenetic marker-driven approach is to sequence random clones, which is likely to produce dramatic insights, especially when conducted on a massive scale. The recent monumental sequencing efforts, which include reconstruction of the genomes of uncultured organisms in a community in acid mine drainage and the Sargasso sea, illustrate the power of large-scale sequencing efforts to enrich our understanding of uncultured communities [49, 50]. These studies have established new linkages between phylogeny and function.

Sequence-driven analysis approach has several advantages. Foremost is that it is not dependent on the expression of cloned genes by foreign hosts. In addition, along with this approach, similar well established screening strategies, such as PCR or colony hybridization, can be employed for different targets [51, 52]. Majority of the sequence-based approaches were performed to retrieve clones containing ribosomal RNA genes for phylogenetic surveys. The advantage of the identification of clones harboring ribosomal RNA genes or other marker genes over large inserts is that sequencing of the DNA surrounding these genes is feasible. This method provides access to genomes of uncultivated soil microorganisms and can yield clues about the physiology of unknown organisms. Eventually, novel biomolecules of industrial interest can be identified [35]. One more promise of this approach is its future use in DNA microarray technology. This technology has been proven useful for the detection of selected genes in the environment, for profiling of metagenomic libraries, and for functional genomic studies of soil microbial communities.

3.5.3 Substrate Induced Gene Expression Analysis (SIGEX)

In an effort to improve the frequency of screening hits, Uchiyama and his colleagues proposed a substrate induced gene expression screening (SIGEX) [41]. This method is based on the finding that genes encoding catabolic pathways are usually clustered in operons that are induced by a relevant substrate and are often controlled by regulatory elements located in the proximity of the catabolic genes. An operon-trap *gfp* (green fluorescent protein) – expression vector was constructed, which allowed shotgun cloning of metagenomic DNA upstream of the *gfp* gene, thereby placing the expression of this gene under the control of promoters that were present in the metagenomic DNA. Clones influencing the expression of *gfp* on addition of the substrate of interest can be isolated by fluorescence-activated cell sorting (FACS). Uchiyama *et al.* constructed a metagenomic library of ground water sample and successfully applied SIGEX to isolate 33 clones induced by benzoate and two clones induced by naphthalene from 1,52,000 clones [41]. In addition, these researchers showed that the enzyme BZO 71-8 P450 from the metagenomic library is novel.

SIGEX has many advantages in metagenome screening. It provides an efficient and economic way of high-throughput screening, because it allows for semi-automation, thereby saving time, labor and expenses. This is particularly important for industrial applications. SIGEX is also advantageous because it can detect catabolic genes for which colorimetric or other on-plate screening methods are not established. Moreover, SIGEX enables the deduction of substrate for an unknown enzyme from the induction substrate used in the SIGEX screening. This helps to increase scientific knowledge about the genetics of previously unknown and hypothetical genes [53]. Nevertheless, the application of SIGEX has certain limitations. First, SIGEX is sensitive to the structure and orientation of genes with desired traits. SIGEX misses catabolic genes that are expressed constitutively. In addition, SIGEX cannot detect any active clones in the direction opposite *gfp*. Moreover, it misses the active clones that have a transcription terminator between catabolic genes and the following *gfp*. Also, SIGEX is not suitable for applying to metagenomic libraries harboring large insert DNA due to the abundance of transcription terminators [54]. Therefore, when SIGEX is applied, these drawbacks should be considered carefully.

4. APPLICATION OF METAGENOMICS – CASE STUDIES

Applications of metagenomics are immense. The heart of this approach is that it has provided us with an access to the genome heterogeneity of both culture-dependent and culture-independent microorganisms. Metagenomics has already opened new avenues of research by enabling unprecedented analysis of genome and evolution in environmental contexts and providing access to far more microbial diversity. The early screening campaigns of metagenomic libraries centered on the cloning of genes encoding phylogenetically conserved molecular traits to explore microbial diversity. However, recent studies are mostly concentrating on retrieving novel bioactive compounds from metagenome. The potential for application of metagenomics to biotechnology seems endless. In fact, the promise of metagenomics as a source of new technology has yet to be fully realized, partly because of challenges in screening for desired activities.

4.1 Ecological Inferences

Metagenomic approaches have recently been made to studies on soil microbial ecology and their influence on field has been considered. These methodologies are now being widely adopted in laboratories around the world. Some spectacular achievements have been made in understanding the structure and dynamics of microbial communities of different ecosystems.

4.2 Biogeochemical Cycles

An exciting potential of metagenomics is that it provides a community-wide assessment of metabolic and biogeochemical functions (Table 5). Analysis of specific functions of individual members of a community can generate information about how organisms share the workload of maintaining the nutrient and energy budgets of the community. One such analysis is the nearly complete sequencing of the metagenome of a community in acid drainage of Richmond mine (California, USA), one of the most extreme environments on earth. The community is dominated by a few bacterial genera, *Leptospirillum, Sulfobacillus, Acidomicrobium* and archaeal species *Ferroplasma* and members of *Thermoplasmatales*. This interesting simple community structure made it possible for Tyson *et al.* to clone total DNA and sequence most of the community with high coverage [49]. This metagenomic sequence substantiated a number of significant

Table 5 *Biogeochemical Cycles*

Carbon	*Nitroge*	*Phosphorus*	*Sulphur*
Carbon is the fundamental element of the organic systems. It is required in photosynthesis, process by which plants convert the sun's light energy into chemcial energy for use in food webs. Carbon is in every part of every human.	Nitrogen is required for construction of amino acids and proteins without which organisms can not exist. Certain plants and bacteria fix nitrogen from the atmosphere and soil and convert it to a form useful for amino acid construction	Phosphorous is required in the conversion of chemical energy into kinetic energy and heat. Its concentrations in the environment affect the rate and growth of organisms. Excess phosphorous can accelerate growth in part of a food web misbalancing the whole food web.	Sulphur is normally present in large quantities in mineral deposits. Its involvement in life cycles is normally relatively small. It is important for use in protein construction. Human activity, releases significant amounts of sulphur, which make environment too acidic for life

hypothesis [23]. As such, the acid mine drainage serves as a model for the analysis of other communities.

The Sargasso sea analysis is another gigantic metagenomic project taken up [50]. Sargasso sea ecosystem is one of the complex ecosystems of earth. Venter group sequenced over 1 billion bp of metagenomic library and claimed the discovery of about 1.2 million new genes. Intriguing inferences are expected to be drawn because of the sheer size of the data set. They placed 7,94,061 genes in a conserved hypothetic protein group many of which are with unknown functions. The next most abundant group contained 69,718 genes apparently involved in energy transduction. Out of these 782 belong to rhodopsin-like photoreceptors. Thus, the vast data available on Sargasso Sea iş a gold mine for further analysis. It is expected to unravel the many intriguing settings of ecosystem function. The sequence data set from the Sargasso Sea provided the means to reassemble a number of genomes with criteria that include the depth of sequencing coverage, oligonucleotide frequencies and similarity to previously sequenced genomes.

4.3 Population Micro-heterogeneity

Metagenomic analysis has revealed several interesting genetic features of microbial communities *hitherto* otherwise unknown. For example, within the population of *Cenarchaeum symbiosum*, a bacterium associated with marine sponge, the rRNA genes were found to be highly conserved, showing > 99.2 percent homogeneity providing an indication that the population comprises a single species. However, the DNA sequences flanking the rRNA genes have exhibited only 87.8% identity [55]. Similarly, Tyson *et al.* reported the genome variation in the species occurring acid mine drainage [49]. Thus, these studies emphasize the importance of investigations on a mixture of strains to obtain a true picture of heterogeneity within the species.

4.4 Functional Diversity and Relations in a Community

Functional diversity is an aspect of the overall microbial diversity in a community and encompasses a range of activities. The relationship between microbial diversity and function in most of the communities is largely unknown, but functional diversity has been assumed to influence ecosystem stability, productivity and resilience towards stress and disturbance. The most intriguing aspect of microbial ecology is that how do the microorganisms establish different types of relations with eukaryotes, compete and communicate with other microorganisms and acquire nutrients and produce energy [23]. Metagenomics has provided insights into each of these areas. Some functional phenomena revealed by metagenomic are given below.

4.4.1 Symbiosis

Many of the bacterial symbionts, because of their obligate dependence on their hosts live in specialized niches, very often in pure or highly enriched cultures. These are ideal candidates for metagenomic analysis. Studies of Morans group on *Buchnera aphidicola*, an endosymbiont of aphids, revealed that this species contains a reduced genome of 564 open reading frames and upon comparison with a reconstructed ancestral genome, about 1900 genes appear to have lost [56]. Interestingly, many of the lost gene functions are contributed by the host that suggests the genome shrinkage is the result of symbiotic life style. Thus elucidation of *B. aphidicola* genome provided insights into the evolution of the symbiosis between the insect and bacterium, the biochemical mutual dependence that they have developed, and the mechanisms of genome shrinkage and rearrangement.

Proteobacterium-tube worm symbiosis is an interesting association in many respects. The bacterium is a member of γ-proteobacteria, as identified by 16S rRNA gene analysis. Hughes *et al.* isolated DNA from the bacterial symbiont and constructed fosmid libraries with an objective to understand the physiology of the bacterium [57]. Subsequently, Robinson *et al.* identified a gene with similarity to Rubisco from the same fosmid library, which has shown highest similarity with the Rubisco from *Rhodospirillum rubrum* [58]. Based on these observations, it was hypothesized that *Rhodospirillum pachyptila*, a tentatively identified bacterium is the actual symbiont of the tube-worm that fixes the carbon to its host. Another interesting conclusion was drawn based on the identification of gene encoding flagellin. The presence of genes for flagella suggested that endosymbiont has a free-living stage in its life cycle and may infect each generation of tube worms rather than being passed maternally [59].

4.4.2 Competition and Communication

Understanding of microbial interactions such as competition, antagonism and mutualism is central to advancing the definition of principles that govern microbial community structure function. Genes for competition and cooperation are difficult to identify based on sequence alone because the expression of their functions is entirely dependent on ecosystem context. Therefore, genomics by itself does not provide a means to test ecological hypothesis, but it can provide the basis for forming hypothesis. Functional genomics coupled with chemical ecology may yield necessary information.

Chemical ecology involves the identification of small molecules with biological activity and proposed ecological function. These compounds can be identified through a variety of methods, including metagenomics.

4.4.3 Phylogeny

Archaea represent a considerable fraction of the prokaryotic world in marine and terrestrial ecosystems. Many novel archaeal lineages that have been detected through molecular phylogenetic approaches have remained elusive because a majority of them are not available in culture. Metagenomic analyses have provided clues about the potential metabolic strategies of several of the uncultivated and abundant archaeal species including marine Crenarchaeota and Methanotrophic Euryarchaeota. These initial studies on archaeal populations also revealed a high degree of genomic variation that suggests the existence of considerable heterogeneity among archaeal strains [60]. Schmeisser *et al.* in a significant study have analyzed the complex metagenome of biofilms which represents a highly diverse microbial community growing on rubber-coated valves within drinking water networks [61]. This analysis revealed significant insights into phylogenetic, catabolic and metabolic abilities of the microbial community. Together with the random snap shot sequences, the authors identified more than 100 kb of contiguous environmental DNA sequences.

4.5 Biotechnological Potentials

Most natural products of economic importance such as enzymes, antibiotics or other pharmaceuticals are derived from cultured soil microorganisms, but the rate of discovery of novel natural products derived from isolated microorganisms has significantly decreased during the last few years [62]. This is mainly due to a high rediscovery frequency, as the same organisms are recovered repeatedly by using standard cultivation techniques. Culture-independent microorganisms, which constitute about 99 percent of the microbial diversity still remains to be untapped reservoir for exploitation of useful novel bioactive compounds. There is an ample demand for novel enzymes, antibiotics and biocatalysts and metagenomics is currently thought to be one of the most likely technologies to provide candidate molecules required [24, 63]. For this reason, metagenomes have attracted the attention of academia and specialized pharmaceutical companies. Metagenomics together with *in vitro* evolution and high-throughput screening technologies, provides the industry with an unprecedented chance to bring biomolecules into industrial application.

Since its introduction, metagenomics has identified a significant number of novel genes encoding for biocatalysts or molecules with high potential for use in pharmaceutical products or production processes. The list of published biocatalysts and bioactive compounds discovered by screening of complex soil derived DNA libraries is still rather small, but grows rapidly.

4.5.1 Enzymes

Screening of large metagenomic libraries has fortuitously revealed numerous open reading frames (ORF), many of them encoding enzymes [66]. Enzyme-focused screens have revealed chitinase, 4-hydroxybutyrate dehydrogenase, lipase, esterase, protease, oxygenase, amylase, DNase, xylanase and polyketide synthase [63]. Most expression screens so far detected activity in simple plate or liquid assays, whereas Daniel and coworkers have performed an elegant gene capture experiment complementing growth-deficient knockout mutants on selective media with metagenomic DNA [28, 45]. Richardson *et al.* described a high performance amylase for starch liquefaction that was generated by combining environmental discovery and laboratory evolution for the identification and optimization of the enzyme [65]. A similar approach was reported by Knietsch *et al.* [51].

4.5.2 Antibiotic Resistance Genes

Antibiotic resistance is a concern for the management of diseases in humans, animals and plants. The intensive research efforts to elucidate mechanisms of resistance have focused on genes derived from a narrow range of environments. Identification of antibiotic resistance genes from uncultured soil microorganisms is believed to contribute to the search for new antibiotics [66, 67]. Riesenfeld *et al.* constructed four libraries that collectively contained 4.1 gigabases of cloned soil DNA [32]. From these and two other previously reported libraries, they identified nine clones expressing resistance to aminoglycoside antibiotics and one expressing tetracycline resistance. With one exception, all the sequences were found to be considerably different from previously reported sequences. The results of these studies emphasized the fact that the vast uncultured community of soil bacteria is a reservoir of antibiotic resistance genes.

4.5.3 Antibiotics

Small molecules discovery by functional metagenomics has concentrated on antibiotics, which are of interest for their pharmaceutical applications as well as their role in ecosystem function. Traditional antibiotic screens for molecules with antimicrobial property have led to the discovery of new antibiotics in metagenomic libraries (Fig. 3). Antibiotic producing clones were detected at a frequency of approximately one producer per 10^4 clones [68, 69]. Terragine, a *Mycobacterium* – inhibiting antibiotic, was discovered from a soil metagenomic clone maintained in *Streptomyces lividans* [47]. Similarly, acyltyrosines were discovered in a clone maintained in *E. coli* [70]. Turbomycin, one of the first antibiotics discovered by metagenomics, was identified by accident in a clone that was showing hemolytic activity [68]. Osburne's group identified structurally related compounds, indirubin and indigo blue in a soil metagenomic library [71]. The same group screened a 5000-member metagenomic library for conserved regions of genes encoding typing I polyketide synthase. The screens yielded 11 new polyketide synthase homologues that contained significant sequence similarity to polyketide synthase genes from cultured of organisms. In addition, these screens revealed two novel compounds, fatty dienic alcohol isomers. Violacein, a purple colored antibiotic, was earlier recorded to be produced by *Chromobacterium violaceum*. However, sequence of metagenomic violacein biosynthetic operon suggested that the clone was derived from an organism other than *C. violaceum*.

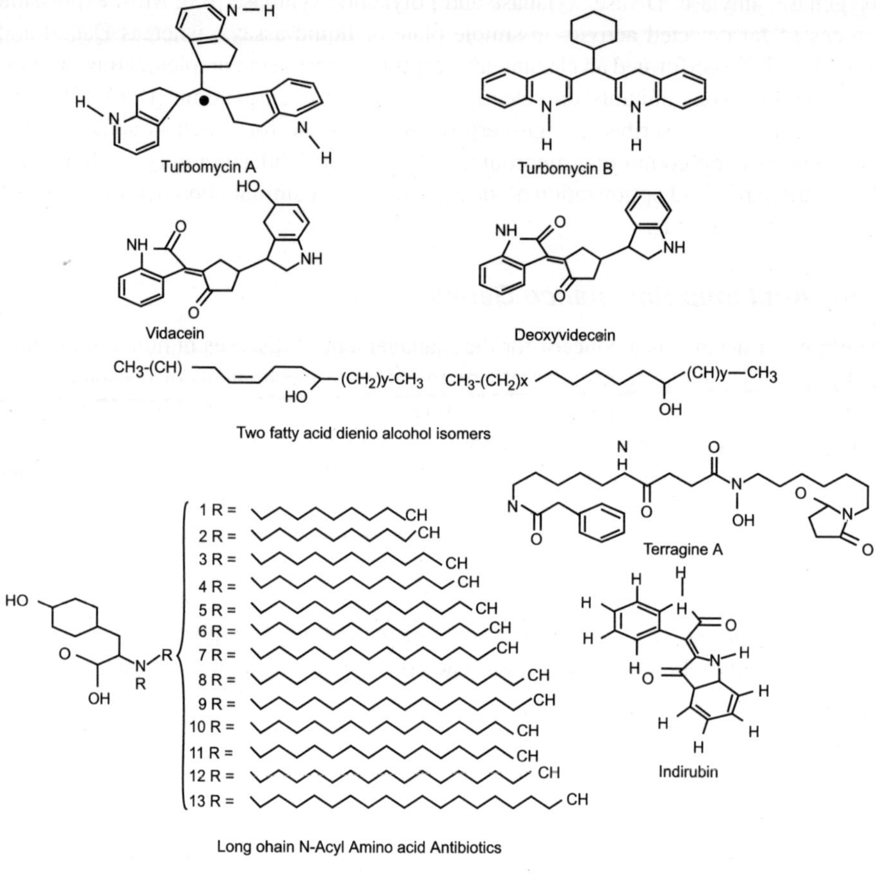

Fig. 3 *Antibiotics discovered in metagenomic libraries*

4.5.4 Anticancer Drugs

The fact that discovery rate of new structural classes of antineoplastic agents from cultured organisms has significantly decreased and the simultaneous mounting evidence that the vast majority of soil microbes are unknown, has heralded a new approach to natural product drug discovery. Uncultured soil species represent a tremendous untapped resource of antineoplastic agents [72]. Actinomycetes are looked upon as potential microorganisms for biologically active microbial products including anticancer drugs. Recently, some bacteria like myxobacteria, chromobacteria, cyanobacteria are also reported to produce antineoplastic agents some of which are in clinical trials [73]. The FDA-approved anticancer drug, pentostatin, is produced by both *Streptomyces* and *Aspergillus*. Clearly, the message is that a range of organisms appear to be the potential sources for the production of antineoplastic agents. Advances in metagenomic methods are likely to bring into light many novel anineoplastic agents from diverse microbes.

5. PERSPECTIVES AND CONCLUSIONS

The bulk of research on biodiversity is being devoted to plant and animal communities. In the 1960s, following the footsteps of plant and animal ecologists, microbiologists began investigating the impact of biodiversity on the function and structure of microbial communities [74]. The potential of heterogeneity and variability of microorganisms is probably larger than for any other group of organisms. However, our knowledge of microbial diversity is limited partly because of our inability to culture a great majority of the organisms. Problems associated with studying microbial diversity arise not only from methodological limitations, but also from a lack of taxonomic knowledge [75].

The diversity of soil microorganisms has been exploited for many years based on the cultivation and isolation of microbial species. Despite of novel cultivation techniques, we could access less than one percent of richness of microbial community. Culture-based probing of biodiversity has inherent limitations. To circumvent some of the problems associated with culturing microorganisms, cultivation-independent molecular approaches, such as DNA reassociation, DNA-DNA and mRNA: DNA hybridization, DNA cloning and sequencing and PCR methods such as denaturing gradient gel electrophoresis (DGGE), temperature gradient gel electrophoresis (TGGE), ribosomal intergenic spacer analysis (RISA) and automated ribosomal intergenic spacer analysis (ARISA) have been made [76]. Although molecular methods have the advantage of obtaining information about non-cultivable organisms, they also have limitations that cannot be ignored. These methods may throw a light on structural diversity but it is not possible to obtain information on functional diversity which is essential for understanding the ecosystem functioning. The best way to study microbial diversity would be to use a variety of tests with different end points and degrees of resolution to obtain the broadest picture possible and the most information regarding the microbial community.

The advent of metagenomics has changed the entire scenario. It has opened up new avenues of research by enabling unprecedented analysis of genome heterogeneity and evolution in environmental contexts and providing access to far more microbial diversity than has been viewed in the petridish. This novel approach, wherein larger fragments of eDNA are the object of study, has received denominations like environmental genomics, ecogenomics. The idea of the genomes of all microbes present in a specific habitat as an entity carries a certain conceptual load. Besides, establishing phylogenetic relationships and throwing a light on ecosystem functioning, the vast potential of the thousands of genomes of microorganisms present in environments such as soils, waters or sediments can be exploited for biotechnology. The obvious potential of the metagenomic library for bioactive compounds has been realized by companies around the world. Different industries are showing keen interest in exploiting the resource of uncultivated organisms that has been identified through large-scale environment genomics. In fact, they are pumping a lot of funds for these projects. Diversa, the largest and most prominent specialist biotech company (San Diego, California, USA) and TerraGen Discovery (Cambridge, Massachusetts, USA) are the first companies to file patents on specific aspects of metagenomic technology in 1996.

Diversa claimed methodologies to isolate and normalize environmental DNA prior to cloning and screening and described a resulting library created from marine picoplankton. TerraGen's claims were focused on the retrieval of environmental DNA sequences from soil to express industrially relevant xylanases [77]. Since then, several small to medium sized enterprises have used uncultivated environmental microbial resources to retrieve relevant enzymes and biocatalysts.

As often the case with many technological revolutions, metagenomics also suffers from many initial problems. These problems, primarily pertain to (i) isolation and purification of metagenomic DNA, (ii) construction of libraries and (iii) screening of libraries. To arrive at reliable information on ecophysiology and biodiversity, unbiased sampling is an important first step in metagenomic studies. It has been realized that with today's technology it would be difficult to obtain complete coverage of all of the genomes in a soil community. Approaches that enrich for a portion of the microbial community or for collection of metagenomic clones will enhance the power of metagenomic analysis to address targeted questions in microbial ecology and to discover new biotechnological applications [24]. However, arguments against enrichments are the loss of microbial diversity and the favoring of the fast-growing and cultivable part of the microbial community. Daniel suggested that as the enrichment steps yield more-biased metagenomic soil libraries, coupling of classical enrichment and modern metagenomic technologies is an efficient approach for the isolation of a number of novel genes conferring the targeted reaction from soil samples [64].

The presence of a cluster of genes in an environment is much more relevant ecologically than the presence of one specific gene and provides much more information about the organism carrying it. Similarly, the genes encoding the biosynthesis of bacterial secondary metabolites are also tightly linked. Size of these gene clusters range from 20 to > 100 kb. Obviously, improved methods in DNA extraction and cloning vectors and surrogate hosts are required to retrieve the contiguous gene clusters. No standard protocols exist for isolating sufficiently purified metagenomic DNA from environmental samples. Improved or alternative protocols for DNA extraction are to be looked into. Heterologous expression system of genes from metagenome requires further improvement. An efficient system other than *E. coli* should be developed. Innovation will be required to identify and overcome the barriers to heterologous gene expression and to detect rare clones efficiently in the immense libraries. Construction of gene libraries in eukaryotic hosts such as *Schizosaccharomyces pombe* which is capable of properly splicing and expressing genes should also be given a priority consideration to retrieve fungal genes. Construction of cDNA libraries would allow the retrieval of biocatalysts derived from eukaryotes.

The most cumbersome step in the metagenomic analysis is the handling of large number of clones and screening of complex libraries. For example, Henne *et al.* had to evaluate 2,86,000 of a small insert library to isolate three novel genes encoding esterases [42]. Therefore, simple and cheap activity based strategies such as the detection of clones exhibiting the targeted phenotype on indicator agar media or through complementation of mutants are needed to be employed. Conventional screening systems such as sequence driven and function driven methods are proving to be costly and time consuming despite the robotic automation. SIGEX, as an alternative, holds promise as a powerful tool for screening genes involved in bioactive molecules of industrial importance. However, the applicability of this technique has yet to be verified against a range of gene expressions. Similarly, the efficiency of searching novel catalysts from metagenome needs an improvement. Venter and colleagues have introduced a new approach to the field, the shotgun sequencing without the need to construct large-insert gene libraries [50]. They have applied this approach to analyze metagenome of marine microbial populations collected *en mass* from the Sargasso sea. In future this approach may become a model to explore the microbial diversity of many other complex ecosystems.

And of course in future, the management and analysis of enormous data generated through this technology is bound to pose a big challenge. Advances that facilitate the management and analysis of large libraries include bioinformatics tools to analyze vast sequence databases and reassemble multiple genomes rapidly and affordable gene chips for library profiling [78] or that readily distinguish clones that are expressing the genes from those clones that are silent.

By and large, metagenomics is all set for a big leap. Coming decades are going to witness a profound research activity in this area and hopefully generation of a vast amount of useful data sufficient enough for biologists and biotechnologists to be obsessive for another half century or so.

Epilogue

India, with its diverse habitats and ecosystems, is a treasure trove of biodiversity. Microbial diversity of Indian ocean, glaciers of Himalayas, desert sands, hot water springs of western Himalayas, sediments of lakes and eusterins, soils of tropical rain forests and many other habitats remained largely unexplored. As rightly pointed out, whatever the little information is available with scientists often working solitarily with almost no interaction [79]. A greater thrust is required to unravel the bacterial and archaeal diversity. At a right time 'Alexander von Humboldt Stiftung" has taken an initiative providing a platform to highlight the richness and applications of microbial diversity in India. Metagenomic technology, which is yet to be employed in the Indian context, has boundless potential for exploring and exploiting the microbial diversity. Needless to say such type of studies spanning over length and breadth of India are the urgent need of the hour, which hopefully unfold rich microbial diversity and also many of the microorganisms of biotechnological importance. The mythological analogy is the spontaneous generation of divine boon granting subjects from the churning of milky ocean. Clearly the message is we need foresight, planning, determination, perseverance and sustained efforts to achieve the results in the direction duly supported by public and private organizations.

Acknowledgments

Author Ajit Varma is thankful to Dr. Alok Srivastava for kindly inviting me to compile this chapter. The encouragement received from Dr. Ashok K Chauhan, Founder President, Amity University is appreciated. Partial financial assistance received from CSIR, New Delhi is acknowledged.

REFERENCES

1. Wall, D. H. and Virginia, R. A. (1999) Controls on soil biodiversity insights from extreme environments. *Appl. Soil Ecol.* **13,** 137-150.
2. Timonen, S., Finlay, R. D., Olsson, S. and Soderstrom, B. (1996) Dynamics of phosphorous translocation in intact ectomycorrhizal system: non-destructive monitoring using a B-scanner. *FEMS Microbiol. Ecol.* **19**, 171-180.
3. Smith, K. P. and Goodman, R. M. (1999) Host variation for interactions with beneficial plant-associated microbes. *Annu. Rev. Phytopathol.* **37**, 473-491.
4. Dodd, J. C., Boddington, C., Rodriguez, A., Gonzalez-Chavez, C. and Mansur, I. (2000) Mycelium of arbuscular mycorrhizal fungi (AMF) from different genera. *Form, Function and Detection, Plant and Soil* **226**, 131-151.
5. Yao, H., He, Z., Wilson, M. J. and Campbell, C. D. (2000) Microbial biomass and community structure in a sequence of soils with increasing fertility and changing land use. *Microb. Ecol.* **40**, 223-237.
6. O'Donnell, A. G., Seasman, M., Macrae, A., Waite, I. and Davies, J. T. (2001) Plant and fertilizers as drivers of change in microbial community structure and function in soils. *Plant Soil* **232**, 135-145.
7. Rodriguez-Valera, F. (2004) Environmental genomics, the big picture? *FEMS Microbiol. Lett.* **231,** 153-158.
8. Rosello-Mora, R. and Amann, R. (2001) The species concept for prokaryotes. *FEMS Microbiol. Rev.* **25**, 39-67.
9. Torsvik, V., Salte, K., Soerheim, R. and Goksoeyr, J. (1990) Comparison of phenotypic diversity and DNA heterogeneity in a population of soil bacteria. *Appl. Environ. Microbiol.* **56**, 776-781.
10. Torsvik, V., Goksoyr, J. and Daae, F. L. (1990) High diversity in DNA of soil bacteria. *Appl. Environ. Microbiol.* **56**, 782-787.
11. Hugenholtz, P., Goebel, B. M. and Pace, N. R. (1998) Impact of culture – Independent studies on the emerging phylogenetic view of bacterial diversity. *J. Bacteriol.* **180,** 4765-4774.
12. Pace, N. R. (1997) A molecular view of microbial diversity and the biosphere. *Science* **276**, 734-740.
13. Rappe,. M. S. and Giovannoni, S. J. (2003) The uncultured microbial majority. *Annu. Rev. Microbiol.* **57**, 369-394.
14. Kellenberger, E. (2001) Exploring the unknown: the silent revolution of microbiology. *EMBO Rep.* **2**, 5-7.
15. Joseph, S. J., Hugenholtz, P., Sangwan, P., Osborne, C. A. and Janssen, P. H. (2003) Laboratory cultivation of widespread and previously uncultured soil bacteria. *Appl. Environ. Microbiol.* **69**, 7210-7215.
16. Wery, N., Gerike, U., Sharman, A., Chaudhuri, J. B., Hough, D. W. and Danson, M. J. (2003) Use of a packed-column bioreactor for isolation of diverse proteases-producing bacteria from antarctic soil. *Appl. Environ. Microbiol.* **69,** 1457-1464.
17. Janssen, P.H., Yates, P.S., Grinton, B.E., Taylor, P.M. and Sait, M., Improved culturability of soil bacteria and isolation in pure culture of novel members of the divisions acidobacteria, actinobacteria, proteobacteria and verrucomicrobia. *Appl. Environ. Microbiol.*, 2002, **68**:2391-2396.
18. Kaeberlein, T., Lweis, K. and Epstein, S. S. (2002) Isolating 'uncultivable' microorganisms in pure culture in a simulated natural environment. *Science* **296**, 1127-1129.

19. Zengler, K., Toledo G., Rappe, M., Elkins, J., Mathur, E. J., Short, J. M. and Keller, M. (2002) Cultivating the uncultured. *Proc. Natl. Acad. Sci. USA* **99**, 15681-15686.

20. de la Torre, J. R., Christianson, L. M., Beja, O., Suzuki, M. T., Karl, D. M., Heidelberg, J. and Delong, E. F. (2003) Proteorhodospin genes are distributed among divergent marine bacterial taxa. *Proc. Natl. Acad. Sci. USA* **100**, 12830-12835.

21. Handelsmann, J., Rondon, M. R., Brady, S. F., Clardy, J. and Goodman, R. M. (1998) Molecular biological access to the chemistry of unknown soil microbes: a new frontier for natural products. *Chem. Biol.* **5**, R245.

22. Randon, M. R., August, P. R., Bettermann, A. D., Brady, S. F., Gross-man, T. H., Liles, M. R., Loiacono, K. A., Lynch, B. A., Mac Neil, I. A., Minor, C., Tiong, C. L., Gilman, M., Osburne, M. S., Clardy, J., Handelsman, J. and Goodman, R. M. (2000) Cloning the soil metagenome : a strategy for accessing the genetic and functional diversity of uncultured microorganisms. *Appl. Environ. Microbiol.* **66**, 2541-2547.

23. Handelsman, J. (2004) Metagenomics: Application of genomics of uncultured microorganisms. *Micro. Molec. .Biol. Rev.* **68,** 559-685.

24. Schloss, P. D. and Handelsman, J. (2003) Biotechnology prospects from metageonomics. *Curr. Opin. Biotechnol.* **14**, 303-310.

25. Healy, F. G., Ray, R. M., Aldrich, H. C., Wilkie, A. C., Ingram, L. O. and Shanmugam, K. T. (1995) Direct isolation of functional genes encoding cellulases from the microbial consortia in a thermophilic, anaerobic digester maintained on lignocellulose. *Appl. Microbiol. Biotechnol.* **43**, 667-674.

26. Stein, J. L., Marsh, T. L., Wu, K. Y., Shizuya, H. and De Long, E. F., Characterization of uncultivated prokaryotes: isolation and analysis of a 40 kilobase-pair genome fragment from a planktonic marine archaeon. *J. Bacteriol.* **178**, 591-599.

27. Handelsman, J., Randon, M. R., Brady, S. F., Clardy, J. and Goodman, R. M. (1998) Molecular biological access to the chemistry of unknown soil microbes: a new frontier for natural products. *Chem. Biol.* **5**, 245 249.

28. Henne, A., Daniel, R., Schmitz, R. A. and Gottschalk, G. (1999) Construction of environmental DNA libraries in *Escherichia coli* and screening for the presence of genes conferring utilization of 4-hydroxybutyrate. *Appl. Environ. Microbiol.* **65**, 3901-3907.

29. Krsek, M. and Wellington, E. M. H. (1999) Comparison of different methods for the isolation and purification of total community DNA from soil. *J. Microbiol. Methods* **39,** 1-16.

30. Berry, A. E., Chiocchini, C., Selby, T., Sosio, M. and Wellington, E. M. H. (2003) Isolation of high molecular weight DNA from soil for cloning into BAC vectors. *FEMS Microbiol Lett.* **223**, 15-20.

31. Courtois, S., Cappellano, C. M., Ball, M., Francou, F. X., Normand, P., Helynck, G., Martinez, A., Kolvek, S. J., Hopke, J., Osburne, M. S., August, P. R., Nalin, R., Guérineau, M., Jeannin, P., Simonet, P. and Pernodet, J. L. (2003) Recombinant environmental libraries provide access to microbial diversity for drug discovery from natural products. *Appl. Environ. Microbiol.* **69**, 49-55.

32. Riesenfeld, C. S., Godman, R. M. and Handelsman, J. (2004) Uncultured soil bacteria area a reservoir of new antibiotic resistance genes. *Environ. Microbiol.* **69,** 981-989.

33. Steffan, R. J., Goksøyr, J., Bej, A. K. and Atlas, R. M. (1998) Recovery of DNA from soils and sediments. *Appl. Environ. Microbiol.* **54**, 2908-2915.

34. Brady, S. F., Chao, C. J., Handelsman, J. and Clardy, J. (2001) Cloning and heterologous expression of a natural product biosynthetic gene cluster from eDNA. *Org. Lett.* **3**, 1981-1984.

35. Quaiser, A., Ochsenreiter, T., Klenk, H. P., Kletzin, A., Treusch A. H., Meurer, G., Eck, J., Sensen, C. W. and Schleper, C. (2002) First insight into the genome of an uncultivated crenarchaeote from soil. *Environ. Microbiol.* **4**, 603-611.

36. Shizuya, H., Birren, B., Kim, U. -J., Mancino, V., Slepak, T., Tachin, Y. and Simon M. (1992) Cloning and stable maintenance of 300-kilobase-pair fragments of human DNA in *Escherichia coli* using an F-factor-based vector. *Proc. Natl. Acad. Sci. USA* **89**, 8794-8797

37. Randon, M. R., Raffel, S. J., Goodman, R. M. and Handelsman, J. (1999) Toward functional genomics in bacteria: analysis of gene expression in *Escherichia coli* from a bacterial artificial chromosome library of *Bacillus cereus*. *Proc. Natl. Acad. Sci. USA* **96,** 6451-6455

38. Sheng, Y., Mancino, V. and Birren, B. (1995) Transformation of *Eschericia coli* with larger DNA molecules by electroporation. *Nucleic acids Res.* **23,** 1990-1996.

39. Sosio, M., Giusino, F. A., Cappellano, C., Bossi, E., Puglia, A. M. and Donadio, S. (2000) Artificial chromosomes for antibiotic-producing actinomycetes. *Nat. Biotechnol.* **18,** 343-345.

40. Handelsmann, J., Liles, M., Mann, D., Riesenfeld, C. and Goodman, R. M. (2002) Cloning the metagenome: culture-independent access to the diversity and functions of the uncultivated microbial world. *Methods Microbiol.* **33**, 241-255.

41. Uchiyama, T., Abe, T., Ikemura, T. and Watanabe, K. (2005) Substrate induced gene expression screening of environmental metagenome libraries for isolation of catabolic genes. *Nat. Biotechnol.* **23**, 88-93.

42. Henne, A., Schmitz, R. A., Bomeke, M., Gottschalk, G. and Daniel, R., Screening of environmental DNA libraries for the presence of genes conferring lipolytic activity on *Escherichia coli*. *Appl. Environ. Microbiol.* **66**, 3113-3116.

43. Handelsman, J. (2005) Sorting out metagenomes. *Nat. Biotechnol.* **23**, 38-39.

44. Entcheva, P., Liebl, W., Johann, A., Hartsch, T. and Streit, W. R. (2001) Direct cloning from environment cultures, a reliable strategy for isolation of complete operons and genes from microbial consortia. *Appl. Environ. Microbiol.* **67,** 89-99.

45. Majernik, A., Gottschalk, G. and Daniel, R. (2001) Screening of environmental DNA libraries for the presence of genes conferring Na^+ (Li^+) / H^+ antiporter activity on *Escherichia coli* : Characterization of the recovered genes and the corresponding gene products. *J. Bacteriol.* **183**, 6645-6653.

46. Diaz-Torres, M. L., McNab, R., Spratt, D. A., Villedieu, A., Hunt, N., Wilson, M. and Mullany, P. (2003) Novel tetra cycline resistance determinant from the orgal metagenome. *Antimicrob Agents Chemother*. **47**, 1430-1432.

47. Wang, G. Y., Graziani, E., Waters, B., Pan, W., Li, X., McDermott, J., Meurer, G., Saxena, G., Anderson, R. J. and Davies, J. (2000) Novel natural products from soil DNA libraries in a streptomycetes host. *Org. Lett.* **2,** 2401-2404.

48. Beja, O., Aravind, L., Kohin, E. V., Suzuki, M. T., Hadd, A., Nguyen, L. P., Jovanovich, S. B., Gates, C. M., Feldman, R. A., Spudich, J. L., Soudich, E. N. and De Long, E. F. (2000) Bacterial rhodopsin : evidence for a new type of phototrophy in the sea. *Science* **289**, 1902-1906.

49. Tyson, G. W., Chapman, J., Hugenholtz, P., Allen, E. E., Ram, R. J., Richardson, P. M., Solovyev, V. V., Rubin, E. M., Rokhsar, D. S. and Banfield, J. F. (2004) Community structure and metabolism through reconstruction of microbial genomes from the environment. *Nature* **428**, 37-43.

50. Venter, J. C., Remington, K., Heidelberg, J. F., Halpern, A. L., Rusch, D., Eisen, J. A., Wu, D., Paulsen, I., Nelson, K. E., Nelson, W., Fouts, D. E., Levy, S., Knap, A. H., Lomas, M. W., Nealson, K., White, O., Peterson, J., Hoffman, J., Parsons, R., Baden-Tillson, H.,

Pfannkoch, C., Rogers, Y. –H. and Smith, H. O. (2004) Environmental genome shotgun sequences of the Sargasso sea. *Science* **304,** 66-74.

51. Knietsch, A., Waschkowitz, T., Bowien, S., Henne, A. and Daniel, R. (2003) Metagenomes of complex microbial consortia derived from different soils as sources for novel genes conferring formation of carbonyls from short-chain polyols on *Escherichia coli. J. Mol. Microbiol. Biotechnol.* **5**, 46-56.
52. Liles, M. R., Manske, B. F., Bintrim, S. B., Handelsmann, J. and Goodman, R. M. (2003) A census of rRNA genes and linked genomic sequences within a soil metagenomic library. *Appl. Environ. Microbiol.* **69**, 2384-2691.
53. Yun, J. and Ryu, S. (2005) Screening for novel enzymes from metagenome and SIGEX, as a way to improve it. *Microb. Cell Factories* **4**, 8-13.
54. Ermolaeva, M. D., Khalak, H. G., White, O., Smith, H. O. and Salzberg, S. L. (2000) Prediction of transcription terminators in bacterial genomes. *J. Mol. Biol.* **301,** 27-33.
55. Schleper, C., Delong, E. F., Preston, C. M., Feldman, R. A., Wu, K. Y. and Swanson, R. V. (1998) Genomic analysis reveals chromosomal variation in natural populations of the uncultured psychrophilic archaeon *Cenarchaeum symbiosum. J. Bacteriol.* **180**, 5003-5009.
56. Abbot, P., Withgott, J. H. and Moran, N. A. (2001) Genetic conflict and conditional altruism in social aphid colonies. *Proc. Natl. Acad. Sci. USA* **98**, 12068-12071.
57. Hughes, D. S., Felbeck, H. and Stein, J. L. (1997) A histidine protein kinase homolog from the endosymbiont of the hydrothermal vent tubeworm *Riftia pachyptila. Appl. Environ. Microbiol.* **63**, 3494-3498.
58. Robinson, J. J., Stein, J. L. and Cavanaugh, C. M. (1998) Cloning and sequencing of a form II ribulose-1,5-biphosphate carboxylase / oxygenase from the bacterial symbiont of the hydrothermal vent tubeworm *Riftia pachyptila. J. Bacteriol.* **180**, 1596-1599.
59. Millikan, D. S., Felbeck, H. and Stein, J. L. (1999) Identification and characterization of a flagellin gene from the endosymbiont of the hydrothermal vent tubeworm *Riftia pachyptila. Appl. Environ. Microbiol.* **65**, 3129-3133.
60. Schleper, C., Jurgens, G. and Jonuscheit, M. (2005) Genomic studies of uncultivated archaea. *Nat. Microbiol.* **3**, 479-488.
61. Schmeisser, C., Stockigt, C., Raasch, C., Wingender, J., Timmis, K. N., Wenderoth, D. F., Flemming, H. C., Liesegang, H., Schmitz, R. A., Jaeger, K. E. and Streit, W. R. (2003) Metagenome survey of biofilms in drinking-water networks. *Appl. Environ. Microbiol.* **69**, 7298-7309.
62. Strohl, W. R. (2000) The role of natural products in modern drug discovery. *Drug Discov. Today* **5**, 39-41.
63. Loreng, P., Liebeton, K., Niehaus, F. and Eck, J. (2002) Screening for novel enzymes for biocatalytic process: accessing the metagenome as a source of novel functional sequence space. *Curr. Opin. Biotechnol.* **13**, 572-577.
64. Daniel, R. (2004) The soil metagenome – a rich resource for the discovery of novel natural products. *Curr. Opin. Biotechnol.* **15**,199-204.
65. Richardson, T. H., Tan, X., Frey, G., Callen, W., Cabell, M., Lam, D., Macomber, J., Short, J. M., Robertson, D. E. and Miller, C. (2002) A novel, high perfo-rmance enzyme for starch liquefaction. Discovery and optimization of a low pH, thermostable a-amylase. *J. Biol Chem.* **277**, 26501-26507.
66. Benveniste, R. and Davies, J. (1973) Aminoglycoside antibiotic-inactivating enzymes in actinomycetes similar to those present in clinical isolates of antibiotic-resistance bacteria. *Proc. Natl. Acad. Sci. USA* **70**, 2276-2280.

67. Anderson, A. S., Clarck, D. J., Gibbons, P. H. and Sigmund, J. M. (2002) The detection of diverse aminoglycoside phosphotransferases within natural populations of actinomycetes. *J. Ind. Microbiol. Biotechnol.* **29**, 60-69.

68. Brady, S. F., Chao, C. J. and Clardy, J. (2002) New natural product families from an environmental DNA (eDNA) gene cluster. *J. Am. Chem. Soc.* **124**, 9968-9969.

69. Gillespie, D. E., Brady, S. F., Bettermann, A. D., Cianciotto, N. P., Liles, M. R., Rondon, M. R., Clardy, J., Goodman, R. M. and Handelsman, J. (2002) Isolation of antibiotics turbomycin A and B from a metagenomic library of soil microbial DNA. *Appl. Environ. Microbiol.* **68**, 4301-4306.

70. Brady, S. F. and Clardy, J. (2000) Long-chain N-acyl amino acid antibiotics isolated from heterologously expressed environmental DNA. *J. Am. Chem. Soc.* **122**, 12903-12904.

71. MacNeil, I. A., Tiong, C. L., August, P. R., Grossman, T. H., Loiacono, K. A. Lynch, B. A., Phillips, T., Narula, S., Sundaramoorthi, R., Tyler, A., Aldredge, T., Long, H., Gilman, M., Holt, D. and Osburne, M. S. (2001) Expression and isolation of antimicrobial small molecules from soil DNA libraries. *J. Mol. Microbiol. Biotechnol.* **3**, 301-308.

72. Pettit, R. K. (2004) Soil DNA libraries for anticancer drug discovery. *Cancer Chemother Pharmacol.* **54**, 1-6.

73. Cragg, G. M. and Newman, D. J. (2000) Antineoplastic agents from natural sources: achievements and future directions. *Expert Opin. Investig. Drugs* **9**, 2783-2797.

74. Swift, M. J. (1974) Species diversity and the structure of microbial communities in terrestrial habitats. *In* Anderson, J. M. and McFayden, A. (eds.), The role of terrestrial and aquatic organisms in decomposition processes, Blackwell Scientific Publications, Oxford, United Kingdom, pp. 185-221.

75. Overeas, L., Jensen, S., Daae, F. L. and Torsvik, V. (1998) Microbial community changes in a perturbed agricultural soil investigated by molecular and physiological approaches. *Appl. Environ. Microbiol.* **64**, 2739-2742.

76. Kirk, J. L., Beaudette, L. A., Hart, M., Moutoglis, P., Klironomos, J. N., Lee, H. and Trevors, J. T. (2004) Methods of studying soil microbial diversity. *J. Microbiol Methods* **58**, 169-188.

77. Radomski, C. C. A., Seow, K. T., Warren, R. A. J. and Yap, W. H. (1998) Method for isolating xylanase gene sequences from soil DNA. Compositions useful in such methods and compositions obtained thereby. *US Patent* US 5, 849, 491.

78. Sebat, J. L., Colwell, F. S. and Crawford, R. L. (2003) Metagenomic profiling: Microarray analysis of an environmental genomic library. *Appl. Environ. Microbiol.* **69**, 4927-2934

79. Johri, B. N., Ganguly, B. N., Goel, S. K., Virdi, J. S., Tripathi, A. K., Jain, R. K., Kamra, D. N. and Ashish Bhatnagar (2005) Microorganism diversity: Strategy and action plan. *Curr. Sci.* **89,** 151-154.

19

ETHICS IN PUBLIC DOMAIN: BIOMEDICAL RESEARCH AND BEYOND

Giridhari Lal Pandit

ABSTRACT

The article covers the issue of ethics in public domain and the challenge that lies in protecting life and human dignity in all the frontiers of research and its applications. Taking the case of biomedical/biogenetic/biotechnological research as one of the frontiers, the paper develops an argument for full participation in the debate by all the stakeholders, among them the professional philosophers and legal luminaries who can impart not just rigor bur teeth to ethics in public domain by coming out with universally binding regulations and enforcement mechanisms. In the absence of such regulations, mankind will be groping in the dark while asking how to cope with the situation where the world is divided against itself even on commonly shared vital issues of moral progress. All the steps taken until now at the UN and at the national legislatures fall short not only of full participation by all the stakeholders but of legally and ethically binding regulations, failing to recognize the complexities and the implications of the challenge. Ethics committees have, no doubt, cropped up world-wide in some fields. The fear of proliferation of unregulated research and its applications leading to undesirable, even unintended consequences, on the one hand, and the paradox of moral progress being excluded by scientific and technological progress, on the other hand, haunt the public policy makers everywhere. Unfortunately, ethics committees have had only a limited role and success (in and outside their institutional boundaries) marked by fragmented approaches to guidelines being formulated in individual countries or regions of the world. In the opinion of the author, the real breakthrough that has yet to come will have to pass the test of moral progress at the frontier defending life and human dignity.

1. THE CRISES OF KNOWLEDGE-PRODUCTION WITHOUT WISDOM

My presentation may be broadly viewed from the perspective of *ethics in public domain* which I like to describe as the frontier of all frontiers of research and its applications. Fierce debates on issues like biotechnology, stem cell research and cloning have shown that the rapid scientific and technological progress can give rise to serious ethical questions. These concerns must be addressed by individuals, policy makers and the scientific community alike (i) to take into account public fears and (ii) to avoid alienating citizens and creating an environment hostile to scientific and technological innovations. Moreover, in the light of efforts to develop in a sustainable way, implications of new technologies for future generations must be thoroughly considered. Thus, the present and future tasks at the greatest frontier before humanity have to do with issues of economic, scientific, technological and moral progress [1, 2]. *The past and present institutional failure universally in almost all fields gives rise to the dilemmas of promoting*: *knowledge* with or without wisdom; *scientific progress* with or without moral progress; *economic and technological progress* with or without proliferating forms of exclusion; *economic prosperity and political freedom* with or without structural violence; *biomedical research* with or without informed access to new technologies and mechanisms; *education* with or without literacy; *democracy* with or without participation; *human rights* with or without access to them; *development* with or without environmental damage and climate change; and *Earth as a family* with or without the human species *Homo sapiens* and so on.

Faced with the challenging task of protecting life, human dignity and values, *ethics in public domain* invites our attention, *first of all*, to the paradox of moral progress being excluded by scientific, technological and economic progress every where in the world. Indeed it invites our attention to the universality of this paradox. One of the basic issues, here, is how far the interdisciplinary courses of studies at the universities worldwide could help in fighting exclusive images of knowledge and value which still dominate our lives and activities. How far could these help in overcoming the menace of forms of exclusion proliferating everywhere and in all fields? *Secondly*, whether we are willing to pursue a model of education and a style of living where besides intelligence, professional success, personal attainments, we also learn to value *wisdom*? In a nutshell, I am referring here to the value of bringing moral progress to bear upon all other forms of human progress so that (i) the universality of the crises of *knowledge without wisdom* can be avoided and (ii) forms of exclusion do not proliferate with progress in knowledge-production, with scientific, technological and economic development. In my view, these questions bring us face to face with future generations as *stakeholders*. Thus, they bring us face to face with the task of interrogating our present policies on education, on societal and state responsibility for the upkeep of health, on quality of life, on environment, on science, technology and development. If we are guided by multiple images of knowledge and value, we can prepare all the *stakeholders* for a future which is open. Since the future is unpredictable, critical reflection on such challenging questions and open problems remains our best guide to possible directions for seeking improvements. And, in any discourse on educational policies at various levels, philosophy and ethics in public domain must assume a fundamental role.

2. STAKEHOLDERS IN MORAL PROGRESS

I will begin by asking certain fundamental questions in order to indicate why, and to what extent, mankind depends upon moral progress for regulating the investments which the state, the individual, the institution, the civil society and corporate industry, each as a *stakeholder*, makes where progress in the production and the uses of scientific knowledge is concerned. First, can moral progress in public domain co-evolve with public policy on investment, on planning, on development, on the beneficial uses of the fruits of scientific progress? In fulfilling the tasks of reaping the fruits of the tree of knowledge in diverse fields, when can we say that we are making moral progress which is commensurate with human dignity and human freedom? Since the United Nations adopted the Universal Declaration of Human Rights (UDHR) in 1948, their enforcement and protection as instruments for protecting human agencies and human beings against cruelty, oppression, abuse and degradation has assumed unquestionable priority and importance among the members of the UNO, in their conduct of international relations and in their treatment of their citizens and in their public policy on development. Yet, their universal recognition, their successes, their prospects and the failures of their universal enforcement tell a story not only of moral progress but of moral failure of mankind at the most difficult frontier of improving the human condition of individual dignity and individual freedom. In terms of the UDHR alone, we can claim, with Michael Ignatieff [3], that the moral intuition that "our species is one, and each of the individuals who compose it is entitled to equal moral consideration" and that "Human rights is the language that systematically embodies this intuition, and to the degree that this intuition gains influence over the conduct of individuals and states, we can say that we are making moral progress". But in the absence of an enforcement mechanism which is commensurate with the UDHR, an ever widening gap between the word and the deed tells how difficult it is to guard all that which is at stake at the frontier of human rights.

In scientific journals and in conferences devoted to discussions on human rights, bioethics, medical ethics and environmental ethics we are not even surprised to find how frequently references are made (i) to ethical dilemmas, but none to the question of how we might make and judge moral progress, given scientific and technological advance; (ii) to contexts of participation, but none to the environment for participation; and (iii) to rules for participation, but none to the social indicators of participation. I think that this kind of situation is not fundamentally different from those talk-shows in which we find people speaking of (a) education as if it was about everything else except education; (b) UDHR as if it was about everything else except human rights and access to them; and (c) science as if it was about everything else except the question of its being in charge of its own progress and its many uses. As to education, let me quote John Raven (1994) who argues "that the areas in which research and innovation are *most* badly needed in our society do *not* have to do with finding better ways of producing goods of one kind or another but with finding better ways of running society itself" [3]. Building up his detailed arguments in his book on how to manage education for effective schooling, he concludes: "In essence, what we have seen in this book is that the failure of the educational system to achieve its main goals is multiply determined and that that failure contributes to the perpetuation of a society which, at least in the short term, requires its educational system to perform functions which are in sharp tension with the educational activities it manifestly needs to undertake" [5].

More specifically, take the example of schools failing to teach poor children to read within their 1st two or three years of schooling. Even when they complete the schooling, they are found functionally illiterate. Thus, at a recent *World Bank*

Conference on Aid Effectiveness for Human and Social Development, participants showed conclusively that completing school is a very practical and often irrelevant achievement. The most likely explanation for this is institutional failure, given children's diverse backgrounds at the school entry point. While better-off children, whose reading habits begin to develop already at home, enter school with a vocabulary of 2000 to 4000 words, poor children entering school, without any such background, have barely a vocabulary of only 600. This raises the question whether schools are meant to churn out students who are without any improvement in learning outcomes on the completion of school. What do the schools then invest on, be it the developed or developing countries? Why does it take a few right thinking and questioning men and women to ask paradoxical questions such as these: Whether schools are not meant for teachers instead of children? Whether hospitals are not meant for doctors instead of patients? Whether Parliaments are not meant for their members instead of their constituencies where people with votes belong? Why does our world have to churn out institutions which breed forms of exclusion and collective crime? Today, there is an urgency to seriously address the questions of moral progress and forms of exclusion. This task assumes increasing importance in our highly globalized world which has made scientific, technological and economic progress but which is torn by structural violence, terrorism, human rights violations, poverty, hunger, refugee problems and so on. So long as moral progress not only lags behind but is even excluded by scientific, technological and economic progress, the two kinds of progress will never be synonymous. As a result, everywhere in society there will be forms of exclusion and structural violence. In many dark areas, the future generations too will have to pay the price on our behalf. This is particularly true in case of environmental damage caused by the present and past human activities. The question is why moral progress of mankind lags far behind scientific and technological progress, always showing a negative correlation.

What I find equally paradoxical is that most of the troubles in our modern world lie in those areas of experience where people already feel strongly connected. Take for example religious fundamentalism which is responsible for terrorism and massive structural violence in many parts of the world. The only explanation for strong connections on which it thrives could be this. Like (the political discourse on) education, or human rights awareness campaigns, or science and technology, fundamentalism too is about everything else except the cause for which the fundamentalists and their known and unknown sponsors seem to fight. Is this kind of situation not really highly paradoxical? Is it not quite reasonable, therefore, to propose the following general systems theoretical strategy? *Participate and reconnect, or participate and rediscover the connections*. This alone should allow us to discover more and more participants, more and more stakeholders, to be empowered and more and more connections than we have thought there are to the situation which we so proudly celebrate as our achievement, or as an object of scientific study. If used as an epistemological strategy, it should allow us to test improvements in the human condition by testing the correlations between the conditions of human dignity, human freedom and human knowledge. Thus, the task here is to engage in those research strategies and developmental models with which we can expect to bring about improvements in the human condition. But at this very level, the most challenging question still remains with us: When can we say that we are on our way to moral recovery and moral progress or on our way to controlling technological progress before we are controlled by it? Is there not a rationality of connecting the conditions of human dignity, human freedom, human knowledge and moral progress, which we can no longer ignore? We can answer this question in the affirmative, because without such rationality it would not be possible to improve even our causal knowledge of a whole diversity of forms of exclusion in society.

3. WHY ETHICS IN PUBLIC DOMAIN

Negligence and breach of duty of care have become a matter of everyday experience for patients being admitted in hospitals for treatment of disease. There are serious *issues* of medical practice relating to standard and quality of health care, informed access to medical care, ethics consultation service, complementary and alternative medicine, medical reform, informed consent, medical malpractice, assisted reproductive technologies, reproductive health technologies. Take the case of biomedical/biogenetic/biotechnological research as one of the *frontiers*. One of the serious issues is how to promote full participation, in the debate on fundamental issues of policy and implementation, by all the *stakeholders*, among them the professional philosophers and legal luminaries who can impart not just rigor bur teeth to ethics in public domain by coming out with universally binding regulations and enforcement mechanisms. In absence of such regulations, mankind will be groping in the dark while asking how to cope with the situation where the world is divided against itself even on commonly shared vital issues of moral progress. All the steps taken until now at the UN and at the national legislatures worldwide fall short not only of full participation by all the *stakeholders* but of legally and ethically binding regulations, failing to recognize the complexities and the implications of the challenge. There is no doubt that ethics committees have cropped up worldwide in some fields. But they have done so under the fear of proliferation of unregulated research and its applications leading to undesirable, even unintended consequences, on the one hand, and under the paradox of moral progress being excluded by scientific and technological progress, on the other hand. In practice, they have had only a limited role and success (in and outside their institutional boundaries) marked by fragmented approaches to guidelines being formulated in individual countries or regions of the world. The real breakthrough that has yet to come will have to pass the test of moral progress at the frontier defending life, human dignity and values.

Why ethics research committees have cropped up worldwide? How far can they help humanity in facing the present as well as future challenges in science, in technology, in biomedical research and other fields? There is no simple answer. One answer is that it has become imperative to bring issues facing mankind in all fields of knowledge production and its application in full public view for proper and thorough debate, before letting them go before the national legislatures, or parliaments, and UN bodies for policy decisions, particularly where universal legal and ethical regulations are concerned. This is all the more necessary in the context of our aiming at a common sustainable future.

The most important yet neglected aspect of the problem is that with every breakthrough in science and technology, particularly in biomedical research, diseases are invented with the same ease with which their diagnoses, tests and treatments are advertised for doing business with the rich. The invention of **stem-cell/ umbilical cord banking** serves as a good example. These are the banks where rich couples can deposit stem-cell rich cord blood that can be stored and used in the future in case their baby is faced with any serious medical disorder or health problem. If the tests (prenatal and postnatal) reveal that their child is healthy, the cord blood can be donated for scientific research. The charges for registration include enrolment fee and annual storage charges up to two decades or more. Umblical cord banking is still at a very nascent stage in India. What is true of this example is also true of assisted reproductive technologies.

Leaving out for the moment the specific issues, let me raise those issues which are general and common to several fields at the same time, for example the issues of "informed consent" in varied research contexts. Pre-implant-testing of the artificially-inseminated embryos (before they are implanted in a woman's body) where couples feel concerned about potential life-threatening genetic or chromosomal defects/diseases in their offspring, stem-cell banking, human genomics, assisted reproductive technologies and reproductive health technologies in biomedical research, even organ procurement and intensive care for children and surgical patients, *all* raise serious issues not only of "informed consent" but of *informed access*. There are also common issues arising from scientific and technological progress, or those connected with freedom of scientific research. For the moment I shall keep them aside. As the medical technology grows at an incredibly fast speed, the treatment options keep increasing too, though not necessarily in an egalitarian way with equal access to all. In this context, it must be remembered that "informed consent", an ethically loaded concept, gets inseparably linked with *informed access*, entailing the intervention by ethics consultation service. Without ethics consultation service, "informed consent" loses its meaning and significance in research. Yet most countries in the world have not been able either to institutionalize ethics consultation service at all or to establish a network of such services as a first step to a well-integrated ethics consultation service.

Again, a very serious issue regarding the very access to new technologies such as assisted reproductive technologies or reproductive health technologies arises in the present context. What is it to have, or nor to have, *access to new technologies* (**ANT**), be it in medicine or outside medicine? Does **ANT** imply just the ability to pay for them? If yes, then new technologies surely breed new forms of rich-poor divide, since evidently not all who are in need of it can pay for them. The moment we ask who can *afford* to benefit from new technology, the old *rich-poor divide* repeats itself. However, the issue of **ANT** goes much deeper than this. **ANT**, where rational, entails knowledge-based access, allowing one to exercise a dignified rational choice. That is, at a deeper level, "informed consent" is crucial to **ANT**, not just one's economic status. While choosing a new technology, one should *know* not only the consequences of such a choice but also the reasons *why* one chooses to have the technology at all.

Thus ethically viewed, **ANT** is an issue fundamentally involving decision-making, given sufficient information or knowledge, on the basis of which one may decide to have or not to have the new technology. **ANT** in this sense gets closely reconnected to "informed consent". The potential beneficiaries must be able to take the right decision whether or not to go for it. Alas! presently, this is universally not the case. This easily results in *medical malpractice*, in manipulation and exploitation. The worst example is the South Korean scandal in which the disgraced scientist unethically procured the eggs for stem-cell research from his own employees.

Universally, what is worse is this: Often, it turns out that the right information about new technologies we are here concerned with is missing just because there is no investment on research on **ANT**, although there is no dearth of investment on producing and selling the new technology to the potential beneficiaries. This raises the question: *Why is it necessary for our world to produce inaccessible technologies*? Is it a step in progress or in regress to do so repeatedly? *The perennial paradox in which we all are involved here is the paradox of producing knowledge and technology without wisdom or moral progress.* Until the present human condition of **ANT** changes, the new technologies will continue to take over, manipulate and exploit human beings at the cost of human dignity and human life. This is particularly true of new technologies connected with bio-medical research.

How does the scenario for ethics in public domain look like in India? In late 1990s, the Indian Council of Medical Research (ICMR) set up a Central Ethics Committee on Human Research (CECHR) with the aim of updating its Ethical Committee's 1980 "Policy Statement on Ethical Considerations Involved in Research on Human Subjects". In its September 1996 meeting, the CECHR set up 5 sub-committees identifying major areas of concern, covering (1) Clinical evaluation of drugs/devices/diagnosis/vaccines/herbal remedies; (2) Epidemiological research; (3) Human genetics research; (4) Transplantation research, including foetal tissue transplantation; (5) Assisted reproductive technologies. In August 1997, the CECHR met to deliberate on their draft guidelines and a Draft Consultative Document. It is difficult to comment how public and interdisciplinary was the public debate intended to be generated region-wise and across the scientific institutions in the country before the final drafting of the guidelines by a drafting committee set up for the purpose. It is difficult to say how many universities, or disciplines, how many professional philosophers of repute, or how many scientists genuinely interested in ethics in public domain were encouraged to participate. Absence of a professional philosopher on all the sub-committees of the CECHR, even on the Sub-Committee for General Principles, is revealing for a people with a civilization rich in moral philosophy going thousands of years back in history. In sharp contrast to this, in Europe, particularly in Germany, ordinary pathways and streets are named after individual scientists and philosophers, celebrating the life of the mind as torch-bearer at the frontier of human life. However, it took CECHR four years to document "Ethical Guidelines for Biomedical Research on Human Subjects" (**EGBRHS**), a publication which ICMR brought out in 2000. The *Foreword* written by the Chairman of the CECHR says: "Biomedical Research has acquired dimensions which are at once exciting and awesome. It raises some delicate and difficult issues of ethics which need to be dealt with sensitivity to human values and with great circumspection. While research which promises to mankind the great blessings of Science should not be stifled by too restrictive an approach, however, great care should be taken to ensure that something does not go out of hand. Therefore, any system of ethical guidelines on research needs to be cognizant of, and informed by, sensitive balance of the risks and benefits". In the last statement which sounds rather overly utilitarian in restricting itself to the decisive balance of the risks and benefits, it is not possible to miss its sole emphasis on ethical guidelines, raising serious doubts about **EGBRHS** (2000) as an effective instrument of protection of life and dignity. Protection of life and dignity goes far beyond utilitarian rationality. If this is recognized, it entails our ability to distinguish between morally and legally binding regulations on research from mere non-binding guidelines. When we read through the pages of **EGBRHS** (2000), we merely find statements laying down guidelines to be followed, not even any mention of the need to monitor how they are to be followed.

The debate on a total ban on reproductive and therapeutic cloning divides the world in a manner which makes it clear what is at stake if no consensus on ethical and legal regulations on research is arrived at. A culture of life and its protection is, as some have argued, at stake. At the UN the debate over a ban on reproductive cloning began in 2001with France and Germany in the lead, with near universal national support. But governments led by the United States and Costa Rica urged the world to back a total ban, including a ban on therapeutic cloning. In all its forms, including the therapeutic embryonic stem cell research in which the human embryos cloned for purpose of obtaining the cells used in the studies are later discarded, human cloning should be banned. The *stakeholders* in this issue seem to be having too many divergent considerations, coming in the way of a universal agreement. On the 18th February 2005, a deeply divided UN General Assembly Committee adopted a nonbinding statement calling on governments to prohibit all forms of human cloning, including techniques used in research on human stem cells, abandoning efforts for stronger action because their divisions were too deep. From the beginning, the debate hinged on whether to outlaw all cloning or permit cloning for research. Nations that sought a total ban always had more votes, but never enough to achieve broad consensus or a binding worldwide treaty. In November 2005, the debate met with failure in arriving at a worldwide treaty to that effect, despite the efforts of the UN legal Committee. The only option left was to work toward the nonbinding political statement, hoping to achieve broad consensus. However, the hope still remains alive at the UN in so far as the present debate is seen as a step toward a universal declaration in the future, provided the national legislatures around the globe work for essential inputs. The declaration adopted in November 2005 also demands that Member states "adopt the measures necessary to prohibit the application of genetic engineering techniques that may be contrary to human dignity" and to take measures to prevent the exploitation of women in the application of life sciences. The debate on embryonic stem cell research will continue. Stem cells are nascent cells which can develop into replacement cells that researchers believe could help treat damaged organs and illnesses. The embryonic stem cell research aims at designing such cells to treat conditions considered incurable by traditional medical means, such as hearing disorders, nervous system failures, diabetes, Parkinson´s disease and glaucoma. Recently news came in that some scientists have developed alternatives like developing stem cells from umbilical cord blood taken from newly born babies, sparing embryonic stem cells. If this should work, it will not only make the morally objectionable embryonic stem cell research avoidable but give the message that it is possible for scientists to learn from ethics in public domain while working harder on safer alternatives for applying the fruits of scientific progress for improving the human condition.

I am here concerned with the question why moral progress of mankind is excluded by scientific and technological progress, always showing a negative correlation. At the global level, state of the art of embryonic stem cell research, of assisted reproductive technologies and reproductive health technologies provide good examples. There are other areas of research and technological development where questions of moral progress arise in so far as investment in areas of development of new and faster technologies is incommensurate with (near absence of) investment on enhanced access not only to new but even to old technologies. As a result, while new technologies reach quickly the privileged few, large masses of people continue to be deprived of even the older and outdated technologies, repeating over and over again the world's oldest divide between the rich and the poor. What is worse is how incommensurate investment on development is with (near absence of) investment on research on the social and other consequences of new technologies. Should scientific, industrial and technological development imitate fashion designing industry in order to reach the chosen few or should it aim at moral progress by improving the human condition of individual dignity and individual freedom on Earth and minimizing structural violence. If a project investing in development, change or reform breeds forms of exclusion, doing good to some by degrading others, it must be given up in favour of an alternative project which enhances the chances for improving the human condition on Earth. I think that this clearly shows us how mankind may make moral progress in the future while engaging in projects of scientific, technological, industrial and economic development. Let me quote Werner Heisenberg (1989:494), warning us about the new dangers of scientific and technological progress as a challenge calling for our response. Concerned as he was with the ambivalence of science, Werner Heisenberg proposed, as a guiding principle, "to consider any special progress in science or technology as a part of the whole, as something that cannot be separated from the general problems of our way of life, our environment, our political behaviour. This obligation - to keep in mind the unavoidable connection or interplay between *all* actions - will set and should set limits to our blind confidence in science and technology, but not necessarily to science and technology itself" [6]. For detailed, discussion, see references [1, 7]. Similarly, Richard Rhodes [8] reminds us of the deeper truth how "Violence and scarcity are connected. At the end of the 20th century, when the world does not lack resources, scarcity is itself a form of violence – structural violence, as it is termed, meaning violence that results from the way power and wealth are distributed. The fact that African-Americans in the United States have an average 4 years' shorter lifespan than whites quantifies one evident manifestation of structural violence. Hunger, malnutrition, extreme poverty, and preventable disease throughout the world are all manifestations of structural violence" [8]. If we don't stop at some point, we can go on adding challenge after challenge which tells how difficult some human beings make it for others to fight at the frontier, defending human dignity and human freedom. Think of the arrogance of some of us who want to conquer nature where the most urgent task of science and technology is to remove our prejudices, to civilize us and to teach us by example how we can learn from our mistakes. Think of corruption in public life. Even the media, "the free press", are not free from it. Particularly in India, the human condition of coastal and rural life as actually lived by poor masses never comes to public attention. Even in times of disaster, whether "natural" such as earthquakes, cyclones or tsunamis or "man-made" such as the uprootment of whole ethnic communities caused by killings and threatenings by organized terrorists, all that we receive are short glimpses only. Why is this so?

4. FREEDOM OF SCIENTIFIC RESEARCH

It is not invariably clear in all kinds of contexts of scientific and technological progress, more so in those of biomedical research, whether the scientist's freedom of scientific research, which civil society more than the state values so much, should automatically extend to their all possible and desirable applications that manifestly aim at the improvement of the human condition. It is as if the freedom of scientific knowledge production was neither synonymous with nor sufficient for the freedom of deciding which *uses* of knowledge to avoid and which to foster for society's, even future generation's, long-term interests. Let us admit that the two kinds of freedom are *asymmetrical* with each other. No limits may be set to the scientist's freedom to search for knowledge, or to fundamental research. But the same does not hold true in the case of society's search for improvement in the human condition, making use of the fruits of the tree of scientific knowledge. The problem is that scientists and policy makers often do not heed this asymmetry in all its moral and social complexity that involves many stakeholders. *Ethics in public domain strikes at the root of this asymmetry*. It allows us to pause and reflect on our own achievements as soon as these are clouded by our arrogance, uncritically extending freedom of scientific research to its most ambitious and wildest applications. The most spectacular example is provided by advances being made by genetics which raises the prospect of future enhancement and control of human characteristics, even extending enhancement to mind-enhancing drugs, bodybuilding and cosmetic surgery. Biomedical research generally *and* human genetics, human reproductive technologies, human embryonic stem cell research and human enhancement technologies particularly, pose serious problems of consensus and principled co-operation in the domain of a public policy dispute. The scenario is one in which many stakeholders, representing diverse kinds of interests and epistemic communities, rooted in different moral traditions and values, are sometimes found genuinely interested in overcoming the tensions of a fragmented approach to resolving complex ethical issues which the technologically and market driven scientific research and experimentation pose. But is consensus on values in the public domain attainable? Within the fast developing technology-driven scientific practice itself, how far can the diverse interests of researchers, scientists, human test-subjects, the potential beneficiaries, technocrats, corporations and popular media be integrated with the culturally fostered public ethical concerns? Some of these ethical concerns find expression in our respect for human life, for the embryo, for traditional ways of human procreation, for a properly regulated human subjects research, for human rights of the human subjects participating in research, for their informed consent, for human dignity and human welfare without commercialization, and for improvement of the human condition without commodification. It is no surprise, if the biomedical researchers often find themselves pulled by the forces of (i) technological progress opening up new possibilities of conquering disease and new horizons of controlling human procreation, even enhancing human characteristics, and (ii) moral progress making it imperative to bring more and more of the complexities of scientific practice under principled ethical scrutiny and debate.

In this article, I am arguing that the old images and distinctions regarding scientific knowledge production must be allowed to dissolve, if ethics in public domain is to empower different stakeholders on an equal basis. There may be something to learn from the example of the highly co-operative and highly instrumentalised extended scientific community in fundamental physics and astronomy. Here too scientific knowledge production involves many stakeholders instead of knowledge simply flowing from the university to industry, as was demanded by the older images. If the methodology of inquiry in the natural and social sciences recognizes *co-operation* at many levels as a necessary condition of knowledge production, we must not forget its role as an important *strategy of inquiry* in this context. Strategies can change over time depending on the complexities of the problem-situation at hand. However, in the social sciences and bio-medical research, strategies of inquiry, of debate and publicity in the media, assume a special significance because the complexities of research cannot be separated from the social responsibility of the stakeholders. Nor can they be separated from the individual interests and rights of those who are being projected as the potential beneficiaries of the new technologies. Thus, they demand situation-to-situation variations in the rules of method. They even demand variations in the priorities of research, in the allocation of funds, in the ethical regulations for obtaining informed consent from the human subjects of research, in the ethical regulations for guaranteeing to all the stakeholders increased access to technologies meant to benefit human beings and in regulating research so that it reflects public values and not just the exclusive interests of researchers and corporations who stand to profit by research. If the methodology lends scientific inquiry its stability, strategy allows it to cope with the increasing complexities of the changing problem-situations and their wider social contexts. While strategies may keep changing with the changes in the researchers' cognitive and practical ends, methodological rules - at least some of them – need not do so over different epochs. However, within one and the same methodology, different, even incompatible strategies may be resorted to if only to realise the cognitive and practical ends of inquiry. If understood as a *strategy of inquiry* in the field of biomedical research, *co-operation* so *internalized* can play a key role in arriving at consensus and in resolving the complex ethical issues and dilemmas in the domain of public policy where there is always a plurality of stakeholders [1, 9-13].

5. EXPERIMENTING WITH HUMAN SUBJECTS: HOW INFORMED CAN INFORMED CONSENT BE?

The single page Consent Form for Normal/Control Samples prescribed at most Centres for biochemical technology under the CSIR within the **EGBRHS** (ICMR 2000, pp. 87-88) provides for a statement of declaration (either signed or thumb impressioned) by the Normal Volunteer/Patient, giving his/her consent freely to participate in the (genetic) study aimed at understanding the role of indoor fungi in Asthma/ various disease processes/ and the like, depending on the nature of the experimental project being undertaken. The signatory then adds "that it has been explained to me that all information on medical history and the analysis of my blood samples will be stored anonymously and will be used as Normal (Volunteer)/Patient samples for molecular research of fungal allergy...". In the same Consent Form the investigator signs, certifying "that the above consent form has been signed in my presence. The purpose for which the sample will be used has been explained to the above named Normal (Volunteer)/ Patient. The individual is free to withdraw as and when he/she feels so inclined. All the results of the investigation will be made available to the participants".

The **EGBRHS** (ICMR 2000, pp. 87-88) are generally rooted in a fairly broad consensus regarding the importance of informed consent, given the correlative expert verbal explanations of the risks involved in experiments on human subjects. In many areas of biomedical/biotechnological/biogenetic research involving human subjects, such as human embryonic research and research in assisted reproductive technology, the ethical guidelines require obtaining informed consent from donors and beneficiaries alike. Generally, each area of research on human subjects demands additional requirements for informed consent, besides the general requirements documented under **EGBRHS**. **EGBRHS** lay down, as a guideline for assisted reproductive technologies, that "After duly counselling the couple/oocyte/semen donor, an informed and written consent should be taken from both the spouses as well as the donor, as the case may be. They should be explained the various risk factors associated with the procedures in simple language and the words that they can understand. These include risks associated with an ovarian hyperstimulation, anaesthetic procedures, and invasive procedures like laparoscopy, aspiration of ovum etc. They should also be explained the possibility of multiple pregnancies, ectopic gestation, increased rate of spontaneous abortion, premature births, higher prenatal and infant mortality as well as growth and developmental problems. They should also be explained that there is no guarantee on the success/failure of the procedure". The human subjects-research ethical guidelines such as these, which in turn determine the format of biomedical and biotechnological research proposals in various areas, raise serious ethical and legal questions as follows. There are serious problems here in correlating the verbal explanations to be given to the declarations of informed consent to be taken. Notice that the kind of standard these set for the explanation of the risks associated with the procedures involved in assisted reproductive technologies only underlines the fact that the procedures in question are best understood as *experiments on human subjects*. The question arises, *first of all*, how explanatory can such explanations be for the experimental human subject. *Secondly*, on whom should really the onus of the declaration that, in a given case, the informed consent is a properly informed consent rest? Should it rest solely on the donors/patients/subjects themselves, who are required to sign the declarations? In many cases of biomedical research, as the subject of experiments and trials, the human subject is used as the source of samples over a period of time. The same human subject is also used as the source of the information that the informed consent is a properly informed consent. *In the context of scientific inquiry, what is the difference between the two procedures of (i) extracting the samples for the tests/experiments and (ii) extracting the whole truth of the informed consent from the same human subject?* For *fulfilling* the presupposition that the verbal explanation given to the experimental human subject is properly correlated with the informed consent taken from her/him, the question arises what should be the form and content of those explanations without which no consent can become informed consent. Based on verbal explanations only, when does consent, in writing, become informed consent? When does explanation, even if it is given in writing, allow it to become informed consent, justifying the declarations made by the patient/human subject? Normally, it has been found that the pattern of verbal explanation remains invariably the same from experiment to experiment, heedless of the background variations among subjects, such as their language, education, literacy, culture, life-style, family life, psychology and the like. Given the asymmetries of information between the scientist-researcher and the experimental human subject (which are similar to the more familiar scenario of doctor-patient asymmetries in all medical practice), ethical dilemmas are bound to proliferate in all biomedical and biotechnological research experimenting with human subjects. Hence the question: How informed can informed consent be qualitatively? When and on what basis can a human subject decide to withdraw from participation in the experiment? Whether she/he should do so just on the basis of verbal explanations which are given at the beginning of research/experiment? Or whether she/he should do so only on the basis of explanations furnished in writing? Or whether she/he should do so only after discovering aspects of research which were being concealed from the outset?

It can be argued that informed consent in the context of experiments on human subjects is a subject far more complex than is generally assumed. Therefore, no simplistic guidelines are going to suffice. Above all, it is an ethically loaded concept. But it is this aspect which has gone unrecognized even in those institutions where ethics committees are in place. I think that declarations of informed consent should have a mandatory provision for relevant explanations in writing. Such declarations should have at least three components. One of these should be signed by the human experimental subjects themselves and the other two by the scientist/researcher and the Ethics Committee who are empowered with the responsibility to clear and monitor the research proposals. But the problem is that at present such committees are generally appointed only to issue clearance of research proposals on the basis of guidelines that are legally and morally non-binding, lacking universal validity, and formally over-simple or ritualistic. In the guidelines, there is no provision requiring the research to be monitored after the clearance has been given. It can be argued that it is imperative to empower such committees with regulatory, adjudicatory and monitoring authority, going beyond their present bureaucratic advisory role. Moreover, it is imperative at all the levels of decision-making to be properly fed with relevant information based on prior research on what kind of knowledge is helpful to all the stakeholders when it comes to actual decision-making in using a new technology. This raises the most fundamental issue of access to new technologies.

5.1 Health Care Reform, Reproductive Health Technologies and Assisted Reproductive Technologies

First, what are assisted reproductive technologies? Assisted reproductive technologies are those technologies which enable the childless couples, given surrogate mothers, to have children, who if left to themselves would never have a baby either because of infertility or because of treatment for diseases like cancer which could leave a patient infertile or vulnerable even after her full recovery from the disease. Let us take an example. Jennifer Rutansky, a young woman of 26 years, from Jacksonville, Florida, had to go for chemotherapy following surgery and radiotherapy to treat her cancer which was diagnosed in August 1997. Seven years later, having recovered, she got married. But she was warned not to get pregnant. The couple found a surrogate mother. Her own frozen eggs and her husband's sperms were used to produce embryos through the *in vitro fertilization* techniques. The embryos were then transferred into the surrogate mother's womb. The first two experiments, carried out at the Florida

Institute for Reproductive Medicine, failed. The third experiment was successful. On October 3, 2005, their biological son was born, thanks to fertility preservation techniques and assisted reproductive technologies. Assisted reproductive technologies go beyond reproductive health technologies that offer reproductive health resources. Reproductive health technologies include drugs, devices, and medical interventions to control reproduction and/or prevent sexually transmitted infection, such as contraceptives and products used to enhance fertility. There has been a race for investing in technologies that are needed to manage the physiological and emotional transitions associated with menopause. But there is hardly any investment in research on how to educate the potential users and practitioners about the uses of these technologies and proper decision-making. The basic question about any such technology is how accessible it is to those who might be most needy or vulnerable to risk. For example, it is not commensurate with the rights of couples to make ethically and epistemologically responsible decisions regarding the use of the reproductive health technologies merely on the basis of advice about the clinically-proven safety and efficacy of such technologies [13]. A proper investment in research on reproductive health decision-making, sexual practices, and norms and values for childbearing and family formation is needed. In the words of Woodsong and Severy, "The concept of sexual health further affirms the human rights of all individuals to access sexual and reproductive health care services and information related to sexuality, including information that can aid in deciding if, when, and how to have children" [14]. Serious debate and research is needed on answering the question what kind of knowledge is necessary for an individual to influence her use of such technologies or to influence her reproductive health decision-making. There is no doubt about one thing: Whatever information and advice are made accessible to the potential users and practitioners of reproductive health services, these should be more accurate, comprehensive and comprehensible than they are actually. In a nutshell, investment in reproductive health technologies should be accompanied by investment in social and behavioural research on the potential users and practitioners of these technologies [14].

Assisted reproductive technologies, on the other hand, are defined as non-coital methods of conception to overcome infertility problems that involve manipulation of eggs and sperms, the most popular among them being *in vitro fertilization*. Used in 70% of all assisted reproductive technology procedures, *in vitro fertilization* uses drugs to stimulate egg production in a woman. The ripened eggs from the ovary are retrieved, in the laboratory, and fertilized with semen. The resulting embryos are then transferred back into the uterus for implantation. Assisted reproductive technologies are not only highly expensive and therefore not accessible to the ordinary poor people but also time consuming. Moreover, they involve multiple injections of drugs. What is worse, they have a modest success rate, with less than 25% of cycles involving fresh, non-donor eggs resulting in a live birth. Besides this, there are reproductive health technologies that include drugs, devices, and medical interventions to control reproduction and/or prevent sexually transmitted infections. The **EGBRHS** (ICMR 2000, p. 86) defines "assisted reproduction" as manipulation of the gametes outside the body and transfer of gametes or embryos into the body. Advances in medicine and biomedical research raise important questions which are rarely asked. This is true of the current medical claims to `cure´

infertility in couples who are desirous of having a child by accessing the newly available techniques of medically assisted procreation. One of the fundamental questions concerns its inevitable consequence of framing disease, where there may be none, by treating infertility as if it was a disease. It has been rightly argued by some critics that this amounts to confusing "*the survival of the species with that of the individual*" [15]. Infertility is framed as a disease in so far as it is in its name that every attempt is made to publicly defend the medically assisted management and manipulation of women's bodies for procreation. Again, contrary to their claims, one might ask whether doctors treat sterility at all. The truth is that with artificial insemination, they manipulate male gametes. And with in vitro fertilization and embryo transfers, they produce embryos. Thus, we can look at the medically assisted procreation as a medical takeover and control of the reproduction of the human species. As a worst case scenario, women, who are treated as the suppliers of oocytes, are completely de-humanized (reminding us of the recent scandal in South Korea). Think of the medical control over fertilization. In vitro fertilization makes it possible to control the genetic inheritance of the embryo. This control extends further to the manipulation of the transmission of the genes of the species at the end of the pregnancy.

We can look at assisted reproductive technologies as the medically assisted human procreation by the medical centres which make experimental use of women's bodies without women themselves having any control over the modalities of assisted reproductive technology. As experimental subjects, women are here caught between their desire for motherhood, along with their willingness to readily 'consent´ to their participation in the new practices in which two things are happening simultaneously: The assisted reproductive technology based laboratory experiments extend to their bodies; *and* in this very process they lose their individual rights, their freedom and dignity. This shows the urgency with which the legal and ethical framework of 'informed consent´ in the context of assisted reproductive technologies, even reproductive health technologies, need to be probed even more deeply and critically than elsewhere in biomedical research.

5.2 Biomedical Research: A Question of Access to New Technologies

In all experimentation using human embryos or foetuses what is at stake is public morality. Hence it is imperative not only to raise ethical questions concerning such research which reflect a diversity of views that must be taken into consideration but to seek consensus through public debate with a view to legislating and, through proper legislation, regulating research wherever it is to be allowed. *It is imperative to arrive at regulations with a legal and moral force instead of mere guidelines which are non-binding and, therefore, not only a waste of time but self-deceit.* In the United Kingdom, it was in 1978 that the world's first baby was born using the technique of *in vitro* fertilization. Until then there was no public debate in sight on the subject of experimentation on fertilization of human eggs in the laboratory simply because most people were ignorant about what was happening in the laboratories. In particular, what attracted attention was (i) the new technique of using couples desirous of having a child as experimental subjects for research, *even extended research* demanded by the new technique trying to understand the very process of fertilization, the structure of the human egg and implantation, while promising to offer them remedial therapy; (ii) the success rate and failure rate of the technique: "removing eggs from a woman's ovary and preparing them for fertilization outside the body, mixing them with male sperm, growing them in a dish for two or three days, in a specially prepared fluid, and finally replacing them in the uterus"; (iii) the morally most repulsive part of the technique which uses live human embryos for research and then destroys them; (iv) the urgent need for proper legislation, which is based on co-operative consensus on questions of public morality, if such research is to be allowed; and (v) the need to study those factors which restrict women's access to assisted reproductive technologies with a view to understanding what can make them acceptable to them. Availability of assisted reproductive technologies *alone* is not going to matter in women's decision-making whether to use or not to use it. Cultural beliefs and cultural values play an important role. There is also the question whether increased access and choice means enhanced quality of life or enhanced health care, particularly when the partners involved have differences of belief regarding the methods and their impact on the quality of their sexual life or intimacy. Recent studies have suggested that cultural, social and economic barriers to women's access to assisted reproductive technologies vary in different parts of the world [12].

Issues of public morality are not decidable by resort to arguments which are based on a calculus of benefits or utilities, although it is true that business economics, or corporate interests, exercise great influence on public policy, even preventing proper regulation against unethical practices in research and in business. Otherwise public morality would look like morality in private space and lead to trivializations highlighted by the dilemma: either no actions need moral justification in terms of universal moral laws or all actions are justifiable in terms of moral considerations by individuals in private space.

6. ENFORCEMENT MECHANISMS: WHAT SHOULD ETHICS COMMITTEES BE AND DO?

The role of ethics committees in the field of bio-medical research is well-recognized universally, both at the national and international levels. But this is not the case when we turn our attention to the policies, strategies and solutions where problems of development-planning, sustainable development and management of energy needs in the midst of environmental change or protection of environment are concerned. If it makes sense to think in terms of guidelines for sustainable-development and environment-related problem-posing and problem-solving, then it also makes sense to widen the scope of environmental discourse by evolving uniform policies with regard to the structure and role of ethics committees in all relevant fields, be it environment-related research and engineering, be it sustainable development-planning, or be it new technologies needed to reduce the risks of environmental/climate change. In the changing relationship between science and its various publics, where criteria other than pure scientific merit have gained acceptance as a basis for the funding of basic research, there is the real but most challenging frontier confronted with the following questions: What is the future of human and ethical values? Who should determine the development of scientific knowledge and its applications? For whom should science be done? Who should be able to tell scientists what methodologies to employ? Whichever way these questions are answered, the structure and role of *ethics committees* at the national and international levels assumes significance in connection with the *enforcement mechanisms* without which all strategies, policies and solutions would have no teeth when it comes to their implementation.

7. WHAT ABOUT INFORMED ACCESS TO NEW TECHNOLOGIES (IANT)?

In a nutshell, let me restate the thesis which I defend in this article and to which all thinking men and women must pay the attention which it deserves. In **§3** above, I deliberately introduced the concept of *informed access* in order to develop the argument that "informed consent", an ethically loaded concept, is inseparable from *informed access*, entailing the intervention by (the nearly absent) *ethics consultation service*. That is to say, without *the ethics consultation service*, "informed consent" loses all its meaning and significance in research. In the biomedical research involving experimentation on human subjects, ethics committees working with ethical guidelines for research emphasize regulations regarding *informed consent* just as a matter of formality. As they stand, these regulations themselves need constantly to be reviewed, if they are to mean anything to human subjects of biomedical experimentation. What is more serious is this. *Be it biotechnology, genomics, medicine or public health, informed consent without informed access makes no sense, particularly in the context of new technologies cropping up at an amazing speed, even before the older technologies are made fully accessible to people.* They not only crop up, they also extend the laboratory of experimentation to those human beings who willingly participate in research but who have no *informed access* to new technologies. Human beings used as the laboratory extensions under new technologies cannot just be taken for granted. Neither does so-called *informed consent* justify such use of human beings. In my view, therefore, there can be no *informed consent* without *informed access.* As soon as there is debate on *informed consent*, there will be an increasing demand for *informed access* to new technologies. That is to say, without *informed access* to new technologies, there can be no meaningful debate on *informed consent.* And without proper investment on research on *access to new technologies*, any investment on new technologies itself becomes questionable. The concepts of informed consent and informed access are, moreover, ethically loaded concepts. Thus, they entail not only *media advocacy* **(16)** but *ethical consultation service*, besides an open debate in public domain, more so in the new technologies regime. Once we recognize this, we have to introduce necessary changes in the entire framework governing theory and practice of biomedical research, medicine and public health initiatives world wide. We have to change the very rules of the game governing them as public good.

By bringing a deep human concern to bear upon the ethical problems arising from the implications of (heavy investments on) science and technology, without proper investment on research on their impact and accessibility, I have highlighted, I hope, the biggest challenge before mankind today: *How to foster a strong tradition of interdisciplinary research across different sciences, including the social sciences and humanities, to explore the complex frontier of human and ethical values with a view to inculcate a concern for them in the educational and research institutions world-wide*? Another fundamental question, closely related to this question, was raised several years ago by Gerard Toulouse, Director of Research at the Ecole Normale Superieure, Paris, France, as follows: *Given scientific progress, how should the responsibility for raising ethical questions be shared between the individual and the community*? [17]. Toulouse's question is primarily addressed to scientific institutions, such as the Royal Swedish Academy of Sciences, the Royal Society in the UK, the US National Academy of Sciences and the German Research Foundation (Deutsche Forschungsgemeinschaft), which play such an important role in the development of science on the one hand and in building better nations on the other. He cites three examples to bring home to us the need for professional bodies and scientific academies to lead the way in fostering ethical debates about science and protecting those individuals who take risks in upholding the values of scientific freedom and social responsibility. The *three* examples are as follows. *First*, in Soviet dissident scientist Andrei Sakharov´s life, which was full of protests against the arms race and in favour of democratic freedom, one already comes across an excellent example of how science and ethics interact. Yet why did the collapse of the Soviet Union lead to the collapse of the scientific establishment there? Moreover, why did the latter fail to protect Sakharov when his life was at great risk? Sakharov was not only ignored, he was indeed exiled to Gorky for seven years. *Secondly*, the American Association for the Advancement of Science (AAAS) has created a prize for scientific freedom and responsibility with the aim of supporting and recognizing courageous individuals as stakeholders in scientific progress. We can say that even more remarkable than this is the fact that as early as 1975, the Report of the Committee on Scientific Freedom and Responsibility of the AAAS initiated a detailed discussion on many of the ethical problems faced by the contemporary society, scientists and engineers [18]. The Report *concluded* by advocating that the AAAS should no longer take a neutral stance with regard to such problems. This step in itself marked a big departure from the dominant tradition of regarding professional scientific institutions as essentially passive organizations which are meant to facilitate communications between scientists by arranging meetings and publishing journals [19]. Toulouse's *third* example concerns the German Research Council: How the Council has set up in 1997 an international commission on self-regulation in science for promoting good scientific practice.

There is no doubt how much we have to learn from each of these different kinds of examples. To these we must, of course, add those further initiatives such as the Ethics Committees that have been set up across Europe in recent years. Particularly noteworthy is the fact that in December 1997 the European Commission set up the European Group on Ethics in Science and New Technologies (EGE) to succeed the Group of Advisers on the Ethical Implications of Biotechnology (GAEIB 1991-1997). The Group is an independent, pluralist and multidisciplinary body which advises the European Commission on ethical aspects in science and new technologies in connection with the preparation and implementation of Community legislation or policies. During its first mandate the EGE (1998-2000) provided opinions on subjects as diverse as human tissue banking, human embryo research, personal health data in the information society, doping in sport and human stem cell research. At a specific request of the President of the Commission, Romano Prodi, the Group also wrote the Report on the Charter on Fundamental Rights related to technological innovation. On 24 April 2001 the Commission has appointed the twelve Members for the period 2001-2004 and amended the EGE guidelines in order to strengthen the role of the Group (EGE 2001-2004). The secretariat of the Group is an integral part of the Group of Policy Advisers. A conference entitled "Research Ethics committees in Europe: facing the future together" was to take place in Brussels on 27-28 January 2005. This event was the first of its kind and should have gathered together around 450 representatives of European research ethics committees (RECs), which evaluate, at local or regional level, any type of research protocols involving human beings. The conference was meant to provide a real opportunity to meet one another to share information, best practices and ideas for future recommendations, initiatives and actions.

Acknowledgment

The author thanks an anonymous referee for valuable comments and suggestions resulting in several improvements in the text. He also thanks Die Alexander von Humboldt-Stiftung for support for research, building upon Pandit (1982, 1991, 1995, 1996, 1998, 1999a, 1999b, 1999c, 2000a, 2000b, 2001a, 2001b, 2001c, 2001d, 2002a, 2002b, 2004a, 2004b, 2004c, 2005a, 2005b, 2005c, 2006a, 2006b, 2007a, 2007b, 2007c, 2007d, 2007e). These references are cited under 'References' and 'Suggested Reading' sections. Earlier versions were presented for discussion at the Humboldt Kolleg Indien "Bio-Nano-Geo Sciences: Addressing Issues of Concern to Mankind" (Institute of Himalayan Bioresource Technology, Palampur, Himanchal Pradesh: 24.-26. März 2006) *and* at the **Seminar für Übersetzen und Dolmetschen** (**University of Heidelberg**) under the Lecture Series „Gesundheit – ein kostbares Gut: Wie können wir es bewahren?" (Summer Semester, 26th June 2006). For questions and comments, formal and informal, the author expresses his thanks to all those who participated

REFERENCES

1. Pandit, G. L. (1995) *Von der Oekologie des Bewusstseins zum Umweltrealismus – Die Wiederentdeckung menschlicher und nicht-menschlicher Interessenssphaeren*. Picus Verlag, Wien.
2. Schwarzburger, H. (1998) Kein blindes technik-vertrauen - Indischer philosoph fordert umweltrealismus und neue ethik - Bildung und Wissenschaft. *Der Tagesspiegel* (Berlin): Sonntag, **7**. Juni 1998, s.44. (Review of [1]).
3. Ignatieff, M. (2001) *Human Rights as Politics and Idolatory*. Ed. Gutmann, A., Princeton University Press, Princeton/Oxford, pp. 3-4.
4. Raven, J. (1994) *Managing Education for Effective Schooling*. Oxford Psychological Press/ Trillium Press, New York/Ontario, p. 96.
5. Raven, J. (1994) *Managing Education for Effective Schooling*. Oxford Psychological Press/ Trillium Press, New York/Ontario, p. 148.
6. Heisenberg, W. (1989) *Gesammelte Werke* Band **V**, Ed. Blum, W., Dürr, H. –P. and Rechenberg, H., Piper, München/Zürich, p. 494.

7. Pandit, G. L. (2000) Environmental actions: Can a part control the whole? In *Problems of Action and Observation. Systemica* **12**, 351-368. Proceedings of the International Conference held in Amsterdam, the Nethzerlands, in April 1997.
8. Rhodes, R. (1995) The new morning of the world. *Am. J. Phys.* **63**, 1070-1073.
9. Pandit, G. L. (1991) *Methodological variance: Essays in epistemological ontology and the methodology of science.* **Boston studies in the philosophy of science Vol. 131**, Springer, Dordrecht/Boston/London.
10. Annas, G. J. (1994) Regulatory models for human embryo cloning: the free market, professional guidelines, and government restrictions. *Kennedy Inst. Ethics J.* **4**, 235-250.
11. Khushf, G. (1997) Embryo research: The ethical geography of the debate. *J. Med. Phil.* **22**, 495-519.
12. Beckman, L. J. and Harvey, S. M. (2005) Current reproductive technologies: Increased access and choice?" *J. Social Issues* **61**, 1-20.
13. Saxena, R. K., Sheoran, A. and Pandit, G. L. (2005) How not to interpret the advances of biotechnology. In *Interpretation and culture: Themes in the philosophy of Michael Krausz* - A special issue of *Philosophy in the contemporary world.* **12.1,** 93-102**.**
14. Woodsong, C. and Severy, L. J. (2005) Generation of knowledge for reproductive health technologies: Constraints on social and behavioural research. *J. Social Issues* **61**, 193-205
15. Rouch, H. (2002) 'Nouvelles techniques de reproduction: De la difference a l'inegalite´ in Marie-Claude Hurtig, Michele Kail and Helene Rouch, eds. *Sexe et genre: De la hierarchie entres les sexes*. CNRS, Paris, pp. 245-56.
16. Pandit, G. L. (2007e) "Ethical Tasks of Media Advocacy in the 21st Century"**,** Philosophie der Informationsgesellschaft / Philosophy of the Information Society, **Beitraege des 30. Internationalen Wittgenstein Symposium** (5. - 11. August **2007**), Band **XV**, edited by Herbert Hrachovec, Alois Pichler and Joseph Wang, pp. **169-173**. Kichberg am Wechsel: Austrian Ludwig Wittgenstein Society.
17. Toulouse, G. (1999) Ethics enters the 21st century. *Physics World* 13-14.
18. Edsall, J. T. (1975) Report of the AAAS committee on scientific freedom and responsibility. *Science* **188**, 687-693.
19. Holton, G. and Blanpied, W. A. (eds.) (1976) Science and its Public: The Changing Relationship. *Boston Studies in the Philosophy of Science*, D. Reidel Publishing Company: Dordrecht/Boston.

SUGGESTED READING

- Chimni, S. (ed.) 2004: *International Refugee Law: A Reader*, Sage Publications: New Delhi/ Thousand Oaks/London 2000, pp. xxix+613; Filippo Osella and Katy Gardner (eds.) *Migration, Modernity and Social Transformation in South Asia*, Sage Publications: New Delhi/Thousand Oaks/London 2004, pp. xlviii+380.
- Pandit, G. L. (1982): *The Structure and Growth of Scientific Knowledge – A Study in the Methodology of Epistemic Appraisal.* **Boston Studies in the Philosophy of Science Vol**.**73** (with an Editorial Preface written by Robert S. Cohen and Marx W. Wartofsky). Dordrecht/ Boston/London: D. Reidel Publishing Company/Springer Verlag: Netherlands.
- Pandit, G. L. 1996: "Bewusstsein und seine Existenzformen: zurueck zum moralisch ungeteilten Universum", in Christoph Hubig und Hans Poser (Hrsg.) *Cognitio humana – Dynamik des Wissens und der Werte* – **XVII.** Deutscher Kongress fuer Philosophie: Workshop-Beitraege **Band I**, Leipzig (**1996**, 34-41).

- Pandit, G. L. 1998: "Review of Huntington, Samuel P. (1997) *The Clash of Civilizations and the Remaking of World Order*", Viking: Penguin India, pp. 367, in *World Affairs* **2** (Jan.-March **1998,** 135-137).
- Pandit, G. L. 1999a: "Weltumfrage zur Situation der Philosophie am Ende des 20.Jahrhunderts" in Raul Fornet-Betancourt (Hrsg.) *Quo vadis, Philosophie? Antworten der Philosophen: Dokumentation einer Weltumfrage*. Wissenschaftsverlag Mainz in Aachen **1999**, pp. 222-227.
- Pandit, G. L. 1999b : "L´impact de la pensee de Sir Karl Popper sur notre comprehension du monde naturel et du monde des homme,,, *Quo vadis: Revue Internationale de Philosophie* **35** (**1999**, 82-94).
- Pandit, G. L. 1999c: "Paradoxes of Progress: How Can Humanity Improve Environmental Thinking and Environmental Action?", in *Politics and the Life Sciences* **18** (September **1999**, 241-242).
- Pandit, G. L. 2000a: ´´Review of Strange, Susan (1998) *The Retreat of the State: The Diffusion of Power in the* World Economy", Cambridge: Cambridge University Press,pp.xvii+218, in *World Affairs* **4**(Jan.- March **2000**, 113-115).
- Pandit, G. L. 2001a: "Review of von Welck, Stephan F. (Ed.) *Crossing Borders, Stretching Boundaries: The Bose-Einstein Lectures on Indo-German Cooperation in Science, Technology and Environment*", Manohar Publishers & Distributors: New Delhi, 2000, pp. 237, in *World Affairs* **5** (October-December, **2001**, 126-130).
- Pandit, G. L. 2001b: "Participate and Reconnect: The Problem of Improving the Human Condition", in Gerard de Zeeuw, Martha Vahl and Ed Mennuti (Eds.) *Problems of Participation and Connection*, a Special Issue of *Systemica* **13** (**2001**, 337-341), the Journal of the Dutch Systems Group. Proceedings of the International Conference held in Amsterdam, the Netherlands, 5-9 April 1999. Lincoln Research Centre.
- Pandit, G. L. 2001c: "Tradition and Innovation in Asian Culture: India", in Raul Fornet-Betancourt (Hrsg.) *Kulturen zwischen Tradition und Innovation: Stehen wir am Ende der traditionellen Kulturen?* IKO – Verlag fuer Interkulturelle Kommunikation: Frankfurt, (**2001**, 48-64). Dokumentation des **III.** Internationalen Kongresses fuer Interkulturelle Philosophie: Aachen, 22.-25.November 1999.
- Pandit, G. L. 2001d: "Leibniz and the Changing Images of Universal Interconnectedness", in Hans Poser et al (Hrsg.) *VII. Internationaler Leibniz-Kongress "NIHIL SINE RATIONE – Mensch, Natur und Technik im Wirken von G. W. Leibniz*", Vortraege 2. Teil. Technische Universitaet Berlin (**2001**, 942-947).
- Pandit, G. L. 2002a: "Heisenberg-Einstein Context Principle and the Dynamic Core-Context of Discovery in Physics", in *Fortschritte der Physik* **50** (**2002**, 461-482). Selections from Proceedings of the International Symposium "100 Years Werner Heisenberg – Works and Impact", organized by the Alexander von Humboldt-Foundation: Bamberg, Germany September 26-30, 2001. Special Editors: D. Papenfuss, D. Luest and W. Schleich.
- Pandit, G. L. 2002b: "Heisenberg-Einstein Context Principle and the Dynamic Core-Context of Discovery in Physics" in D. Papenfuss, D. Luest, W. P. Schleich (eds.) *100 Years Werner Heisenberg - Works and Impact.* Wiley-VCH **2002**, pp. 32-53.
- Pandit, G. L. 2004a: "Review of Stevis, D. & Assetto, V. J. (Eds.) *The International Political Economy of the Environment – Critical Perspectives*", Lynne Rienner Publishers, Boulder/ London, 2001, pp.viii+309, in *World Affairs* **8** (July-September **2004**,131-135).
- Pandit, G. L. 2004b: "Ecologies of Environmental Nesting", in Karimo Bakhtiyor (ed.) *Proceedings of Humboldt-Kolleg (July 6-10,* **2004***) Uzbekistan – Use of Geographic Information Systems and Simulation Models for Research and Decision Support in Central Asian River*

Basins. Tashkent, 2004, p. 154 (an International conference sponsored by the Alexander von Humboldt-Stiftung, Germany and co-sponsored by the Humboldt-Society of Uzbekistan).

- Pandit, G. L. 2004c: "Review of Maude Barlow & Tony Clarke, Blue Gold: The Battle Against Corporate Theft of the World´s Water", Earthscan: London, 2002, pp.xviii+ 278, in *World Affairs* **8**, October-December **2004** (119-122).
- Pandit, G. L. 2005a**:** "World´s Refugees: What Humanity Owes to Them" in *World Affairs,* ***The Journal of International Issues*** **9**, Jan-March **2005** (148-156). a Review Essay based upon on Dagmar Bernstorf and Hubertus von Welck (Eds.) *Exile as Challenge: The Tibetan Diaspora*", Orient Longman 2003, pp.vii+488; B.
- Pandit, G. L. 2005b: "Consciousness in Search of its Ecology", in D. P. Chattopadhayaya (ed.) *Self, Society and Sciences: Theoretical and Historical Perspectives* (**History of Science, Philosophy and Culture in Indian Civilization** Vol. **XI**/part **2**), Centre for Studies in Civilizations: New Delhi, **2005**, pp.70-76.
- Pandit, G. L. 2006a: "Leibniz´s Weltbild of Universal Interconnectedness," in **VIII**. International Leibniz-Kongreß: Unity in Plurality: **2006a** (769-779).Gottfried-Wilhelm-Leibniz Gesellschaft E. V. Hannover, Germany/University of Hannover.
- Pandit, G. L. 2006b: "Exclusion and Uprootment in Kashmir: From a Political Tragedy to an Ecological Catastrophe" in *World Affairs, The Journal of International Issues* 10, **Autumn** 2006b **(150-169).**
- Pandit, G. L. 2007a: "Emergence of the Human Interests Studies: The Environmental Dimension", in Martha Vahl, Gerard de Zeeuw and Ed Mennuti (Eds.) *Problems of Individual Emergence*, a Special Volume of *Systemica* **14** (295-312), the Journal of the Dutch Systems Group, ISSN 0167-9961, Lincoln Research Centre UK, 2007 (Proceedings of the International Conference held in Amsterdam, April 2001).
- Pandit, G. L. 2007b: "Human Interest Studies within Environmental Realism" in Dhankar, R. S. (ed.) *Environment and Quality of Life: Canada and India*. Allied Publishers: New Delhi 2007 (forthcoming).
- Pandit, G. L. 2007c "Exclusion and Uprootment in Lalla´s Kashmir: A Political Tragedy or an Ecological Catastrophe" in Amy Colin (ed.) *Marginalization: Dynamics of Injustice and Discrimination* (**Forms and Dynamics of Exclusion** Vol. **I**). Editions UNESCO (forthcoming)
- Pandit, G. L. 2007d: "Consciousness: Cosmic Dimension", in R. C. Pradhan (ed.) *G. C. Pande: Philosopher, Historian and the Literary Genius*, Indian Council of Philosophical Research: New Delhi, 2007, pp. 172-181 (in press).

DETAILS OF CONTRIBUTORS

Natural Nano-Machines of Life: A Biological Route to Nanotechnology

Debashish Chowdhury
Department of Physics, Indian Institute of Technology,
Kanpur, Kanpur 208016, India.
debch@iitk.ac.in

Abstract

An eukaryotic cell is filled with machines, each typically a few tens of nanometers long, which have specific tasks and operate in a well coordinated manner. I present examples of linear and rotary motors as well as pumps that drive various processes essential for sustaining life. Then, I summarize the methods of studying the materials and mechanisms of these machines emphasizing the crucial differences between the natural nano-machines and their macroscopic counterparts. Finally, I highlight some potential applications of the fundamental research on molecular machines in biomedical science and nano-technology.

Author Details

Professor Debashish Chowdhury studied at the St. Xavier's College Ranchi (B.Sc. with Honors in Physics and Distinction in Mathematics), Indian Institute of Technology, Kharagpur (M.Sc. in Physics, 1980) and Indian Institute of Technology, Kanpur (Ph.D., 1984). After his post-doctoral research in Cologne (as an Alexander von Humboldt Fellow) and Philadelphia, USA (as a senior research associate) he joined Jawaharlal Nehru University, New Delhi in 1987 as an Assistant Professor in the School of Physical Sciences. He served as an Associate Professor of Physics at IIT, Kanpur, from 1993 and was promoted to a full Professor in May, 1997.

Professor Chowdhury's research area is Statistical and Biological Physics. His first book was published in North America by the Princeton University Press (Princeton series in Physics edited by A. S. Wightman and P.W. Anderson, NL) and elsewhere by World Scientific Publishing Co., Singapore in 1986. His second book (with Prof. D. Stauffer) was published in 2000 by Wiley-VCH, Germany, a section of John-Wiley. He has so far published more than 120 papers in leading international refereed journals some of which have been cited more than 500 times. Prof. Chowdhury is a member of the Board of Editors of European Physical Journal B which is co-published by Springer (Germany), EDP Sciences (jointly owned by the French Physical Society, French Chemical Society and French Society of Applied and Industrial Mathematics) and the Italian Physical Society. Prof. Chowdhury is a recipient of the Young Scientist Medal of the Indian. National Science Academy (1989). He has been a (elected) Fellow of the Indian Academy of Sciences Bangalore since January 2004.

Hybrid Bionanodevices for Engineering and Healthcare

Lalit M. Bharadwaj and R. P. Bajpi

Biomolecular Electronics & Nanotechnology Division
Central Scientific Instruments Organization (CSIO),
Sector 30, Chandigarh-160030, India.
lalitmbharadwaj@hotmail.com
Guru Jambeshwar University,
Hisar, Haryana, India

Abstract

Development of biomoleculear and nanodevices for engineering and medical applications has been described in this article. These include: biomolecular electronics, disease diagnosis and targeted drug delivery based on molecular devices working in living systems. The group is taking multi-disciplinary integrated approach to various R & D activities related to broad areas such as DNA nanowire, Bio-MEMS for disease diagnosis, biomolecular motors for nano electronic switching & nanorobotics, etc. Studies on Carbon Nanotubes (CNTs) interfaced with biomolecules is targeted towards new sensor development.

Author Details

Dr. R. P. Bajpai did his doctorate from Indian Institute of Technology, Delhi and D.Sc. from Hokkaido University, Japan. He was a Humboldt Fellow in the Physikalisches Institute für Technische Universität in Clausthal Zellerfield, Germany. He has been a Visiting Professor/Scientist to a number of academic institutions in India, Japan, Switzerland and USA. Dr. Bajpai was Director of CSIO, a CSIR Laboratory in Chandigarh, before taking over his present assignment as Vice Chancellor of Guru Jambeshwar University, Hisar. Dr. Bajpai works in the area of micro-electronics, nanotechnology, biomolecular electronics, etc. and has so far authored two books and has published close to 180 articles in national and international journals. Dr. Lalit M Bharadwaj is Deputy Director and Head, Biomolecular Electronics and Nanotechnology, at Central Scientific Instruments Organisation (CSIO), Chandigarh, India. He has a basic degree in Physical Chemistry and wide multidisciplinary research experience in the area of semiconductors , biophysics, nanotechnology, biomolecular electronics, etc. He has 78 publications in international journals and was awarded the 16th IETE Hari Ramji Toshniwal Gold Medal Award in recognition of his contribution in the field of nanotechnology.

Pharmacoscintigraphic Evaluation of Nanoparticulate Drug Delivery Systems

Rakesh Kumar Sharma[1], Sheetal Soni,[1,2] Anil Kumar Babbar[1] and Amarnath Maitra[2]

(1) *Division of Radiopharmaceuticals, Nanotechnology and Stem Cell Research, Institute of Nuclear Medicine and Allied Sciences, Brig. SK Mazumdar Road, Delhi 110 054, India*

(2) *Department of Chemistry, University of Delhi, Delhi 110 007, India*

rksharmadrl@yahoo.com

Abstract

Drug delivery to the brain is often restricted by the blood-brain barrier (BBB), which regulates the passage of substances including drugs, between peripheral circulation and CNS. One approach to overcome this barrier has been using appropriately modified polymeric nanoparticles. Pharmacoscintigraphic is the application of nuclear medicine techniques for tracing the radiolabeled ingredient of the active component of drug/device or the formulation excipient. We have used surface modified chitosan nanoparticles as brain delivery carriers and evaluated their efficacy by pharmacoscintigraphy. The chitosan nanoparticles were radiolabeled with Technetium-99m, and the *in vivo* brain uptake profiles were studied in Swiss Albino Strain 'A' mice. The biodistribution of the preparation in the brains of New Zealand rabbits was also studied with gamma scintigraphy. Using Polysorbate-80 (1% w/w) coated nanoparticles, we have demonstrated a nearly five-fold increase in the brain accumulation compared with uncoated nanoparticles. Our results indicate that surface modification of chitosan nanoparticles by coating with polysorbate-80 is effective in drug delivery through BBB.

Author Details

Born on 16th July 1958, Dr. Rakesh Kumar Sharma has more than 25 years of rich Research & Development experience in multidisciplinary fields. He has also been actively involved in Research, Planning, Co-ordination & Project Management, Teaching, Human Resource Management and Scientific Administration. Dr. Sharma started his career as Defence Research and Development Organisation (DRDO) Scientist in 1983 after a brief tenure in teaching and research. He is presently holding the post of Additional Director at Defence Research Laboratory, Tezpur, Assam and heading the Division of Phytochemistry. Earlier he had worked at Defence Research and Development Establishment (DRDE), Gwalior for two years and at the Institute of Nuclear Medicine and Allied Sciences (INMAS), Delhi for over two decades. In research, his interests include revitalization of traditional medicines for development of modifiers of radiation response, Medical informatics; Development of SPECT/PET radiopharmaceuticals for nuclear scintigraphy & therapy, Smart Magnetopharmaceuticals for MRI, Cancer Research, Stem Cell Research, development of Nuclear medicine and Nuclear Magnetic Resonance techniques for biopharmaceutics studies of NDDS, and Radiation Biology & Radiation Protection.

During 1991, Dr. Sharma was bestowed GSF Guest Scientist award to work at the Institute of Biophysics, University of Frankfurt, Frankfurt/M, Germany for a period of 4 months. He had been a Visiting Scientist at the Institute of Biophysical Chemistry, University of Frankfurt,

Germany for three months in 1992. He was given the prestigious Young Scientist Award instituted by Brig. SK Mazumdar Memorial Trust for the year 1993 and Laboratory Scientist of the Year Award 2001 instituted by DRDO. During 2002-03 he worked at Max-Planck-Institute for Biological Cybernetics at Tuebingen, Germany to pursue his interest in Molecular Magnetic Resonance Imaging (MRI). He is also recipient of INMAS Team Leadership Awards consecutively for the years 2003 and 2004, Chandra Kanta Dandiya Prize in Pharmacology (2004) instituted by Prof Prem Chand Dandiya Endowment Trust and Jaipur Prize of the Indian Pharmacological Society (2005). One of his most outstanding contributions has been in the field of MR Spectroscopy of intact cells and mouse tumor model. He was instrumental in promoting the introduction of post-graduate course in Radiation Sciences at Institute of Nuclear Medicine and Allied Sciences in 2004. He also holds position of Adjunct Professor in the Faculty of Pharmacy at Jamia Hamdard (Hamdard University). Dr. Rakesh Kumar Sharma is a venerated member of the pharmacy profession and has been intimately involved in the working of several professional, statutory bodies. He is life member of various professional societies including IPGA, IHPA, and Indian Society for Radiation Biology, The Photodynamic Group of India, Medical Library Association of India, Society of Nuclear Medicine (India), National Magnetic Resonance Society of India and Indian Science Congress Association. Recently, Dr. Sharma was elected as a Fellow of the Institution of Chemists, Kolkata. He was a member of the National Task Force on Biomedical Devices Constituted by Indian Council of Medical Research, whose recommendations have been accepted by Ministry of Health and Family Welfare, Govt. of India. Dr. Sharma is Coordinator of the Core Groups on Chemical (Industrial) Disaster Management, Chemical (Terrorism) Disaster Management and a Member of Core Group on Nuclear Disaster Management appointed by National Disaster Management Authority, Govt. of India, New Delhi.

Dr. Sharma has 7 patents and 150 publications to his credit besides contributing 17 chapters in books and editing five books/monographs. Four technologies developed by him have been successfully transferred to Industry and are at production or final stages of trials.

Nano Materials of Carbon –Fullerenes and Nanotubes

V. K. Jindal
Department of Physics, Panjab University
Chandigarh 160014, India.
jindal@pu.ac.in

Abstract

The purpose of this paper is to highlight the significance of nonmaterials and nanotechnology as a general awareness. I also aim to pay special attention to fullerenes and carbon nanotubes, whose availability and discovery after 1985 laid the foundation for rapid growth of interest and progress in the field of nanotechnology. After introducing these materials in reasonable detail, I also bring out some methods of production of fullerenes and carbon nanotubes used by students at our University. Further, a theoretical methodology employed by our research team to understand the special properties of fullerenes and carbon nanotubes is briefly given with some references for those who wish to look into details. Still, the primary aim of this article remains to introduce fullerenes and carbon nanotubes.

Author Details

Dr. V.K. Jindal is a Professor at Department of Physics, Centre for Advanced Study at Panjab University, Chandigarh. He is also Coordinator, M.Tech Program in Nanoscience & Nanotechnology, Centre for Emerging Areas in Science & Technology at Panjab University.

He has contributed significantly in the area of Anharmonicity of Solids. As a Humboldt Fellow and as BMFT (Ministry of Research and Technology of Germany) Research Fellow, he established computer programmes out of which the ones used in calculating phonon life times and line shifts for molecular crystals became verywell known because of their usefulness in correlating theory and experiments. His work in the area of molecular crystals was recognised by the Third World Academy of Sciences (TWAS) in ICTP when they offered him grant to work in Florence in Italy. As a Fulbright Foundation Fellow and on support of University of Illinois, Urbana-Champaign, USA, he collaborated on applied problem to understand how materials break under shock pressure and how to control and prevent them from breaking. His involvement in research has been in both theoretical and experimental work of significance. He has conducted and collaborated in performing Neutron Inelastic Scattering Experiments at ILL, Grenoble, France; at KFA, Juelich, Germany and at HMI, Berlin, Germany several times. His experimental work also comprises of Small Angle Neutron Scattering Experiments at ILL, Grenoble. All this work has been very well recognised and cited. He has kept on working in current topics of research interests. Important deviations from normal condensed matter being micellar solutions for anisotropic study, shock pressure effects for material study, fullerene solids, and bio-physical materials. Presently, he is heading a research group which devotes effort to understanding theoretical methodology for nanomaterials, especially carbon nanotubes, experimenting with ion irradiation on these, and also attending to their production techniques.

Synthesis and Characterization of Nanosized Materials for Lithium Rechargeable Batteries

S. Selvasekarapandian, T. Savitha, J. Malathi, R. Mangalam and C. S. Ramya
Solid State and Radiation Physics Lab,
Bharathiar University, Coimbatore -46, India
sekarapandian@yahoo.com

Abstract

Nanosized cathode materials based on Phospho olivine, $LiNiPO_4$ and $LiCoPO_4$ and nano composite polymer electrolyte, PVAc:DMF:$LiClO_4$:TiO_2 have been synthesized for Lithium rechargeable batteries. Cathode materials have been characterized by Thermogravimetric/ Differential Thermal Analysis, X-Ray Diffraction, Laser Granulometry and Electrical conductivity studies and composite polymer electrolytes by XRD, and Electrical conductivity studies.

Author Details

Dr. S. Selvasekarapandian, retired Professor of Physics and Professor and Head (i/c) of Nanotechnology, joined Bharathiar University in the year 1989. He has, to his credit, a total of 38 years of teaching experience for undergraduate and post graduate students and 30 years of research experience. He received his M.Sc and Ph.D from Indian Institute of Technology, Madras. His field of research includes thermoluminescence and its applications, super ionic conductors, environ-mental radiation survey and Laser- Raman study of ferroelectric and biomolecules. He was a visiting professor in Multidisciplinary Research Institute of Tohoku University, Sendai, Japan during 2003. He has attended 24 International conferences in various countries like USA, Canada, Germany, France, Greece, Japan, Korea, Australia, New Zealand, Malaysia, Singapore, etc. and presented his work. He has edited 11 volumes in different fields and also referees for leading international journals like Journal of Material Science, Journal of Radio Analytical and Nuclear Chemistry, Journal of Hazarded Materials, Journal of Applied Polymer Science, Material Letters, Physics B. Prof. Selvasekarapandian was responsible for the computerization of Bharathiar University Library when he was acting as officer in-charge for three years from 1998-2001. At present he is the executive member of Asian Society for Solid State Ionics and president of Indian Society for Radiation physics Coimbatore chapter and president of Luminescence Society of India since 2005.

Clay Nano-composites as Environmental Friendly Catalysts for Green Chemical Synthesis

Vasundhara Singh
Punjab Engineering College, Sector 12, Chandigarh 160012, Punjab, India.
vasun7@yahoo.co.in

Abstract

Catalyst is one of the fundamental pillars of 'Green Chemistry', which is defined as the design of chemical products and processes that reduce or eliminate the use and generation of hazardous substances. Heterogeneous catalysis in particular addresses the goals of 'Green Chemistry' by providing the ease of separation of products and catalyst, recyclability, increased yields and selectivity, thereby eliminating the need for separation through distillation or extraction or other unit operations. Another class of liquid–liquid biphasic catalysis using ionic liquids as green solvents due to their non volatility and easy recylability has been exploiter in synthesis. The development of new methodologies based on modified forms of clay and clay nano-composite materials give an insight both in the studies towards the behavior and synthetic potential of clays in conjunction with ultrasonic and microwaves and their applications in natural product synthesis and biologically active compounds.

Author Details

Dr. Vasundhara Singh has done her Ph.D in 1992 from Panjab University, Chandigarh in the field of Synthetic Organic Chemistry. She has done her Post Doctoral Research from Stanford Research Institute, USA and she has got DAAD fellowship and worked at University of Regensburg, Germany in the area of asymmetric catalysis. She has worked as visiting scientist at University of Rennes 1, Rennes, France funded by CNRS, France. She has got Career Award for Young Teacher's (CAYT) from AICTE, New Delhi for her contribution in the field of Green Chemistry in 2003. She has previously held position of Assistant Prof at Thapar Institute of Engg. & Tech., Patiala and Reader at Punjabi University, Patiala. Presently, she is Assistant Prof. at PEC, Chandigarh. She has 32 research papers in national and international Journals besides several papers presented at national and International Journals. She has guided two Ph.D students and has five PhD students presently working in her group. She has completed various research projects funded by UGC, DST, AICTE, New Delhi and is presently working on a joint Indo French project funded by CEFIPRA, New Delhi with University of Rennes, France. Her research interests are in Natural Product Synthesis, Green Chemistry, Catalysis, Ionic liquids and photocatalytic Solar detoxification of organic pollutants.

Micronutrients Disorders – Challenge to Sustain Food Grain Production

Upkar S. Sadana
Department of Soils, Punjab Agricultural University, Ludhiana-141004, Punjab, India.
upkar_pau@yahoo.com

Abstract

Micronutrient deficiencies have become one of the major constraints in sustaining agricultural production in India. Analysis of more than two lakh soil samples in different states of the country revealed the predominance of Zn deficiency in 48 per cent samples followed by Fe and Mn deficiency in 12 and 5 per cent samples, respectively. Coarse textured, alkali, flood plain and calcareous soils, which have relatively high pH and/or low organic matter content, are more prone to micronutrient deficiencies. As such it is not advisable to use micronutrient mixtures for combating Zn, Fe or Mn deficiency in crops. Organic manures are not only a good source of micronutrients but have also been found to enhance their availability in soil through chelation and reduction in pH. Judicious and efficient management of costly micronutrient fertilizers holds the key to maintain sustainable agricultural production. Zinc has received the major focus of attention of the micronutrient scientists due to its wide spread deficiency all over the country. Soil application of zinc sulphate is the most commonly used method to correct Zn deficiency in different crops. Severe Fe and Mn deficiency is difficult to manage with soil application due to the oxidation of soil-applied iron and manganese sulphate at high soil pH. Foliar application of Fe and Mn is an immediate effective measure to combat their deficiency; however, Fe and Mn have to be applied repeatedly through foliar application. Other approach could be sowing of nutrient-efficient crops that can grow well on soils low in micronutrients. For nutrients such as Fe, Mn and Zn which generally have an extremely low solubility in soils of arid and semi-arid regions, the response to nutrient deficiency stress is mainly biochemical. The response may involve changes in rhizosphere pH, redox potential, root exudates and microbial activities. Identification and /or evolving of micronutrient efficient crop cultivars having either low micronutrient requirement for potential yields or capable of mining micronutrients from the less available pools should be given priority.

Author Details

Dr. Upkar S. Sadana graduated from College of Agricultural, Punjab Agricultural University, Ludhiana in 1974 and received his M.Sc. (Soil Science) in 1977 and Ph.D. (Soil Science) in 1981 from the same University. He was recipient of University merit scholarship in B.Sc, Indian Council of Agricultural Research (ICAR) Junior Research Fellowship in M.Sc. and ICAR Senior Research Fellowship in Ph.D.

He started his research career in 1982 as Assistant Soil Chemist in the Department of Soils, P.A.U., Ludhiana in the scheme "All India Coordinated Research Project on Micro- and Secondary-Nutrients and Pollutant Elements in Soils and Plants". Since 1997 he is working as Professor of Soils in the Department of Soils, P.A.U., Ludhiana in the scheme "Establishment of Advanced Center of Agricultural Research and Education on Soil Fertility and Plant Nutrition". As Alexander von Humboldt (AvH) Stiftung

Post doctoral Fellow, Dr. Sadana has worked at Institute of Plant Nutrition, Technical University München, Freising-Weihenstephan, Germany from July 1992 to December 1993 and at Institut für Agrikulturchemie der Georg-August-Universität, Göttingen, Germany from September to October 1998, June to August 2000, June to August 2001, September 2002, September to October 2004, June toJuly 2005, June 2006. Dr. Sadana has made significant and technology oriented contribution in the field of chemical equilibria of micronutrients, modeling nutrients dynamics in the rhizosphere and efficient management of micronutrients deficiencies in rice-wheat cropping system. He has developed a simple method to study the oxidizing power of rice roots. This method screen rice varieties suitable for Fe toxic and Fe deficient soils and can help find means and ways to increase rice productivity on problematic soils. In recognition of his research contributions Indian Science Congress Association (ISCA) bestowed on him Pran Vohra Award-1989 for Young Scientist. He was also awarded S.N. Ranade Memorial Award for excellence in Micronutrients Research for the year 2000-01. Dr. Sadana has visited Australia, Japan, Germany, France, UK, Pakistan, China, Turkey, Thailand, and Spain to participate in International conferences. He has so for published 51 research papers.

Using Plants for Remediation of Metal Polluted Soils

G.K. Psaras
Section of Plant Biology, Department of Biology, University of Patras, GR 265 00, Patras, Greece.
g.k.psaras@upatras.gr

Abstract

Organic compounds and inorganic elements constitute the key pollutants of soil. Efforts to clean up the soil using a 'green' technology are on. It has been shown that some plants have the capability of eliminating toxic metals from soil. Metal phytoextraction, either for phytomining, or for the so-called phytoremediation of contaminated soil involves the growth of hyperaccumulating plants to concentrate the metal. The metal is extracted from the harvested crop. There are three points that need to be kept in kind. The first is the discovery of plants that can act as hyperaccumulators. Second is the identification of mechanisms by which the plant handles the toxic metals. The third point is to deal with ecological repercussions of hyperaccumulation. We have used scanning electron microscopy, X-ray microanalysis and atomic absorption spectroscopy to study metal hyperaccumulation.

Author Details

B.Sc. in Natural Sciences, University of Athens, Greece, 1973
B.Sc. in Biology, University of Athens, 1978
Ph.D. Degree, University of Athens, 1980
AvHumboldt Research scholarship 1988-1989, 2005

Teaching

Plant Anatomy and Morphology

Research

Functional plant anatomy. Light, epi-fluorescence and electron microscopy/ microanalysis. 50 Scientific research papers in peer-reviewed Journals

Scientific Film

Scheuerlein R. und G.K. Psaras. 1994. Statolithensedimentation bei *Alocasia odora* (Araceae). Scientific film in collaboration with the Institut für den wissenschaftlichen Film, Göttingen: C 1857.

Books

1. **Dictionary of Botanical terms**. D.K. Christodoulakis, G.K. Psaras, Patras University Press, Patras, 2001. ISBN 960-130-051-6. In Greek
2. **Plant Morphology**. G.K. Psaras, A. Tiniakou, G. Kamari. Patras University Press, Patras, 2002. Patras. In Greek. **Part I: Introduction**. ISBN 960-530-057-5, 960-530-058-3. **Part II: Laboratory manual**. ISBN 960-530-057-5, 960-530-058-1
3. **Atlas of Plant Anatomy**. G.K. Psaras. 2002. Stamoulis, Athens, ISBN 969-351- 415-2. In Greek.
4. **Plant Physiology**. Translation in Greek of the book **Plants** (.ed. I. Ridge.). Transleted by Y. Manetas (ed.), G. Grammatikopoulos, A. Kyparissis, E. Levizou, Y. Petropoulou, G.K. Psaras. IWN, Athens, ISBN 960-411-522-7.

Pollution in the Red-Bloomed Himalayan Dal Lake of Kashmir

Shafiq-ur-Rehman

Division of Environmental Sciences, Sher-e-Kashmir University of Agricultural Sciences and Technology of Kashmir (SKUASTK), Shalimar Campus, G P O Box 56, Srinagar-190 001, Jammu and Kashmir, India.

shafiqurrehman@rediffmail.com

Abstract

Pollution is as old as civilization itself. The problem of pollution of water was not as serious in the past as it is now. Due to human interference, pollution of natural water resources by sewage or plant nutrients causes problems throughout the world. In recent years, the Dal Lake of the Himalayan Kashmir Valley has suffered with formation of a rare phenomenon of red-bloom of a new species discovered as Euglena shafiqii. In a series of examination performed on the pollution scenario of the Dal Lake, we also studied the values of toxic heavy metals (Fe, Cd, Cr, Cu, Mn, Ni and Zn) along with other physical and chemical features like temperature, hydrogen ion concentration, dissolved oxygen, Talkalinity, nitrate-N, ammonium-N, phosphate-P, chloride, sodium and potassium in the waters affected by the red-bloom in different regions of the Dal Lake receiving anthropogenic waste effluents. Most of the overloaded chemical features were found high in the lake. Moreover, the levels of these events were much higher in the basins of the bloom-affected waters. The nutrients have been found important factor for the periodicity and aggregation of Euglena shafiqii, since during the bloom period these nutrients were reduced, thus believed to be utilized by the organisms for growth. The contents of Fe were high in the lake waters. Heavy metals concentrations were higher in the bloom-affected regions which receive wastes from hotels and settlements. Amongst heavy metals Mn levels were highest followed by Zn, Cr, Ni, Cu and Cd in waters at the same regions.

Author Details

Dr. Shafiq-ur-Rehman got early education at his home town Khaga (Fatehpur, U.P.). For higher education he went to the Aligarh Muslim University (A.M.U.) and obtained B.Sc. (Honours) in Chemistry, and M.Sc. in Analytical Chemistry. His hard struggle instigated multidisciplinary research in the A.M.U. and started his research work in the Department of Pharmacology, Jawaherlal Nehru Medical College, in the fields of Biochemical toxicology, Behavioural Sciences, Neurochemistry, Neuropharmacology and related disciplines for his M.Phil. and Ph.D. degrees. He received the prestigious Alexander-von-Humboldt Fellow and conducted advanced research in the Institute of Psychiatry of the University of Munich during 1986-88. After return from Germany, he joined Department of Biochemistry, University of Kashmir at Srinagar as lecturer in 1989, and Div. of Environmental Sciences at SKUAST, Shalimar, Srinagar as Asstt. Professor in 1991. He is working as Assoc. Professor in the same division and is also as Incharge of the Modern Toxicology Laboratory at the Division of Entomology at Shalimar. He established that heavy metal lead causes lipid peroxidative damage to brain and stereo types complex behaviours. He developed a simple and rapid HPLC method of phospholipids analysis in biological membranes. He discovered a new species of Euglena shafiqii that caused Red-Bloom in the famous Dal Lake of Kashmir in 1991. He is working on the novel mode of pesticides toxicity in nervous system, discovering new molecules of pesticidal values and residual analysis of pesticides.

He is a member of several national and international technical societies, including Founder member of the International Society for Trace Element Research in Humans, and Indian Academy of Neurosciences; Life member of Indian Agricultural Society; Member of the International Brain Research Organization. He is Scientific Advisor of a leading J&K NGO (The Kashmir Environmental Protection Coordination Organization, KEPCO®).

Production of Radioisotopes for Positron Emission Tomography

Alok Srivastava
Department of Chemistry, Panjab University, Chandigarh-160014, India.
alok@pu.ac.in

Abstract

Radioisotopes find application in fields as diversified as medicine, agriculture, food preservation, water resource management and environmental studies but it is the medical application which constitutes the most significant part. Radioisotopes are finding increasing applications in the form of diagnostic as well as therapeutic radiopharma-ceuticals. The production of radioisotopes for application in positron emission radiography (PET) is briefly reviewed from the scientific and technological point of view.

Author Details

Dr. Alok Srivastava is a nuclear chemist. He did his doctorate in 1985 from the Radiochemistry Division of Bhabha Atomic Research Centre in the field of Low Energy Nuclear Fission. After completing his doctoral studies he became a Humboldt Fellow in the Institute of Nuclear Chemistry, University of Mainz, Germany in the period 1986-88 and worked in the Laboratory of Dr. Fritz Strassmann, the co-discoverer of Nuclear Fission.

After the completion of Humboldt assignment he became Consultant to European Community and worked in the Institute for Reference Materials previously known as Central Bureau for Nuclear Measurements in Geel, Belgium in the period 1989-90 where he was involved in preparation of standard radioactive sources and development of scanners for non destructive assay of radioactive waste. He then moved to USA to work with Prof. W.D.Loveland and the Nobel Laureate Prof. G.T.Seaborg in the period 1991-93 as Visiting Faculty at Oregon State Univeristy , Corvallis and University of California, Berkeley where he was involved with the study of medium energy heavy ion reactions. After his return from USA he worked as a teaching faculty in the North Eastern Hill University and Allahabad University and presently he is teaching in the Chemistry Department of Panjab University.

He was invited to work as DAAD Professor of Nuclear Chemistry in Aachen University of Applied Sciences, Juelich Division in the period 2002-04 to design and implement an International Academic Program on Application of Nuclear Sciences. He continues to be a Visiting Faculty there till date. Dr.Srivastava works in the area of application of Nuclear Techniques in Material Science and Environmental Science besides Radiation Biology and Nuclear Reactions.

He has published nearly 50 papers in International Journals and about 110 in different National and International Conferences. His scientific contributions have fetched him awards from Indian Association of Nuclear Chemists and Department of Atomic Energy. He has been Visiting Scientist in different laboratories in Germany, France, Japan, Spain, Brazil, Bangladesh and USA and is presently working as Consultant to Defense Research and Development Organization and Indo Tibetan Border Police.

Study of Radiotracer Applications in Bio-Nano-Geo Sciences at Saha Institute of Nuclear Physics

Susanta Lahiri and Kamalika Roy
Chemical Sciences Division, Saha Institute of Nuclear Physics,
1/AF Bidhannagar, Kolkata 700064, India.
susanta.lahiri@saha.ac.in

Abstract

The radiochemistry group of Saha Institute of Nuclear Physics, Kolkata, is actively engaged in the study of different biological and geological problems with the main emphasis being on metal bio-accumulation by different bio-reagents like yeast, algae, mushroom, biopolymers etc. The studies show that certain bioreagents can be efficiently used for preconcentration of some metals. In the field of geo- sciences efforts have been made to develop separation techniques for naturally occurring stable isotopes which are isobaric to some cosmogenic radionuclides. This state of art separations is a step forward in the direction of solving long standing astronomical puzzles. We have also reported for the first time synthesis of radioactive gold nano particles with green (environmentally safe) chemical reagents.

Author Details

Prof. Susanta Lahiri has born at Darjeeling, India on 1st August 1961. He did his M.Sc from the University of Burdwan and Ph. D from Saha Institute of Nuclear Physics (University of Calcutta). He joined as faculty member in The University of Burdwan in 1994 and shifted to Saha Institute of Nuclear Physics in 1997 and became Professor in 2004. Professor Lahiri has published about 110 papers in International peer reviewed journals, and also published more than 125 reports in International and National conferences. Professor Lahiri is the member of editorial board of 'Journal of Radioanalytical and Nuclear Chemistry' since July 2005. He is also the permanent member of International Board of Biomonitoring. Professor Lahiri has visited many countries across the world in connection with his academic activities.

The versatile research interest of Professor Lahiri includes production and separation of no-carrier-added clinically important radionuclides by heavy ion and light ion induced nuclear reactions, bio-accumulation of radionuclides, AMS related chemistry of astrophysical interest, Search for new bio-reagents for green chemistry, etc. Professor Lahiri along with his colleague Dr. Dalia Nayak introduced the concept of "tracer packets", which gives unique opportunity to study the closely associated elements together in some biological/physical systems.

He along with his colleague Dr. Kamalika Roy reported the synthesis of radioactive nano particle for the first. He was awarded twice the associate membership of Third World Academy of Sciences (TWAS). He was also awarded DST-DAAD and INSA-DFG visiting Fellowships to Technical University of Munich, Germany. He has guided six students for their Ph. D degree.

Arsenicosis and the Arsenic-Selenium Connection

Julian E. Spallholz and L. Mallory Boylan
Texas Tech University, Lubbock, Texas 79409, USA
julian.spallholz@ttu.edu

Abstract

Arsenic contamination of the domestic drinking water from shallow aquifers in Bangladesh available to villagers often exceeds the newest WHO standards of >10 ng As/L and the Bangladesh standard of >50 ng As/L. An estimated 9.2 million shallow tube wells in Bangladesh delivers water to 97% of the rural population with an estimated population of 57 million people at risk of arsenicosis. The contamination of drinking water by arsenic extends to W. Bengal, India and Nepal. The same shallow aquifers used for domestic water is also used to irrigate food crops, particularly rice. Irrigation adds arsenic to soils and increases the exposure of the population to additional arsenic via the foods consumed. Selenium, an essential trace mineral found in soils, is absorbed by plants entering the human food chain. We suggested that a low dietary intake of selenium may be contributing to the arsenicosis problem in Bangladesh. Selenium is a natural antidote to arsenic by accelerating its excretion and sequestration by complexation. It is also an antioxidant as a component of the enzyme glutathione peroxidase that may counteract the prooxidant effects of arsenic that contributes to arsenicosis and cancer. Analysis of agricultural soil samples from Bangladesh by flurometry, ICPAES and NAA shows them to be very high in arsenic and biologically deficient in selenium. A low dietary intake of selenium from soils via foods maybe contributing to arsenicosis in the Ganges-Brahamaputa delta.

Author Details

Julian E. Spallholz, born in Boston, Massachusetts in 1943 received his BS degree in Biological Sciences from Colorado State University in 1965, MS in Biochemistry from Colorado State University in 1968 and a doctoral degree from The University of Hawaii in Biochemistry-Biophysics in 1971. From 1971-1974 he was a post-doctoral Research Associate at Colorado State University. At Colorado State University Julian was introduced to selenium research. In 1974 he joined the research laboratory of the late Klaus Schwarz, the person responsible for the 1957 discovery of the biological requirement of selenium. Upon Dr. Schwarz' death in 1978, Julian spent a brief time at the Department of Chemistry, State University of New York at Albany prior to joining the Department of Nutritional Sciences at Texas Tech University, Lubbock Texas in August of that year as Associate Professor. Promoted to Professor in 1982, he has spent most of his research time interested in the nutritional and toxicological attributes of selenium and its compounds. His 1994 review published in Free Radical Biology and Medicine, On the Nature of Selenium Toxicity and Carcinostatic Activity, outlined the chemistry of why selenium and its many compounds were or were not toxic and carcinostatic and predicted selenium toxicity might be used for pharmaceutical applications.

Julian is the author or co-author of 65 peer reviewed publications, over 150 abstracts and numerous chapters in various Proceedings. Julian helped organize and edit both the 2nd and 3rd Proceedings of the International Selenium Symposia held in Lubbock, Texas, 1980 and Beijing, China, 1984. He has contributed to several US and Europeans patents on selenium chemistry and has been the recipient of College and University Research Nominations and Awards. Most recently he has remained interested in the relationship between arsenicosis and the role of dietary selenium in the amelioration of arsenicosis and associated cancers. His other current interest extends to the synthesis of selenium containing small molecules for the treatment of cancer, drug resistant bacteria, bacterial biofilms and viral infections.

He is married to Dr. L. Mallory Boylan, Associate Professor of Nutrition. They make their home in Lubbock, Texas, have two Siamese cats and like to travel frequently.

Biogeochemistry of Lacustrine Sediment: an Archive of Palaeoclimate Variability

Brijraj K. Das
Centre of Advanced Study in Geology,
Panjab University, Chandigarh 160004, India.
bkd_geochem@hotmail.com

Abstract

The Indian monsoon is one major weather systems on the earth and, as evident by the unusually high rainfall in recent years, variations in the intensity have widespread socio-economic impact. Though the dedicated efforts by the scientists have been made over the centuries, a comprehensive picture of monsoon variability, especially on societally relevant time scales, has not emerged. This is because monsoon system is very complicated and to understand it and to predict, high resolution, short term data are required. Scientists have identified such archives while working on continents and obtained encouraging results from the Himalaya on seasonal to decadal scale resolution. It is realized that multiple 'proxy data' can be used to infer past monsoon variability and this is necessary in the assessment and prediction of future climate change.

Since organic matter is precipitated from water as particles of various sizes, its distribution is greatly affected by hydrodynamic conditions. One of the most general distribution features of organic matter in sediments is that its concentration rises as the particle size of sediments decreases. The latter is largely dependent upon source rock, distance of transport, intensity of rainfall and erosion etc. Therefore, the biogeochemical investigation of lake sediments particularly on well dated core sample will unravel the history of climate variability in the past which may correspond to global climate phenomena on a large scale. The biogeochemical studies carried out on lake sediment core samples of Himalya have shown good results on Holocene palaeoclimate changes. A few cases are cited in this paper.

Author Details

Prof. Das is a Humbolet Fellow and is working as a research scientist at the Centre of Advanced Study in Geology, Panjab University, India. His recent studies are in the field of environmental geochemical pollution of water and sediments of freshwater lakes and wetland of Himalayan and some peninsular Indian Lakes. In addition to this, sediment chemistry (inorganic, organic and isotopic) of well dated long core sediment segments and organic constituents of high resolution are being carried out as an archive providing potential data for reconstruction of paleoclimatic history during Holocene (10,000 yrs). The aim is to correlate this data with the marine sediments and ocean trap sediments data to find its relevance in the present day climatic variability. He has carried out a lot of collaborative work with scientists in Germany in related fields. He is presently the Uice President of the Humboldt Academy of Chandigarh.

Engineering Interventions for Management of Marginally Saline Groundwaters

S.K. Kamra
Central Soil Salinity Research Institute, Karnal- 132001, Haryana, India.
skkamra@cssri.ernet.in

Abstract

Indiscriminate pumping is causing groundwater salinization and deterioration of pumped water quality due to upconing of saline water from lower zones in several coastal and irrigated regions of India. Under such conditions, it is imperative to selectively skim thin layers of fresh water accumulated over the saline groundwater through specially designed skimming structures and by enhancing groundwater recharge. In this paper, field appraisal of a few new technologies for skimming and recharging of fresh water in saline groundwater regions of Haryana, Andhra Pradesh, Gujarat and Tamil Nadu is presented. A clear perception emerging out of these studies is the need and importance of groundwater recharge in saline groundwater regions, either alone or in combination with skimming systems. Further individual farmer based skimming and recharge structures seem to have social viability and better chances of success and sustenance than the community/government-operated and maintained systems.

Author Details

Born in Abohar, Punjab, India on 30th December 1954, Sushil Kumar Kamra obtained his B.Tech and Ph.D degrees in Agricultural Engineering from Punjab Agricultural University, Ludhiana, India in 1976 and 1989 respectively. Starting his research career as a Junior Scientist in 1977, Dr. Kamra regularly advanced to different positions at Central Soil Salinity Research Institute, Karnal where he is currently working as Principal Scientist (Groundwater and Drainage). His scientific capabilities and interests include subsurface drainage, groundwater recharge, shallow groundwater structures and hydro- salinity modeling. Dr.Kamra was a visiting FAO Fellow, Cornell University, Ithaca, USA during 1989 and an Alexander von Humboldt Fellow, University of Kiel, Kiel, Germany during 1995- 1997. Dr. S.K. Kamra has presented research papers in more than 10 international conferences/ symposia in India and abroad (Pakistan, The Netherlands, Slovenia, Italy) and has also undertaken a number of assignments/ working visits to USA, Germany, France, Switzerland and Australia. He has been/ is a reviewing referee for International Journals (Water Resources Research, Agricultural Water Management, Irrigation Science, Hydrological Processes, J. Porous Media, Environment Modeling and Software) as well as about 10 national journals.

Dr. S.K. Kamra has made significant contributions in developing the subsurface drainage technology for amelioration of waterlogged saline irrigated lands in India. For areas without natural outlets, evaporation pond technology propagated by him can provide interim solutions for management of saline drainage effluent. These studies paved the way for commissioning of large- scale mechanically- installed subsurface drainage projects in farmers' fields initially in Haryana and Rajasthan and later in Gujarat, Andhra Pradesh, Punjab and Karnataka. Dr. Kamra is deeply committed to academics and basic research on water and solute dynamics in soilgroundwater system. He has developed a 2-dimensional numerical model that provides decision support on environmentally optimal subsurface drainage designs and prediction of long term salinity of drainage water. The scope of the model was extended for pesticides leaching to groundwater during his two year Alexander von Humboldt Fellowship in Germany. Recently Dr. Kamra successfully completed a multi- location research project on groundwater skimming and recharging of fresh water in saline groundwater regions of Haryana, Gujarat, Andhra Pradesh and Tamil Nadu. The impressive accomplishments of the project are paving the way for commissioning of large groundwater skimming and recharging projects in these four states.

River Response to Climate Change: Why, Where and How?

Rajiv Sinha
Engineering Geosciences Group, Department of Civil Engineering,
Indian Institute of Technology Kanpur, Kanpur 208016, India.
rsinha@iitk.ac.in

Abstract

Recent geological history of the earth is known for rapid and frequent climate changes. There is a serious concern about the nature and consequences of climate change as it has an enormous impact on agriculture, water resources and ecological system. There is also a widespread recognition that human activities are altering climatic and environmental conditions over very short time scales. Most analyses of these changes depend on the time series data restricted to a few decades or a century at the most and therefore the causative factors of these changes remain unresolved. A complete understanding of the processes and variability of climate changes would therefore require the use of "proxy data" for tracking the past climate changes for a much longer time scale. As a corollary, future climate change would also be a function of natural factors such as earth's orbital motion, insolation and anthropogenic factors such as greenhouse gas emissions. The assessment and prediction of future climate change would therefore be strongly linked to the understanding of past climates. The response of river systems to ongoing and future climate change has become a subject of intensive research. River landforms and deposits provide one of the most readily available continental records to study environmental change. The Indian subcontinent is host to several large river systems such as the Ganga, Brahmaputra, Narmada and Cauvery. The Gangetic plains of India represent the world's largest area of modern alluvial sedimentation and support a large population of 250 million people. Natural environmental changes therefore have a large impact on the social and economic development of the region. Natural environmental changes have been augmented by the human-caused changes manifested in terms of land cover changes, increased sediment yield in rivers, increased flood frequency and a general degradation of the eco-system.

Author Details

Rajiv Sinha is a professor in geosciences at the Indian Institute of Technology, Kanpur. He obtained his PhD in sedimentology from the Department of Earth Sciences, University of Cambridge, UK in 1992. He joined IIT Kanpur in 1994 as a faculty and since then has been consistently working on various aspects of Quaternary Geology of India. Rajiv Sinha's major interest is to understand the response of river and delta systems to climate change during the Quaternary period. Along with his doctoral students at IIT Kanpur, he integrated the use of remote sensing and hydrological analysis for producing one of the best documented accounts of avulsive river systems in the world. The new approaches to flood analysis have contributed significantly to finding long-term solution to recurring floods in the region.

Dr. Sinha was awarded Alexander von Humboldt fellowship in 2000 to work at the Institute of Mineralogy and Geochemistry at Karlsruhe University in Germany. During his stay at Karlsruhe, he worked on the on the Sambhar playa sediments from the Thar Desert and generated an extensive data set on evaporite mineralogy, geochemistry, stable isotopes to unravel the chemical evolution of the Sambhar lake. Using an integrated mineralogical and geochemical approach, he also reconstructed the changes in salinity, lake level and the underlying climatic changes over the last 25 thousands years. This work has led to several publications in international and national journals.

Dr. Rajiv Sinha is also a recipient of AICTE Career Award in 1998, National Mineral Award in 2002, and S.S. Merh Award in 2006 for his contributions in Quaternary geology. He is currently leading several national and international projects at IIT Kanpur related to river response to climate change with a very clear emphasis on the Ganga plains.

Recession Rate of Tributary Glaciers of Gangotri Glacier System, Garhwal Himalaya-India

B.S. Kotlia, Rakesh K. Domka, M. S. Miral, L. M. Joshi and K. Kumar
Department of Geology, the Durham,
Kumaun University, Nainital, India
bskotlia@yahoo.com

Abstract

The recession rate of the tributary glaciers of the Gangotri glacier system, e.g. Thelu and Raktvarna glaciers in the Garhwal Himalaya has been calculated by GPS and toposheet (Survey of India toposheet no. 53 N/1, 1962, scale 1:50,000). In the last 42 years, the Thelu glacier has receded about 1288 m at an approximate rate of 30.66 m/year. Similarly, the Raktvarna has receded 1585 m with 37.73 m of annual recession. At present, the Chaturangi glacier is found 293 m away from the main trunk of the Gangotri glacier. Evidences of geomorphic landforms, both endogenic and exogenic in the valley also support the recession of glaciers in the recent past.

Author Details

Prof. B. S. Kotlia is a UGC Research Scientist at the Kumaon University, Nainital. He has worked in Germany as an Alexander von Humboldt Research Fellow and is also a fellow of Geological Society of India. His research interests include Climatic changes and quaternary micromammalian fossils in Indian Himalaya (Higher Himalaya; Garbyang, Milam, Spiti, Ladakh: Lesser Himalaya; Kumaun, Garhwal)

Molecular Technologies: Application in Cancer Diagnosis for Improved Treatment Strategy

A. Raghu Ram Rao
Pharmaceutical Chemistry Division,
University Institute of Pharmaceutical Sciences,
Panjab University, Chandigarh 160014, India.
rrakkinepally@pu.ac.in

Abstract

Molecular technology is devoted to the study of synthesis, properties and applications of structures and materials with at least one critical dimension less than the scale of approximately 100 nm. Cancer is one of the most significant causes of morbidity and mortality in the world. Clinical outcome of cancer diagnosis is strongly related to the stage at which the malignancy is detected and therefore, early screening has become desirable. Here comes the role of various molecular technology tools for the early diagnosis of cancer. Various molecular technology tools used for cancer diagnosis are Nanocantilevers, Quantum Dots, Nanoshells, Dendrimers, Nanopores, Nanotubes, Genomic Profiling, Protein Profiling, and Protein Microarray. All these tools are based upon their interaction at the molecular level in the body and size plays an important role in their mode of action. Molecular technology holds a promising future in its hands for the cancer diagnosis and treatment and for many other diseases.

Author Details

Dr. A. Raghu Rama Rao is a Professor of Medicinal Chemistry at Panjab University, India. He has been associated with teaching and research for the last 22 years and has won many awards, including the DAAD fellowship (1991), Senior DAAD fellowship (2002, 2005), AICTE Career award for young teachers (1995), etc. He has 65 research papers to his credit and has guided 10 Ph.D. candidates and numerous Masters' students. His areas of interest include antiasthma agents, anticancer agents and analytical profiles of drug substances.

Metagenomics: A Splendid Novel Approach to Explore and Exploit Microbial Diversity

Pavan K. Pindi[a], N. Kishore[a], S. Ram Reddy[a], S. M. Reddy[a], S. Chatterjee[b], I. Sherameti[c], R. Oelmueller[c] and A. Varma[b]

(a) *Department of Microbiology, Kakatiya University, Warangal 506 009, India*
(b) *Amity Institute of Microbial Sciences, Amity University U. P. Noida, 201303, India*
(c) *Institute of General Botany, Department of Environmental Sciences, Fridrisch Schiller University, Dornburger Strasse Jena 159 07743, Germany*
ajitvarma@aihmr.amity.edu

Abstract

Microbial diversity is one of the greatest knowledge gaps in the biological sciences and remains largely unexplored and unexploited. Despite the improvements in cultivation methods, there has not been any significant change in our perception of microbial diversity. Molecular phylogenetic methods have recently provided a means for identifying the types of organisms that occur in microbial communities without the need for cultivation. However, they do not provide any idea about their ecophysiological functions. Metagenomics, the genomic analysis of a population of microorganisms that provides an access to the physiology and genetics of uncultured organisms, has emerged as a powerful technique in recent times. Metagenomic analysis involves isolating DNA from an environmental sample, cloning the DNA into a suitable vector, transforming the clones into a host bacterium and screening the resultant transformants. Plasmid, cosmid, fosmid and BACs as cloning vectors and *E. coli* as the surrogate host are employed to construct metagenomic libraries. Three types of screening methods, viz. sequence-based, function based and SIGEX are used to screen the libraries for the desired clones. Metagenomics has unlimited applications in basic biology and biotechnology. The long term goal of metagenomic analysis is to reconstruct the genomes of uncultured organisms, understand the functional dynamics in natural communities and also to exploit the culture-independent microorganisms for novel bioreactive compounds. Already, spectacular results have been achieved that exposed only the tip of iceberg. To date, metagenomics has uncovered a variety of novel genes that range from small genes coding for enzymes to complex gene clusters encoding antibiotics. Thus it has the potential to substantially impact white and red biotechnologies. Obviously, metagenomes have attracted the attention of academia and specialized pharmaceutical companies. Notwithstanding certain teething problems, it is sufficiently powerful to answer many intricate biological questions and yield products for solving real world problems.

Author Details

Prof. Dr. Ajit Varma is Director of Amity Institute of Microbial Sciences & Vice Chairman (International), Amity Science, Technology and Innovation Foundation. After completing his Ph.D. in 1964, he started his career as a microbiologist with IARI, New Delhi. He joined School of Life Sciences, Jawaharlal Nehru University, New Delhi, in 1975 and became a Professor in 1985, a position which he held till 2004. He has guided 42 Ph.D. students till date and has published over 273 research articles, in national and international journals as well as many review articles and chapters in books Dr. Varma's research interest is mycorrhizal fungi and their biotechnological applications, an area in which he has published 24 books. Dr. Varma is the recipient of a large number of international scientific fellowships including Alexander von Humboldt fellowship, DAAD fellowship, Commonwealth fellowship, RAISA Fellowship for innovative research in biotechnology (Italy) and many more. Prof. Varma is an elected member of the National Academy of Agricultural Sciences, India.

Ethics in Public Domain: Biomedical Research and Beyond

Giridhari Lal Pandit
*University of Delhi South Campus, 205-Arts Faculty Building,
Department of Philosophy, Benito Juarez Road, New Delhi – 110021, India.*
panditgl@yahoo.co.in

Abstract

Natural scientists have to make themselves aware of the responsibility that on one hand; their work can contribute to survival; while on the other, it may also lead to the destruction of life on the earth. The dangerous consequences of physical, chemical and biological researches and product developments as well as the technical and political possibilities the checking of the dangers through the stimulation life-saving strategies.

Author Details

Born and educated in Kashmir, Giridhari Lal Pandit is Fellow of the Alexander von Humboldt-Stiftung and Professor of Philosophy at the University of Delhi South Campus since 1988. Since 1985 Pandit has been regularly visiting leading Universities in Germany. Pandit teaches and works in the fields of methodology of science, epistemology, analytic philosophy, ethics in public domain, and philosophy of ecology. He has been a Visiting Professor at several Universities/Institutes including University of Poitiers, France, Maison des Science de L'home de Paris and Srinakharinwirot University, Bangkok. Pandit was a permanent member of Faculty of Philosophy at Visva-Bharati (founded by Nobel Laureate Rabindra Nath Tagore), Santiniketan, West Bengal, during 1971-1974. He taught at Delhi University as Lecturer during 1969-1971 and 1974-1979 and as Reader in Philosophy of Science and Methodology during1980-1987. In June 2005, he has been re-nominated as Member of "Institutional Human Ethics Committee for Human Subjects" at the Institute of Genomics & Integrative Biology (Formerly Centre for Biochemical Technology), Council of Scientific & Industrial Research under Govt. of India, Delhi University Campus, Mall Road, Near Jubilee Hall, Delhi – 110007, India.

Pandit has authored two volumes in the Boston Studies in the Philosophy of Science, Vol. 73 and Vol. 131, entitled respectively: The Structure and Growth of Scientific Knowledge: A Study in the Methodology of Epistemic Appraisal (Springer Verlag/D. Reidel: Dordrecht/Boston/London 1982), Methodological Variance: Essays in Epistemological Ontology and the Methodology of Science (Springer Verlag/Kluwer: Dordrecht/Boston/London 1991). In German, he has published a book entitled Von der Oekologie des Bewusstseins zum Umweltrealismus: Die Wiederentdeckung menschlicher und nicht-enschlicher Interessenssphaeren (Picus Verlag: Wien 1995). In addition he has published numerous essays and articles in philosophy of science and philosophy of ecology from Europe/USA/India.

Research interests with key concepts: Fundamental physical theory and cosmology; methodology of theory-problem interactive systems; dynamic corecontexts of theory development and theory testing; aim and structure of physical theory; the resolving power of physical theory; philosophy of ecology-environmental nesting and environmental realism; environmental change. *Focus of Current Research:* Besides focusing on philosophy of ecology, human interests-studies, and ethics in public domain/ biomedical research, Pandit's current research in the field of philosophy of science builds on the work already done and aims at developing a methodological model of theory-development and theory-unification in physics.